T0234661

The last ten years have shown a dramatic revolution in our understanding of early animal development. *From egg to embryo* describes the results of this revolution and explains how the body plan of an embryo emerges from the fertilised egg.

The book starts with a critical discussion of embryological concepts and provides a new reference glossary. It goes on to explain in simple terms the mathematics of cell states, morphogen gradients and threshold responses. The experimental evidence for the mechanism of regional specification in *Xenopus*, molluscs, annelids, ascidians as well as *Caenorhabditis*, the mouse, the chick and *Drosophila* is then discussed. Progress with *Drosophila* has been particularly impressive and the *Drosophila* chapter provides a clear guide to the subject and also includes a new table of developmentally important genes.

Throughout, the emphasis is on conceptual clarity and unity: bringing together the mathematical models, embryological experiments and molecular biology into a single, comprehensible, coherent account. The book is a critical work, designed to evaluate theories and experiments and to stimulate thought by the readers, be they undergraduates, postgraduates or research workers.

FROM EGG TO EMBRYO

Developmental and cell biology series

SERIES EDITORS
Dr P.W. Barlow, *Long Ashton Research Station, Bristol*
Dr D. Bray, *King's College, London*
Dr P.B. Green, *Dept of Biology, Stanford University*
Dr J.M.W. Slack, *ICRF Developmental Biology Unit, Oxford*

The aim of the series is to present relatively short critical accounts of areas of developmental and cell biology where sufficient information has accumulated to allow a considered distillation of the subject. The fine structure of the cells, embryology, morphology, physiology, genetics, biochemistry and biophysics are subjects within the scope of the series. The books are intended to interest and instruct advanced undergraduates and graduate students and to make an important contribution to teaching cell and developmental biology. At the same time, they should be of value to biologists who, while not working directly in the area of a particular volume's subject matter, wish to keep abreast of developments relative to their particular interests.

BOOKS IN THE SERIES

R. Maksymowych *Analysis of leaf development*
L. Roberts *Cytodifferentiation in plants: xylogenesis as a model system*
P. Sengel *Morphogenesis of skin*
A. McLaren *Mammalian chimaeras*
E. Roosen-Runge *The process of spermatogenesis in animals*
F. D'Amato *Nuclear cytology in relation to development*
P. Nieuwkoop & L. Sutasurya *Primordial germ cells in the chordates*
J. Vasiliev & I. Gelfand *Neoplastic and normal cells in culture*
R. Chaleff *Genetics of higher plants*
P. Nieuwkoop & L. Sutasurya *Primordial germ cells in the invertebrates*
K. Sauer *The biology of Physarum*
N. Le Douarin *The neural crest*
J.M.W. Slack *From egg to embryo: determinative events in early development*
M.H. Kaufman *Early mammalian development: parthenogenic studies*
V.Y. Brodsky & I.V. Uryvaeva *Genome multiplication in growth and development*
P. Nieuwkoop, A.G. Johnen & B. Albers *The epigenetic nature of early chordate development*
V. Raghavan *Embryogenesis in angiosperms: a developmental and experimental study*
C.J. Epstein *The consequences of chromosome imbalance: principles, mechanisms, and models*
L. Saxen *Organogenesis of the kidney*
V. Raghaven *Developmental biology of fern gametophytes*
R. Maksymowych *Analysis of growth and development in Xanthium*
B. John *Meiosis*
J. Bard *Morphogenesis: the cellular and molecular processes of developmental anatomy*
R. Wall *This side up: spatial determination in the early development of animals*
T. Sachs *Pattern formation in plant tissues*

FROM EGG TO EMBRYO
REGIONAL SPECIFICATION IN EARLY DEVELOPMENT

SECOND EDITION

J.M.W. SLACK

*Principal Scientist, Imperial Cancer
Research Fund Developmental Biology
Unit, University of Oxford*

CAMBRIDGE
UNIVERSITY PRESS

CAMBRIDGE UNIVERSITY PRESS
Cambridge, New York, Melbourne, Madrid, Cape Town, Singapore, São Paulo

Cambridge University Press
The Edinburgh Building, Cambridge CB2 2RU, UK

Published in the United States of America by Cambridge University Press, New York

www.cambridge.org
Information on this title: www.cambridge.org/9780521401081

First published 1983
First paperback edition 1984
Reprinted 1985
Second edition 1991
Reprinted 1992, 1994, 1997

A catalogue record for this publication is available from the British Library

Library of Congress Cataloguing in Publication data

Slack, J.M.W. (Jonathan Michael Wyndham)
From egg to embryo : regional specification in early development /
J.M.W. Slack. – 2nd ed.
     p.   cm. – (Developmental and cell biology series)
Includes bibliographical references and index.
ISBN 0-521-40108-9. – ISBN 0-521-40943-8 (pb)
1. Embryology.   I. Title.   II. Series.
QL955.S54   1991
591.3′3–dc20   90-2670 CIP

ISBN-13 978-0-521-40108-1 hardback
ISBN-10 0-521-40108-9 hardback

ISBN-13 978-0-521-40943-8 paperback
ISBN-10 0-521-40943-8 paperback

Transferred to digital printing 2006

To Janet, Rebecca and Pippa

Contents

Preface to the second edition   *page* xv
Preface to the first edition   xix

1 Regional specification in animal development   1

Why are we interested in embryos?   2
Homology and universality   3
The developmental hierarchy   4

2 The concepts of experimental embryology   9

Normal development   9
   The fate map   10
   Growth and stem cells   16
Forms of developmental commitment   18
   The hierarchy of commitment   19
   Lineage   20
   Regulation and mosaicism   21
Acquisition of commitment   25
   The embryonic field   25
   Cytoplasmic determinants   26
   Induction   27
Homeosis and the epigenetic coding   30
Glossary   31

3 Theoretical embryology   34

Cell states   34
Dynamical systems theory   35
   The state space   37
   A model for cellular specification   40
Gradient models for embryonic induction and regulation   43
Diffusion of a morphogen   43
   The local source-dispersed sink (LSDS) model   45

Symmetry breaking processes   52
  Double gradient models   53
Thresholds and memory   57
  DNA methylation   59
  Homeogenetic induction   60
Repeating patterns   61
  Reaction–diffusion models   61
  Mechanical instabilities   64
  A clock model   65

4 Hierarchies of developmental decisions   67

Xenopus and other amphibians   67
Normal development   68
  Molecular markers   73
  Fate maps   75
Regionalization within the egg   80
  The dorsoventral polarization   81
Regional organization during cleavage and blastula stages   85
  States of commitment during cleavage   85
  States of commitment in the blastula   87
Early inductive interactions   93
  Dorsalization   96
  Mechanism of mesoderm induction   98
  The organizer   101
  Specification of the anteroposterior pattern   104
Neural induction   106
  Regional specificity of neural induction   107
Later events   110
  Somitogenesis   110
  The endoderm   111
Summary of regional specification in the early amphibian
  embryo   111
Sea urchins   112
Normal development   112
  Molecular markers   115
  Fate map   117
Regional organization   119
  Unfertilized egg   119
  Blastomere isolation and fusion   120
  Morula   121
  Micromeres   123
  Vegetalization   124
  Oral–aboral axis   125
Sea urchins and amphibians   126

5   Development with a small cell number 128

Molluscs and other spirally cleaving forms 129
Normal development and fate map 129
   Other species and annelids 133
   Role of the egg cytoplasm 136
Regional organization 137
   Egg cytoplasm 137
   Lobe removal 138
   Blastomere ablation and isolation 140
   Determination of the D lineage in equally cleaving forms 141
   Segmentation in the leech 142
Ascidians 146
Normal development 146
   Fate map 150
Regional organization 151
   Determination of the muscle lineage 152
   Other possible determinants 156
Caenorhabditis elegans 157
Normal development 158
   Developmental genetics 160
Regional organization 163
   Cytoplasmic localization 163
   Laser ablation 164
   Induction of the pharynx 165
General remarks on the mosaic embryos 165
   An invariant cell lineage 166
   A visible regionalization of the fertilized egg cytoplasm 167
   Mosaic behaviour of defect embryos and isolated blastomeres 167
   Cytoplasmic transfer 168

6   Models for Man: the mouse and the chick 171

The mouse 171
Normal development of the mouse 172
   Fashionable genes 176
   Fate map 178
   Chimaerism and mosaicism 181
Specification of early blastomeres 182
   Isolated blastomeres and giant embryos 182
   Specification of the ICM and trophectoderm 183
   Nuclear transplantation and imprinting 185
Later developmental decisions 187
   Formation of primitive endoderm 187
   Mural and polar trophectoderm 187
   Visceral and parietal endoderm 188

Egg cylinder ectoderm   188
Embryonic stem cells and teratocarcinoma   189
   ES cells   189
   Teratocarcinoma   191
Transgenic mice   192
How typical is the mouse?   194
The chick   195
Normal development   196
   Fate map   200
Regional organization   201
   AP polarity and fragmentation of the blastoderm   201
   Induction   205
   Segmentation   209
A common program for vertebrate development?   211

7 The breakthrough   213

Introduction to insects   214
Normal development   215
   Fate map   218
   Cellular commitment   219
   Pole plasm   220
Drosophila developmental genetics   221
   Identification of relevant genes   222
   Types of mutation   224
   Mosaic analysis   225
   Cloning of genes   227
Overview of the developmental program of *Drosophila*   229
The dorsoventral pattern   230
   Maternal control   231
   Zygotic control   234
   Lessons from the dorsoventral system   235
The anteroposterior system   236
   Maternal information: the anterior system   236
   Maternal information: the posterior system   237
   Maternal information: the terminal system   239
   The gap genes   240
   The pair-rule system   243
   Regulation of pair-rule gene expression   246
   The segment polarity genes   248
   Homeotic selector genes   254
   The anteroposterior body pattern   259
Results on other species   260
   Descriptive embryology   260
   Experimental studies   262

What *Drosophila* has taught us   265
Drosophila gene table   266

8   What does it all mean?   278

Learning from *Drosophila*   279

Appendix: How to write a program for development 282

Nomenclature   282
The program   284
Posterior and dorsal regionalization   286
Gastrulation   288
Dorsoventral interpretation   288
The growth zone engine   289
Anteroposterior specification   291
The body plan complete   295

References   297
Index   322

Preface to the second edition

The first edition of *From Egg to Embryo* was written in 1981–82 which was before the techniques of molecular biology had made a serious impact on the problem of regional specification. Over recent years there has been an absolute explosion of new molecular data, led by the work on *Drosophila*, but with *Xenopus*, the sea urchin and the mouse also participating. I felt that the right time for a second edition would be when the *Drosophila* work showed some signs of stabilizing, which it now seems to be doing.

The aims of the first edition were two-fold: to explain the theoretical basis of the regional specification problem and to summarize, in a compact form, the essential experimental data relating to early embryos. The molecular biology explosion has imposed a third goal on the second edition, which is to give the reader a map for navigating through a world of molecular detail. There are now so many potentially interesting genes being studied in so many experimental systems that it is essential to keep the logical and the dynamical categories of explanation to the forefront in order to decide what to remember and what to put in the database.

From Egg to Embryo is written for anyone who is interested in embryos and so does not conform to any particular course of study. However, the first edition enjoyed some success as an aid to undergraduate and post-graduate teaching and I hope the second edition will be found even more useful. I make no apology for not using the fashionable 'concept subheadings' or 'theme boxes'. Such a dogmatic approach to education may have its appropriate applications, but this is not one of them. *From Egg to Embryo* is a *critical* work, in which assumptions are questioned and problems are faced.

Readers familiar with the first edition will find very substantial changes in the second. The entire text has been rewritten with some simplification and clarification where necessary, and the arrangement of the book has been changed somewhat. There is still an initial discussion of the concepts of experimental embryology, so vital for a clear exposition of the subject but so frequently confused and misunderstood. After this comes a theoretical chapter. It now comes before, and not after, the data because morphogen gradients, bistable circuits and thresholds have become facts rather than hypotheses. The central role which these models have in our field is symptomatic of the difference there must be between an explanation of regional specification and an explanation of how a particular gene is turned on, even though similar techniques may now be used in the two types of study. Then comes the main part of the book which consists of a brief account of the principal experiments which have built up our ideas about early development. Unlike most other books about embryology the treatment is organism by organism rather than stage by stage. Instead of jumping from cleavage in sea urchins, to gastrulation in amphibia, to segmentation in *Drosophila*, I have tried to enable the reader to see clearly the strengths and weaknesses of each experimental system. This should emphasize the truth that there is no ideal organism for this type of work, and that something can be learned from all of them, although it must be admitted that *Drosophila* has contributed more than its fair share in recent years. Organisms are not introduced in a taxonomic sequence, as might seem natural to a zoologist. They are presented in an order which introduces each layer of the conceptual framework in a manageable way. So we start with creatures whose suitability for micromanipulation has enabled the construction of reasonable fate and specification maps and proceed through the exemplars of cytoplasmic localization and cell lineage to the models for Man. *Drosophila* now comes at the end because the progress has been so fantastic that to put it earlier would inevitably lead to an anticlimax. In recognition of the fact that we now really do understand what we are dealing with, there is an Appendix which tries to explore what a real developmental program would look like and what it could be used for.

An information explosion all too easily means an explosion in the number of pages and references. Because so many people told me how much they liked the *shortness* of the first edition I have fiercely resisted this inflationary tendency. Inevitably this has meant that some old topics have had to be omitted to make room for the new ones and one casualty of this has been the section on the ctenophores, together with some of the theoretical material which has not turned out to be so useful. I have also had to be very ruthless about references. In this edition the references are intended purely as a guide to further reading. Where possible I have cited a review instead of an original paper and I have not been able to cite references to indicate priority or antiquity. This means that many excellent, innovative and influential papers have been omitted and I do apologize to their authors for this. Finally, I have also resisted the temptation to include material which does not deal strictly with *early animal development*. Although topics such as the development of the vertebrate limb, *Drosophila* imaginal discs, or the slime mould, all involve *bona fide* questions of regional specification, I felt the focus must be kept concentrated on the original subject area otherwise the length would become unmanageable.

Once again I have been helped by several scientists who have kindly reviewed sections for me and I would like to thank Rosa Beddington, Richard Gardner, Chris Graham, Phil Ingham, David Ish Horowicz, Steve Kearsey, Julian Lewis, Jim Smith, John Sulston, David Tannahill, Jo Van den Biggelaar and Fred Wilt for this most valuble service. I should also like to thank the authors who provided me with copies of their photographs, so important to convey the beauty and visual impact of animal embryos.

Jonathan Slack
Oxford 1990.

Preface to the first edition

This book is an enquiry into the mechanisms by which the spatial organization of an animal emerges from a fertilized egg. It is intended for all students, teachers and research workers who are interested in embryos.

It is divided into three parts. The first two chapters introduce the problem of regional specification and attempt to define the meanings of embryological terms which are used in the remainder of the book. This is necessary because terms such as 'induction', 'regulation' or 'polarity' are often used but rarely defined and many controversies have arisen as a result of unnecessary misunderstandings.

The next four chapters give an overview of the experimental evidence which bears on the processes of cellular commitment from the time of fertilization to the formation of the general body plan. The animal types considered are those on which most experimental work has been done: amphibians, insects, other selected invertebrates, the mouse and the chick. This is a general survey rather than a detailed review but sufficient references are provided to enable interested readers to pursue the topics in greater depth.

The last four chapters attempt to generalize the problems and to investigate the extent to which they have been solved by the theorists and the model-builders. In particular several 'gradient' models are examined and assessed in terms of the experimental evidence. This section has been written principally for non-mathematical readers although a few differential equations are provided for those who are interested. Many of the models are relevant to late development and to regeneration as well as to early development, but the focus of the book has been kept on early development because this provides for maximum conceptual unity without undue length.

I should like to thank the series editor, Chris Wylie, for the invitation to write the book; my wife Janet who expertly drew all the diagrams for Part II; Brenda Marriott for undertaking the lion's share of the typing; Richard Gardner, John Gerhart, Chris Graham, Brigid Hogan, Klaus Sander, Jim Smith and Dennis Summerbell for commenting on portions of the manuscript; and finally the many authors who have provided me with copies of their photographs.

1

Regional specification in animal development

This book is about how an egg becomes an animal. Attention will be concentrated on early development because this is the time at which the important events are happening. As everyone knows, the human gestation period is about nine months long but it is not so commonly appreciated that the basic body plan of the embryo becomes established during the very short period from one to four weeks after fertilization. During this time an apparently homogeneous group of cells, the inner cell mass of the blastocyst, become transformed into a miniature animal consisting of central nervous system, notochord, lateral mesoderm, somites, pharyngeal arches, integument and gut. All of these parts contain specific types of cell and all lie in the correct positions relative to one another. In later development there is a good deal of growth and of histological differentiation of the organs and the specifically human, rather than the general vertebrate, characteristics of the organism become established. However, all this takes place on the framework of the basic body plan which was laid down in early development. As Wolpert has emphasized, it is not birth, marriage or death, but gastrulation which is truly the most important time in your life.

The core problem of early development is that of *regional specification*, also often called the problem of spatial organization or of pattern formation. Regional specification is the process whereby cells in different regions of the embryo become switched onto different pathways of development. It is the mechanisms by which this is achieved with which we are concerned in this book. Regional specification should not be confused either with cell differentiation or with morphogenesis which are processes posing us with important but distinct problems. *Cell differentiation* means the synthesis by a population of cells of types of protein different from those made by their ancestors, and different from those made by other cells in the embryo at the same time. How this is achieved is fundamentally a problem of the mechanism of control of gene expression, still only partially understood. Of course genes are turned on and off in the course of regional specification, but as we shall see the process has its own specificity in terms of explanatory categories, and cannot be reduced to gene expression alone.

Morphogenesis means the creation of form, and it is useful to reserve this term for the processes of cell and tissue movement that shape the embryo. Again gene products are obviously necessary for any type of cellular behaviour, but an explanation of shape changes during development will have a substantial mechanical and physical component, and so needs explanatory concepts not found in molecular biology. Both cell differentiation and morphogenesis are of immense importance and without them there would be no development. However, they are not extensively discussed in this book because they are distinct problems from that of regional specification, and to a large extent they are consequences of those processes of regional specification which occur in the earliest stages of development. The distinction between the three fundamental types of developmental process: regional specification, cell differentiation and morphogenesis was first made by Waddington and has been extensively popularized by Wolpert. This book hopefully represents a continuation of the tradition established by these two great embryologists.

WHY ARE WE INTERESTED IN EMBRYOS?

Most people claim at least some sort of interest in how embryos work, presumably as it stirs a sense of knowing one's own origins. In this regard, they share the concern of the medical research scientist that the object of study should be to some degree a model for human development. Naturally, ethical considerations dictate that most studies in experimental embryology must be carried out on animals other than the human. The conventional view is that relevance decreases with taxonomic distance from Man, so mammalian embryos are usually felt to be the best models. However, viviparity and other problems pose serious technical difficulties and, as we shall see, much experimental embryology has concerned itself with invertebrates and non-mammalian vertebrates. It is therefore an issue of some importance to know how similar are the mechanisms for regional specification in different types of animal.

Before considering this we should not forget that there is also another class of person who is interested in embryos. These are the zoologists who are interested in animals in their own right rather than as models for something else. For them, embryonic life is part of the reproductive process which is concerned with reshuffling the genes of the population for the next generation. They are particularly concerned with the *differences* between developmental strategies in so far as each is supposed to be perfectly attuned to the lifestyle of the species concerned. Variations such as the different yolk content of eggs of related species or the different arrangements of embryonic membranes in different mammalian groups are a source of annoyance to the medical scientist, who would prefer everything to be the same, but a source of delight to the zoologist. The lesson is that animals in general, or embryos in particular, possess two sorts of attribute:

pristine characters which are intimately involved in the formation of a certain type of body plan and which will have remained essentially unchanged during the evolutionary history of the group concerned, and *adaptive* characters to do with reproductive strategy in a particular ecological niche. Adaptive characters may be the stuff of Darwinian evolution but are not fundamental to our present concerns.

HOMOLOGY AND UNIVERSALITY

Zoologists have traditionally used the word *homology* to refer to a pristine type of similarity and based their judgement on morphological characters which had not been greatly altered by adaptive evolution. By contrast, adaptive similarity stemming from a different evolutionary past is referred to as *analogy*. On the level of the developmental program, homology has a precise meaning since it expresses a set of developmental decisions that have been made by the part in question. If the sequence of decisions was the same for a particular part in two species, then those parts are homologous. It is difficult however to deal with differences. If the sequence of decisions is similar but not identical, then is it correct to speak of a degree of homology? Until a number of real developmental programs are understood as well as is *Drosophila* today it will not be possible to produce a real measure of incomplete homology.

Homology also has a precise significance when it comes to molecules. In general, two molecules having the same biochemical function, such as ribosomal RNA in Man and yeast, have many residues in the primary sequence the same. If many organisms are considered then an evolutionary tree can be drawn up based on sequence similarity which is very similar to trees derived from comparative anatomy or from the fossil record. In fact, molecular trees are better because they can reach back in time beyond the limits of morphological similarity and be used to address questions of phylogeny. Conversely, a developmental biologist can use sequence similarity to clone a gene from one organism using a probe from another. It is usually assumed that, if the similarity of sequence falls within the normal range for the taxa concerned, then the new clone is a homologue of the old and has the same biochemical function.

For many years developmental biologists have wondered whether the mechanism of regional specification was pristine and universal, like the genetic code and oxidative metabolism; or parochial and adaptive, like coat colour or body size. Until recently this debate was unencumbered by data, but now we have a plethora of molecular homologies which have tended to make popular opinion swing far to the universal side. The majority of cases are genes involved in regional specification which were discovered in the fruit fly, *Drosophila*, and later found to have homologues in other animals, often also being expressed during early development. There are also the cytokines (growth factors) and oncogenes, initially discovered during work on mammalian cells, which were subsequently found in many diverse contexts.

When assessing the significance of all this it is important to distinguish the *biochemical* function of a protein, from its *logical* function in a developmental program. For example, a tyrosine kinase centre or a DNA binding site can be recognized by inspection of the primary sequence of a gene or protein. This has turned out to be a good, although probably not infallible, guide to the function of the molecule concerned. So if the protein possesses a tyrosine kinase sequence, then it is probable that it actually functions in this capacity and phosphorylates tyrosine residues *in vivo*. But it by no means follows from this that two related proteins which are both tyrosine kinases and which are expressed during early development of, say, sea urchins and *Drosophila*, necessarily have homologous developmental functions. This is because what matters from the point of view of the developmental program is the connection between one element and the next. If the kinases are activated by different means, or if their substrates are different then it is likely that the programs are different, and if the programs are different then the homology of a single component does not necessarily have any more significance than the presence of the same electronic components in two different human artefacts. For example, a TV set and a computer may share a number of electronic components that have exactly the same elementary functions, but, because they are wired up differently, the resulting machines are entirely different.

THE DEVELOPMENTAL HIERARCHY

It is very important to emphasize that even the basic body plan is not specified all at once but is formed as a result of a *hierarchy* of developmental decisions. This statement will be justified more fully by the experimental results reviewed in later chapters, but for the moment consider an embryological lineage such as that shown in Fig. 1.1. This shows the provenance of different regions of the vertebrate body deduced mainly from the experimental embryology of the amphibia. The diagram comprises a number of subdivisions of different multicellular regions. Without considering for the moment the mechanism of these decisions or the exact correctness of the diagram, we note that the familiar histological cell types are to be found only on the right-hand side. These are preceded by a series of states of commitment, each of which is able to make a choice between a restricted set of alternatives and each of which has arisen from a subdivision of an earlier and less committed rudiment. For example, for cells to be competent to form the lens of the eye, they must have already decided that they are ectoderm (rather than mesoderm or endoderm), and epidermis (rather than neural plate or neural crest). So each decision is made by cells already in a

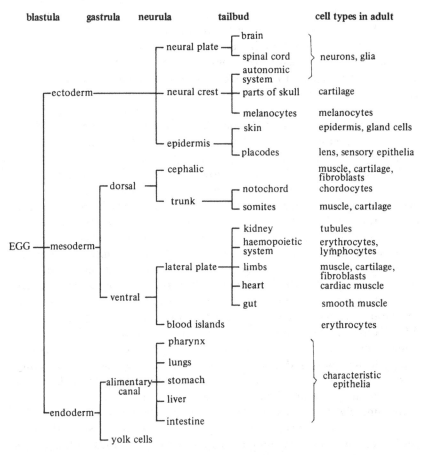

Fig. 1.1. Formation of the basic body plan in a vertebrate (excluding extraembryonic regions). By the early tailbud stage the embryo consists of a mosaic of regions determined to form the principal organs and structures of the body. This body plan is built up as a result of a hierarchy of decisions, and several further decisions will, in most cases, be taken before the cells differentiate into the terminal cell types shown on the right side. It should be noted that some cell types, such as cartilage, arise from more than one lineage. Further details will be found in Chapters 4 and 6.

particular state of commitment, it is made among a small number of alternatives and the outcomes are not in general terminally differentiated cell types but new states of commitment whose potency is further restricted. Most embryologists have always shared the conviction that the basic body plan arises from such a hierarchy of decisions because it is very difficult to make any sense of the data unless we assume the existence of commitments for broad subdivisions of the body at early developmental stages. Unfortunately, the early states are largely covert and with the exception of the three 'germ layers', whose existence was deduced in the nineteenth century from descriptive embryology, there are no suitable names for the committed states. We are thus driven to refer either to positions ('anterior mesoderm', 'posterior mesoderm') or to the future organ ('limb field', 'eye field'), and this makes embryological terminology rather obscure and confusing to the outsider. Fortunately in *Drosophila* we now have a fairly detailed knowledge of the genes whose activity defines these early regions and so can name them after the gene in question: for example the *hunchback*, *Krüppel* and *knirps* domains (see Chapter 7). This makes much of the mystery evaporate and aids general acceptance of the concept of transient states of commitment.

Only two general processes are known which account for the events of regional subdivision: *cytoplasmic localization* and *induction*, and the examples which are discussed in this book are listed in Table 1.1. Cytoplasmic localization has been studied mainly in invertebrate embryos with a small total cell number (see Chapter 5). By means which are still not well understood, regulatory molecules become differentially distributed in the cytoplasm of a cell. Then, in the course of an asymmetrical cell division the two daughter cells inherit different materials and consequently enter different states of commitment. Induction has been studied mainly in vertebrate embryos. Here it appears that the competent tissue becomes differentially determined in response to the concentration of a chemical signal from another region of the embryo. The distinction between cytoplasmic localization and induction is one which has been made using classical techniques of experimental embryology and it is clear that both processes are widespread and probably occur in all types of embryo. Unfortunately the classical techniques are limited when it comes to further analysing the mechanisms of these processes. However, since the first edition of this book the combination of modern methods with classical understanding of the problem has finally enabled some cytoplasmic determinants and some inducing factors to be identified, particularly in *Drosophila* (Chapter 7) and *Xenopus* (Chapter 4).

In the 1970s there was a sharp polarization between molecular biologists and experimental embryologists. At that time, molecular biology had not yet contributed anything to our knowledge of regional specification and it was hoped by the experimental embryologists that their own theoretical models could be tested directly by grafting experiments. As it has turned out, models cannot be tested by grafting (for reasons see below) and in the late 1980s molecular biology has finally begun to produce the goods. The result is that there is now a much higher degree of consensus about how to proceed. It is agreed that molecular techniques have great power, but also that they have to be used to answer the right questions. The questions are precisely those posed over the years by experimental embryologists.

Table 1.1. (a) Cytoplasmic localization

Organism	Position	Stage of localization	Parts specified	Structures formed later
Xenopus	vegetal hemisphere	oogenesis	ventral mesoderm inducing zone	endoderm
	dorsovegetal region	post-fertilization	dorsal mesoderm inducing zone	endoderm
	extreme vegetal	oogenesis	germ cells	germ cells
Sea urchin	vegetal	oogenesis	vegetal centre	micromeres
Ilyanassa	polar lobes	oogenesis	D blastomere	somatoblast, mesentoblast
Leech	pole plasms	post-fertilization	teloblasts	most of ecto and mesoderm
Ascidian	posterovegetal	post-fertilization	posterior parts	muscle, mesenchyme
Caenorhabditis	posterior		asymmetric cleavage	germ cells
	posterior		E lineage	intestine
Mouse	outer surface	8 cell	polar cells	trophectoderm
Drosophila	anterior pole	oogenesis	anterior centre (*bcd* mRNA)	secretion *bcd* protein
	posterior pole	oogenesis	posterior centre (*nos* mRNA)	destruction maternal *hb*
	posterior pole	oogenesis	pole cells	germ cells
	termini	egg	terminal centres (*tor* activation)	formation of termini
	ventral side	egg	ventral midline (*Tl* activation)	nuclear localization of *dl*

(b) Induction

Organism	Signalling region	Responding region	Stage	Outcome
Xenopus	dorsovegetal quadrant	animal hemisphere	blastula	formation of organizer
	rest of vegetal hemisphere	animal hemisphere	blastula	formation of ventral mesoderm
	dorsoequatorial	marginal zone	gastrula	subdivision of mesoderm
	archenteron roof	ectoderm	gastrula	formation of neural plate
Sea urchin	vegetal	animal hemisphere	blastula	formation of oral arms
Ilyanassa	D macromere	animal micromeres	morula	formation of eyes, statocysts
Ascidian	A4.1 blastomeres or progeny	A4.2 blastomeres or progeny		formation of neural tube
Caenorhabditis	EMS cell or progeny	AB cell or progeny	4–28 cells	formation of pharynx
Mouse	inner cell mass	trophectoderm	blastocyst	formation of polar trophectoderm
Chick	hypoblast	epiblast	blastoderm	formation of primitive streak
	Hensen's node	mesoderm	primitive streak	subdivision of mesoderm
	Hensen's node	ectoderm	primitive streak	formation of neural plate
Drosophila	anterior centre	other non-pole nuclei	syncytial blastoderm	activation *hb* and other events
	en/wg regions of parasegment	wg/en regions	germ band elongation	mutual maintenance en and wg

2

The concepts of experimental embryology

As we shall see, the techniques of molecular biology and genetics are now essential in the investigation of regional specification. But, historically it was the experimental embryologists who gave most thought to mechanism and who formulated the basic conceptual framework which enables us to discuss the problems rationally. In this Chapter an attempt will be made to present a vocabulary, based on classical embryology but supplemented with more recent concepts, which can serve as the currency for the arguments in the remainder of the book. Although an attempt has been made to conform as much as possible to common usage, the exigencies of precise definition ensure that some readers will find words used in different senses from those which they favour.

NORMAL DEVELOPMENT

Normal development means the course of development which a typical embryo follows when it is free from experimental disturbance. Most species used for laboratory work have 'normal tables' which break down the course of development into a number of standard stages. The existence of these tables emphasizes the predictability of normal development and enables investigators in different parts of the world to standardize their procedures. Normal development must not be confused with pathways of development which give a normal outcome. For example, when the first two blastomeres of an embryo are separated and each develops into a complete animal the outcome may be normal but the pathway is not normal with respect either to the regions of the egg which become each part of the adult or to the absolute dimensions of parts before compensatory growth has occurred. *Descriptive embryology* necessarily deals with normal development. Although it is not very fashionable and sometimes not very interesting, it really is necessary to understand the normal development of a given organism before attempting to understand the experimental perturbations.

Maternal and zygotic

Features of development are referred to as *maternal* if they are due to components which exist in the egg, having been accumulated during

9

oogenesis. The usual context of this is the *maternal effect gene* where mutations are apparent by their effects on the embryo but the embryonic phenotype corresponds not to its own genotype but to that of the mother.

Features of development are said to be *zygotic* if they are due to components newly synthesized by the embryo itself after fertilization. We speak of 'onset of zygotic transcription' referring to the fact that for many embryo types the zygotic genome does not become active for some time after fertilization. In general, maternal features dominate the embryo during its earliest developmental stages and zygotic features thereafter.

The fate map

A *fate map* is a diagram which shows what will become of each region of the embryo in the course of normal development: where it will move, how it will change shape, and what structures it will turn into. The fate map will change from stage to stage because of morphogenetic movements and growth, and so a series of fate maps will depict the trajectory of each volume element from the fertilized egg to the adult. While it is clear that a series of fate maps must exist for the stages of each individual embryo, it is only worth knowing a fate map if it is the same for different individuals and therefore has predictive value. It is in this sense that we say a fate map 'exists' for a certain type of embryo. Two conditions have to be satisfied for this. First, the final anatomy must itself be constant between individuals. This is true for most animals, but not for creatures such as colonial hydroids or most plants. Secondly, there must be no random mixing of cells for the stages in question. If there is some mixing then a cell in a certain position in one embryo may become something quite different from the cell in the same position in the next embryo. In the complete absence of mixing, a fate map can, in principle, have unlimited precision. Not only will there exist a family tree showing the lineage of every cell in the adult body, but each volume element of the original egg cytoplasm will end up in predictable cells of the adult. This situation is approached by some invertebrates in which the adult cell number is relatively small, and we shall return to them in Chapter 5. For most embryos, however, there is reason to think that there is usually a little local mixing of similar cells and therefore the fate maps cannot be quite this precise. None the less the fate map is an absolutely fundamental concept in embryology and the interpretation of nearly all experiments concerned with early developmental decisions depends on knowledge of the fate map.

Although there are a number of fate maps in the literature based on histological observation alone, it is rarely possible to be confident about the results from this sort of data and so in addition some means has to be found of labelling specific regions of the embryo. In this book the term *label* will be used for marks applied to embryos by experimenters, including genetic labels when a heritable cellular property is utilized solely for this

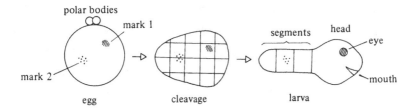

Fig. 2.1. The principle of fate mapping. In the absence of random cell mixing, marks placed on the egg will label particular regions of the larva, in this case the eye and the second abdominal segment.

purpose. Some authors also use the term *marker* but it should be remembered that this is also often used for substances characteristic of a particular cell type: a particular protein or antigenic determinant may be referred to as a marker of neural crest cells, or muscle cells or the anterior half of the neural plate. Cell labels are important not only for fate mapping but for any application in which cell origin needs to be monitored, for example grafts or tissue combinations *in vitro*. A labelling procedure used for fate mapping must be effective enough to ensure accurate identification of labelled cells at a later time, but must also be gentle enough not to disturb normal development. In principle the procedure is very simple: a patch of label applied at time t_1 can be located at time t_2 and we can deduce retrospectively that the t_1 patch was the prospective region for the t_2 patch (Fig. 2.1). It is essential to note that prospective regions are not necessarily *committed* to become the structures which they later form in normal development. An ill-founded belief that they are has caused endless confusion in the past and has sometimes, unfortunately, led to the term 'fate map' being used for maps of potency or states of determination; distinct concepts which will be discussed below.

If there is no mixing then the boundaries of prospective regions will abut one another as shown in Fig. 2.2(*a*). If there is a little mixing, the boundaries will overlap because they include all the cells that have some probability of being incorporated into the structure concerned. Because the map is a statistical average of many individual embryos, the prospective regions on the map are in this case larger than the actual prospective regions in particular individuals (Fig. 2.2(*b*)). In the extreme case where there is total randomization of position there is no fate map at all since any cell, or region of the egg, has a non-zero probability of ending up in any structure whatever (Fig. 2.2(*c*)).

A variety of experimental techniques has been used to construct fate maps and some of the results will be described in later Chapters. An ideal label is one which can be applied at will to any region, at any stage, which is readily visible at all later stages, and which does not perturb normal development. Needless to say no perfect label exists and all methods are to

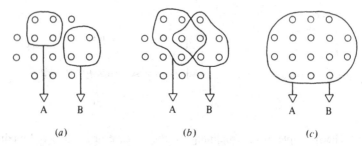

Fig. 2.2. Dependence of the 'primordial cell number' on the amount of mixing which occurs after labelling. (*a*) If there is no mixing at all, then the prospective regions are the same size as the later structures A and B. (*b*) If there is some mixing, then the prospective regions are larger than the later structures because they include all cells which have some probability of contributing to them. (*c*) If there is complete randomization of position then any cell can contribute to any structure. In this example it is assumed that there is no overall growth.

some extent compromises between these qualities. The following list indicates some of the more popular ones:

1. Some eggs possess visible regional differences in their cytoplasm which can be traced through to some of the daughter cells (eg, Conklin,1905*a*). These are useful but leave the rest of the embryo unlabelled.
2. 'Vital stains' are dyes which can be applied to parts of embryos without damaging them (eg, Vogt, 1929). Nile blue and neutral red have been extensively used in amphibian and chick embryology and many text-book fate maps depend on these data. However, they have a limited resolution because of their tendency to spread and fade.
3. Grafts of tissue labelled with tritiated thymidine can be useful if the signal is not diluted too much by cell division (eg, Nicolet, 1971). The labelled nuclei can be visualized in histological sections by autoradiography.
4. For non-growing embryos such as amphibians and most marine inverte-brates, high molecular weight tracers have become very popular in recent years (Stent & Weisblat, 1982). These can be injected into individual blastomeres and are retained in the clonal progeny because they are too large to diffuse to neighbouring cells through gap junctions. The enzyme horseradish peroxidase has often been used since early embryos frequently have no endogenous peroxidase. Also, various fluorescent conjugates of dextrans have found application (eg, Gimlich & Braun, 1985).
5. In mammalian and insect embryology, genetic labels have been popular since applied labels are rapidly diluted by growth. The labels used are usually mutations affecting pigmentation or non-essential enzymes (eg, Gardner, 1978; Janning, 1978). The main problem with them is that they

are rarely genuinely cell-autonomous; either being expressed only in some cell types, or expression being sensitive to a cell's environment, or both.

6. Cytological labels may involve ploidy, chromosomal peculiarities, nucleolar number, or, more recently, DNA sequences visualized by *in situ* hybridization (eg, Thomson & Solter, 1988). They are usually cell autonomous but may sometimes be difficult to visualize in sections.

The same types of label are used to identify cell populations in experiments in which the normal fates of cells are altered in grafts or *in vitro* tissue combinations. Here the requirements for normal behaviour of labelled cells are somewhat less stringent, so, for example, chick–quail combinations are often used because of the Feulgen positive heterochromatic mass associated with quail nucleoli (Le Douarin, 1982). These are less suitable for fate mapping because cells of the different species have somewhat different adhesivity, motility and rates of development.

Clonal analysis

Clonal analysis is a form of fate mapping in which a single cell is labelled and the position and cell types of its progeny identified at a later stage. The principles of fate mapping apply in so far as the labelled cell is judged to be the prospective region for all its progeny. The special interest of clonal analysis lies in the deductions which it enables us to make about determination. The full complexities of 'determination' will be discussed below but for the moment we can accept that a determined cell is one that is irreversibly committed by virtue of its intrinsic character to develop into part of a given structure. Now, since a single cell may be determined to form one of two structures A and B, or neither, but not both, it follows that, if a clone overlaps both structures, the precursor cell cannot have been determined at the time it was labelled. In other words *no clonal restriction means no determination*. The negative form of this principle is crucial because the converse is not true. Clonal restriction by no means implies determination, it simply means that the labelled cell is not in the prospective region of the structure to which it does not contribute (Fig. 2.3). The frequent failure to understand this is due entirely to a failure to understand exactly what is meant by a fate map, a difficulty which should not now be shared by readers of this book.

A particularly clear example of clonal analysis has been provided by the work of Kimmel's group on the zebrafish. Like many fish embryos this species does not have a fate map during the early stages because of extensive cell mixing, and this is shown by the extensive dispersal of clones labelled at the cleavage stages. If a clone is labelled by injection at the mid-blastula stage the progeny will form a mixture of cell types such as neurons, muscle

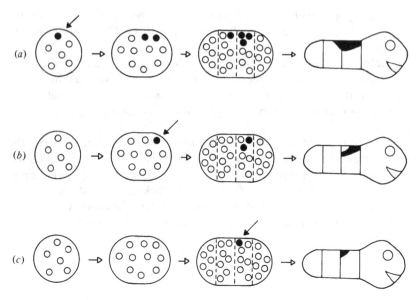

Fig. 2.3. No clonal restriction means no determination but the converse is not true. In (*a*) a cell is labelled in the early embryo and its progeny span the boundary between two segments in the adult. In (*b*) a cell is labelled prior to determination but, because of its position, its progeny do not span the segment boundary. In (*c*) a cell is labelled after determination and its progeny are also confined to one segment.

cells and epidermis. However, when an individual clone is followed it is found that after the onset of gastrulation the progeny of a single cell will only become a single cell type. This allows us to deduce that determination for these cell types cannot have occurred earlier than the beginning of gastrulation (Kimmel & Warga, 1986 and Fig. 2.4).

Other things being equal, clones arising from cells labelled early in development will end up larger than those labelled later because they undergo more cell division. This fact is sometimes used to estimate 'primordial cell numbers' on the grounds that if there is a total of n cells in the prospective region at the time of labelling, then the labelled clone will subsequently occupy about $1/n$ of the structure. This type of calculation assumes that there is no differential growth, either of different parts of the structure or arising from the labelling itself. The procedure is valid if clones are induced *after determination* since the determined rudiment will possess a countable number of cells at each stage, but if they are induced before determination there is a difficulty. When the late embryo is examined, the clone will either be confined to the structure in question or it may overlap other structures. If it is confined then its size will underestimate the size of the prospective region on the fate map since some of the other cells of the prospective region will have contributed some progeny to different struc-

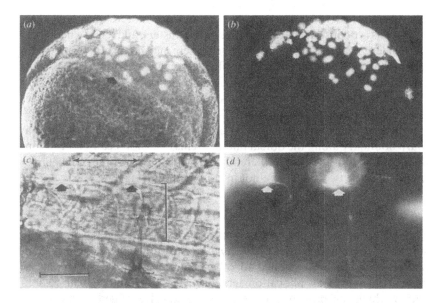

Fig. 2.4. Clonal analysis in the zebrafish, phase/fluorescence and fluorescence views. (*a*), (*b*) a clone at the gastrula stage stemming from a blastomere injected with a fluorescent lineage label at the 64-cell stage. (*c*), (*d*) labelled differentiated motorneurons at a later stage. (Photos kindly provided by Dr C. Kimmel.)

tures. If the clone is not confined then the proportion of the structure labelled will lead to an overestimate of the size of the prospective region since some of the labelled cell's own progeny have gone elsewhere. Estimates of primordial cell numbers therefore need to be treated with some care.

Allocation

What, then is the significance of clonal restriction without determination? From the individual cell's point of view, there is none since its intrinsic character has not changed. But in embryology we often need categories of thought appropriate to a larger scale than that of the single cell, and one such is the term *allocation*, which refers to clonal restriction within a cell population regardless of the state of commitment of the cells. The term is used mainly by mammalian embryologists, although there is no reason why it should not also be applied elsewhere. On some occasions allocation may be a complete artefact based simply on the fact that cells are dividing and the chance of a late, small, clone spanning a boundary is less than that of an early, large, clone. However, it is probably more usually due to some change in the movements or spatial relationships of cells. For example, if cells are condensing into clusters then the chance of a single clone spanning adjacent clusters will progressively reduce, since more clones will arise where cells are densest, near the centres of the clusters, and fewer where they are sparsest, near the edges. Alternatively, there may be absolute barriers to cell mixing at an early stage, such as physical separation or the presence of extracellular structures. If these barriers are removed later on, after cell movement has ceased, then sharp boundaries will be seen between the labelled clone and its surroundings. For example, clones in the vertebrate neural tube which were labelled before neural tube closure often give a sharp boundary along the dorsal midline. This does not mean that cells on the right and left sides were differently determined at the time of labelling, simply that the two sides were not in contact at that stage.

Compartments

The concept of the developmental *compartment* arose from clonal labelling studies on the imaginal discs of *Drosophila* (Garcia Bellido *et al.*, 1979) and has on occasion been applied to other situations as well. Originally a compartment was an allocation territory, that is a multicellular region within which the progeny of every cell remained confined. Particularly intriguing was the existence of a compartment boundary separating the anterior and posterior part of each imaginal disc, because this corresponded neither to any obvious barrier to cell mixing, nor to any difference in terminal differentiation. It did, however, correspond to the domain of action of the homeotic gene *engrailed*, normally on in the posterior and off in the anterior (for more about homeotic genes, see below). Furthermore, posterior clones defective in *engrailed* expression could cross the boundary. Because of this the lineage restriction found in compartments generally was thought to arise from analagous differences in the expression of homeotic genes. Since these genes encode the results of early developmental decisions this makes the compartment the same as the determined region of older parlance. It is still too early to know whether the term 'compartment' is going to establish itself in non-*Drosophila* embryology, but those who encounter it need to be careful whether it is being used in the original sense of an allocation territory, or the later sense of a determined region. In the former case the clonal restriction could depend on some barrier to mixing of otherwise similar cells, in the latter case mixing is inhibited as a consequence of the different states of determination.

Growth and stem cells

In early embryos the cell divisions are usually cleavage divisions which repeatedly reduce the size of the cells, often called *blastomeres* at this stage. True growth usually starts at a later stage and is dependent on the existence of an external food supply from a yolk mass or placenta.

Stem cells, like compartments, mean different things to different people.

In the biology of adult tissues the term 'stem cell' tends to be used for the cells with indefinite or very long self-renewal capability which maintain tissues such as the blood, skin or intestinal epithelium. The ability to do this implies that the population of stem cells can both reproduce itself and generate the differentiated cell types of the tissue concerned, at the correct rate and in the correct proportions. It is often thought that the stem cells must regularly undergo asymmetric divisions, where one daughter remains a stem cell while the other is destined to differentiate and eventually to die. However, it is still not really known to what extent determination is linked in this way to cell division rather than being a probabalistic event or being controlled by microenvironmental differences.

In certain invertebrate embryos in which developmental decisions are taken at a small cell number (Chapter 5) many cells undergo visible asymmetrical divisions, probably because they develop a cytoplasmic asymmetry prior to their division such that regulatory molecules inherited by the two daughter cells evoke different patterns of gene activity in their nuclei. Sometimes, as in the leech, such cells can bud off a long sequence of progeny by repeated asymmetrical division. These have also been called stem cells. They share with the adult tissue case the twin properties of maintenance of the stem cell with generation of differentiated progeny, although here the process is rigid, deterministic and often organized in space, and the period of maintenance and self renewal is limited.

So long as this difference is understood, it would seem reasonable to maintain the definition: *stem cells are cells which, individually or as a population, can both produce differentiated progeny and reproduce themselves.* Although inner cell mass or primitive ectoderm cells of the early mouse embryo are often loosely referred to as stem cells, they do not obviously conform to this definition both because they do not maintain themselves in the same state for very long, and also because the nature of their differentiated progeny changes with developmental time. In such cases there is no evidence of any intrinsic capacity for asymmetric division of the cells concerned and therefore there seems no need to call them 'stem cells' rather than 'cells'.

In fate map terms a stem cell is the prospective region for all of its progeny so any regional organization which is produced in the program of cell divisions from a stem cell will not be represented on the fate map of earlier stages. For this reason the use of clonal analysis can enable us to exclude a mechanism whereby a structure is produced by the division of a single stem cell. If labelled clones are induced at random at a stage before the stem cell starts dividing then the cell must either belong to a clone or not, and the resulting structure will accordingly either be completely marked or completely unmarked. If the structure is partially marked then it must be formed from at least two of the cells present in the embryo at the time of labelling.

FORMS OF DEVELOPMENTAL COMMITMENT

We have seen that a fate map conveys no information about the intrinsic character or the commitment of the prospective regions, and that clonal analysis can do so only in the negative sense that no clonal restriction means no determination. Particularly in *Drosophila* it has recently begun to be possible to infer the commitment of parts of the embryo by direct examination of which regulatory genes are being expressed. But this is true only to a very limited degree for other animal types and because of this our definitions remain operational ones based on particular types of experiment. In this book, three types of commitment will be distinguished: specification, determination and potency. Other authors may use different words but will usually have the same three ideas in their minds.

Specification

A cell or tissue explant is *specified* to become a particular structure if it will develop autonomously into that structure after isolation from the embryo. The medium into which the explant is isolated must be 'neutral' with respect to the developmental pathway followed. The nearest approach to this situation is found in amphibian embryos where tissue explants will develop and differentiate *in vitro* in buffered salt solutions, and in a number of invertebrate embryos where it is possible to obtain quite advanced development of parts of the body from isolated blastomeres. The specification of a region need not be the same as its fate in normal development. For example, the prospective neural plate of an amphibian blastula will differentiate not into a neuroepithelium but into epidermis when cultured in isolation. In order to form neuroepithelium it needs to receive an inductive signal from the mesoderm (see Chapter 4). In principle it is possible to construct a 'specification map' of the embryo by explanting tissue from each region and combining the results. As we shall see later, if the specification map is the same as the fate map, we refer to the region concerned as a 'mosaic'. If the specification map differs from the fate map, we refer to the system as 'regulative'.

Determination

A *determined* region of tissue will also develop autonomously in isolation but differs in that its commitment is *irreversible* with respect to the range of environments present in the embryo. In other words it will continue to develop autonomously after grafting to any other region of the embryo. A very large number of embryological experiments consist of grafting a piece of tissue from one place to another and asking whether it develops in accordance with its new position or its old position. In the former case it is not determined, although it may have been specified, and in the latter case it is determined. A series of such grafts performed at different stages usually show a time at which the tissue becomes determined. For example, the prospective neural plate of an amphibian embryo becomes determined to form neural plate during gastrulation. After this stage, grafts of this tissue to other regions of the embryo will always form neuroepithelium. At the earlier blastula stage it is not determined. It will form neural plate if left *in situ*, but if grafted to a ventral position it will form epidermis or if grafted to a vegetal position it will form mesodermal structures. In the older German literature which some readers of this book may shortly feel impelled to consult, development according to new position was referred to as *ortsgemäss*, and development according to old position as *herkunftsgemäss*. These elegant terms were also used by some English and American authors but have unfortunately now been replaced by the more confusing 'lineage versus position' dichotomy which is discussed below.

Potency

The *potency* of a region of tissue is the total of all the things into which it can develop if put in the appropriate environment. This is not easy to measure experimentally because one can never be sure that all possible environments have been tested. In practice, it usually refers to experiments in which groups of cells are inserted into embryos in such a way that they become well mixed and exposed to the full range of environments present in the embryo at the stage in question. So if the tissue is already determined then its potency is by definition restricted to only one outcome. Mammalian embryologists, who specialize in such experiments, often refer to the 'restriction of potency' rather than to determination. If the tissue can form more than one thing it is called 'pluripotent' and if it can form everything it is called 'totipotent'. It is important to distinguish this sort of totipotency in which cells can contribute to all tissues following implantation into an embryo, from the behaviour shown by isolated early blastomeres which can often generate a complete embryo on their own. This is because the isolated blastomere is not only totipotent but also contains necessary regional inhomogeneities and can produce for itself the appropriate range of environments which its progeny cells require for their regional subdivision.

The hierarchy of commitment

Specification, determination and potency can be defined only for the particular level of the hierarchy of decisions corresponding to the developmental stage in question and this can cause difficulty when the observables are the product of subsequent decisions. For example, the first decision in Fig. 1.1 is between the three germ layers, but we can only know that a region of tissue is committed to form mesoderm if it later produces characteristic mesodermal cell types such as muscle, mesenchyme, kidney, etc. Which of these are produced and how they are arranged will depend on the ability of the graft or explant to generate internal regional subdivisions, as well as on its overall state of commitment. These later decisions may of course be altered independently of the state of commitment, and so, for example, the experimenter can be left unsure of whether his tissue was originally committed to form mesoderm, or somite, or muscle, each of these states being a different level in the hierarchy of decisions. The way out of this difficulty is to find molecular markers characteristic of the states in question rather than of their subsequent derivatives, and as we shall see, this has so far only really been done for *Drosophila*.

The above definitions have been given in terms of regions of tissue. They can also be applied to single cells or even to nuclei. In fact, the states of commitment of nuclei, single cells, and tissues are often not the same. Single cells may appear less committed than the tissues from which they come simply because the environment in the centre of a graft is mainly the product of the graft cells themselves, and it may indeed be the case that some states of commitment require some particular three-dimensional arrangement of cells, or particular extracellular materials, and cannot be maintained at the single cell level.

When a *cell* is said to be determined this carries with it the implication that the state is clonally inherited, since it must be maintained by each progeny cell independently of the environment. A cell can thus acquire a particular state of determination in two distinct ways: either by inheritance from the parent cell or by responding to the appropriate external signals.

Lineage

This alternative is often formulated as 'lineage versus position', which is a single cell version of the old dichotomy between *ortsgemäss* and *herkunfts-gemäss* behaviour of grafts. Unfortunately the term *lineage* has acquired no less than five different meanings and so it is often not at all clear what authors are thinking of in a particular case. These may be listed as follows:

1. Descriptive embryology shows that some animal species, such as molluscs or nematodes, have a pattern of cleavages which is invariant or nearly so between individuals. Other groups, such as vertebrates or insects do not. In the former case it is possible to construct a diagram showing the family tree of every cell in the body, often called a *cell lineage*. This is a necessary but not a sufficient condition for regional specification by means of localized cytoplasmic determinants.
2. By definition a determined cell will pass its state of commitment to its progeny. This is often described as acquisition of the commitment in

question by *lineage*, but says nothing about how the determination arose in the first place.

3. There are several well-attested cases in which cells become committed by inheritance of cytoplasmic determinants. As noted under 1. this can only happen if the pattern of cleavages which parcels up the cytoplasm is invariant between individuals. It is sometimes called a *lineage mechanism* for regional specification.

4. Another possible *lineage mechanism,* which is actually a quite different sort of mechanism and for which there are not yet any proven cases, would depend on some chromosome modification such as DNA methylation with a subsequent directed segregation of the modified homologue in a stem cell division. The difference between this and 2. is that it brings about the *de novo* acquisition of the determination and not merely its subsequent inheritance on cell division.

5. Particularly in the case of the nematode *Caenorhabditis* a description of the cleavage types and directions of a particular cell are often the only way of knowing what sort of cell it is. So here the *lineage* is being used as an observable character or cell marker, although the legitimacy of this is questionable since the lineage may still be affected by unknown environmental effects from neighbouring cells.

Given the present confusion it is very doubtful whether agreement will ever be reached on a single meaning for 'lineage'. In this book it will only be used in the expression 'cell lineage' meaning a diagram of the genealogy of cells in an embryo (1. above). Note that a cell lineage in this sense can only be described for organisms that have an invariant cleavage pattern and not for those in which variable cleavages or random cell movements occur. Also it should be remembered that a cell lineage contains less information than a fate map at single cell resolution because the cell lineage gives only the genealogy while the fate map gives both the genealogy and the spatial relationship between parts at different stages.

Regulation and mosaicism

Regulation and *mosaicism* are old embryological terms and are defined here in such a way that they relate most directly to the presumed mechanisms of regional specification. So mosaic behaviour following experimental interference implies that cells do what they do in normal development while regulative behaviour implies that they do something different. For this reason, regulation and mosaicism should only be defined for situations in which the fate map is known. It should be emphasized that both types of behaviour can occur in a given type of embryo, an early regulative phase usually preceding a later mosaic phase.

Mosaicism is fairly straightforward. An embryo, or part of an embryo, is mosaic when the map of specified regions coincides exactly with the fate map. So if it is cut into pieces, each part develops in the same way as it would in normal development. There is usually a mosaic stage which immediately precedes overt differentiation, and in such cases the significance of this mosaicism is no greater than that of the final anatomy since differentiation may already be under way but not yet visible down the microscope. In fact, immunohistochemistry and *in situ* hybridization now often make it possible to detect the onset of differentiation much earlier than light microscopy. Mosaicism is more interesting when it is detected in very early embryos, as will be discussed in Chapter 5, because this is clearly long before the onset of terminal differentiation. This is often regarded as evidence for the existence in the egg of cytoplasmic determinants which control the specification of parts although it is, in fact, only consistent with this possibility and does not by any means prove it.

In general, *regulation* can be defined as the complement of mosaicism, so that an embryo or organ rudiment is regulative if the state of specification of its isolated parts does not correspond to the fate map. This idea is, however, too vague for practical purposes and it is necessary for us to consider several types of regulative behaviour since each has a somewhat different implications for the nature of the underlying mechanisms. These are twinning, fusion, defect regulation and inductive reprogramming.

Twinning means the production of two complete animals from an embryo divided into two parts. As we shall see, this is quite a common though not universal property of embryos at the two-cell stage. This is true, for example, of the mouse, the frog, the sea urchin and the starfish. Complete twinning from more advanced stages containing thousands of cells can occur in certain short germ insects, the chick and, probably, humans. In all these cases the principal body axes are half normal size at the time of formation. In embryos capable of growth the size of each twin approaches the normal during foetal or larval life. The *proportions* of parts within the miniature primary axes are, however, normal at all times. The phenomenon of twinning reveals several features of normal development which are somewhat obscured by the simple designation of the embryos as 'regulative'. Firstly, it excludes any model for regional specification which is based solely on the localization of determinants in the egg cytoplasm. Any such determinants have to be shared out in such a way that they occupy the same relative positions in the two fragments and from simple geometrical considerations this means that they must be symmetrically disposed with respect to the plane of division (Fig. 2.5). It therefore suggests that most of the regional subdivision arises after fertilization as a result of interactions between different parts of the embryo. Secondly, it shows that these interactions are able to accommodate a change of scale in the pattern. Boundaries which would be formed 100 microns apart in the normal embryo will be formed 100/cube root of 2 = 79 microns apart in the twins.

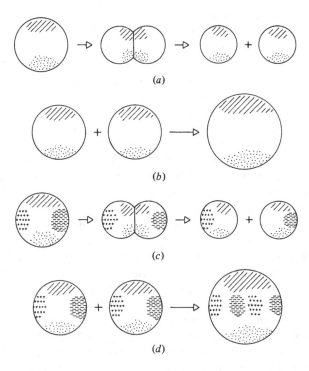

Fig. 2.5. Embryonic regulation and cytoplasmic determinants. (*a*) and (*c*) show blastomere isolation experiments and (*b*) and (*d*) show the fusion of two eggs to form a giant embryo. In either case, individuals with a normally proportioned anatomy can only arise if the determinants are symmetrically disposed along the plane of separation or joining. In (*a*) and (*b*) this is the case and in (*c*) and (*d*) it is not.

This is quite an important constraint on possible signalling mechanisms, and we shall return to it in Chapter 3. Thirdly, it shows that the final size of each structure does not depend on the size of the original primordium but on some mechanism which can stop growth when a certain absolute size has been reached. This is a problem of growth control in late development and adult life and is therefore outside the scope of this book, although it is no less significant for that.

The property of forming normally proportioned half-size patterns is also possessed by some organ rudiments, for example this is true of the eye rudiment in amphibian neurulae and the limb rudiment in amphibian tailbud stages. The implications for the mechanism of regional specification are the same as in the whole embryo except that for an organ rudiment there is no need to assume any subdivision at all at early stages since the signals which eventually bring it about may be emitted by the surrounding tissues.

The converse of twinning is *fusion* of two or more embryos to give one giant embryo. This has similar implications for mechanism because any cytoplasmic determinants have to be symmetrically disposed with regard to the fusion plane (Fig. 2.5). In the fused embryo the parts of the body plan exceed normal size at the time of formation and growth is relatively slowed during late development to bring them back to normal size. This phenomenon has been extensively investigated in mouse embryos where productions of tetraparental mice by fusion of morulae has become a routine procedure. It is further discussed in Chapter 6.

Many embryo types are not capable of producing twins but there are few if any which are not capable of some degree of *defect regulation*. This means that there is at least some region of the embryo which can be removed at an early stage without disturbing the pattern or the proportions of the body plan. Defect regulation tells us about the regions of the embryo which must be present in order to obtain a complete pattern. For example in the sea urchin morula, quite a lot of tissue can be removed from the equatorial region without resulting in the production of defects while only a little can be removed from the terminal regions. In the amphibian blastula ventral tissue can be removed but dorsal tissue cannot. It is natural to identify the regions whose presence is obligatory with the signalling centres or organizers of the early embryo.

A fourth type of phenomenon which should be called regulation, but usually is not, is *inductive reprogramming*. If a signalling centre is grafted to an abnormal position it can cause the surrounding tissue to follow a pathway of development which does not correspond to the fate map. This is often not regarded as regulation because the outcome is not a normal embryo, but at the cellular level the processes are probably the same as those following fragmentation or fusion.

Regulation has been discussed here as though it was always perfect. This is not, in fact, the case; more often than not it is probably imperfect. For example, twins may sometimes be lopsided, the side of origin being bigger than the regulated side. There are few careful studies of proportions in regulated embryos and those that there are often reveal imperfections in embryos which are superficially normal (Cooke & Webber, 1985). However, this is not of great significance. So long as there is some adjustment of the fate map in response to the operation the conclusions reached here hold good. We do not necessarily know whether a failure to regulate completely is an intrinsic inability of the mechanisms to do so or simply because it does not have enough time before the secondary embryonic interactions commence.

Finally, it is important not to confuse embryonic regulation as discussed here with regeneration in adult or larval animals. Regeneration involves the re-establishment of regional differentiations or determinations in newly formed replacement parts while regulation involves the re-establishment of a fate map on a partial domain of uncommitted tissue. Regeneration phenomena can be classified into those involving long-range interactions with extensive cell movements (morphallaxis) and short-range interactions with extensive growth (epimorphosis) (Morgan, 1901). Embryonic regulation by contrast involves no cell movement and no growth, but simply a re-adjustment of the fate map on a domain which is not yet determined.

ACQUISITION OF COMMITMENT

The embryonic field

The term *field* has two clearly distinguishable usages in embryology which we might term the physical and the agricultural forms. In physics a field is a variable, scalar or vector, which varies in space. By analogy with this we could refer to an embryonic field and mean the equations which define the values in different parts of the embryo of the key biochemical variables which control regional specification. We shall come across some hypothetical equations of this type in later Chapters. Usually, however, the agricultural analogy is what people have in mind when they refer to a field; in other words it is the area of tissue within which a certain process, such as an inductive interaction, occurs. So, for example, the 'limb field' is that part of the mesodermal mantle in which interactions occur around the time of gastrulation which result in the formation of a determined limb rudiment. It follows that it is not proper to refer to a limb field at a stage before the relevant interactions commence. There is no limb field in the fertilized egg, simply a prospective region for the limbs. The agricultural usage will be adhered to in this book.

Largely because of the confusion of meanings the term 'field' has been thought to have mystical or vitalistic connotations and has not been very popular in recent years. Modern reductionist biologists have introduced two other terms to denote an area of tissue within which a process occurs. The first is *polyclone*, which emerged from clonal labelling studies in *Drosophila*, and refers to the group of cells making up a developmental compartment (see above and Crick & Lawrence, 1975). The other is the *equivalence group*, introduced to describe a group of cells with a common competence in the nematode *Caenorhabditis* (Kimble, 1981). These are both an improvement on 'field' but have not yet caught on outside their respective organism-based communities.

Polarity and axes

Like so many of our concepts 'polarity' is commonly used in two senses when applied to tissues. The most common meaning and that which will be adhered to below, is regional difference in state of commitment. So when we refer to 'onset of dorsoventral polarity' we mean the establishment of two or more distinct states of commitment along the line from the dorsal to the ventral extremum of the embryo. But polarity is often applied to single cells to indicate that they are visibly different at the two ends, for example the basal and apical surfaces of epithelial cells. It is therefore possible to describe a tissue as polarized if it is composed of polar cells with a preferred orientation, even though there is no regional heterogeneity.

According to Stent (1984) an *axis* is 'an actual or virtual linear component of a structure about which rotation is of some practical significance'. Thus it is correct to speak of the line joining the animal and vegetal poles of an egg as the 'egg axis' since in most cases unfertilized eggs are rotationally symmetrical about this line. It is not, however, correct to speak of 'determination of the dorsoventral axis' when what is meant is the acquisition of a series of states of commitment along the line in question. In this work the terms 'axis' and 'plane' will be used for morphological description alone and should not be taken to be intrinsic components of the embryo.

Cytoplasmic determinants

A *cytoplasmic determinant* is a substance or substances, located in part of an egg or blastomere, which guarantees the assumption of a particular state of commitment by the cells which inherit it during cleavage. In the past the existence of cytoplasmic determinants has often been deduced from quite inadequate evidence such as an invariant cell lineage, the occurrence of mosaic behaviour, or the presence in the egg of visible regional differences. All of these things are consistent with the presence of determinants but do not by any means prove it. The only really satisfactory proof is to transfer cytoplasm from one place to another by microinjection and show that cells inheriting the ectopic cytoplasm become structures normally formed by the egg region from which the cytoplasm came. This has now been done in several cases of which the most spectacular are the anterior and posterior systems of the *Drosophila* egg (see Chapter 7).

Although cytoplasmic determinants certainly exist, it is still unclear in many cases just what they represent at the biochemical level. At one extreme they might be regulatory molecules which are responsible for the derepression of specific genes at a certain time in development. This is the position taken by Davidson in his book *Gene Activity in Early Development* and is borne out by the anterior system of *Drosophila* in which the determinant is localized mRNA from the *bicoid* gene. Alternatively they may be metabolic states, such as the activation of the *torso* product in the termini of the *Drosophila* egg. They might even be a quantitative bias in the overall metabolism or cytoarchitecture which is only the beginning of a complex chain of causation leading some time later to differential gene expression, something which seems possible for the dorsoventral polarity of the amphibian egg. Even where determinants are of a specific character it does not seem likely that they would consist of mRNA coding for terminal differentiation products. If this were the case it would involve an extraordinary degree of 'molecular preformationism' in the egg and would not be compatible with any degree of regulative behaviour. It seems more probable that determinants are responsible for the establishment of the first two or three distinctly specified regions and that subsequent regionalization depends on interactions between these. Again, this view has been borne out by studies in *Drosophila* where all the determinants have early functions (see Table 1.1). It should be noted that the significant cytoplasmic differences which we call determinants are not only established during oogenesis, but also sometimes after fertilization or within early blastomeres after one or more cleavages.

Induction

The existence of regulation in a given case implies that the mechanism of regional specification must involve intercellular interactions. Usually regulation will only occur if certain regions are present, and these often turn out to be the signalling centres or organizers. The signals which control the regionalization of the early embryo are one type of *inductive signal*, and the ability to respond to inductive signals generally is called *competence*.

To give a concrete example of this: there is excellent evidence that in the early amphibian embryo the mesoderm is formed from the animal cap tissue in response to one or more signals from the vegetal region (see Chapter 4). The remainder of the animal cap becomes ectoderm, as does the whole animal cap in isolation. This interaction can occur between small pieces of the blastula cultured in combination and so by using pieces taken from embryos of different stages it has been possible to show that the interaction occurs during the blastula stages. This type of interaction is called an *instructive* (or *directive*) induction (Saxen *et al.*, 1976) because the responding tissue has a choice before it (mesoderm or ectoderm?) and in normal development the interaction results in an increase in complexity of the embryo. The range of choices open to the competent tissue is a property of its state of determination. In this case once the mesoderm is formed it is competent to become somite, kidney, mesenchyme or blood cells in response to further interactions but it is no longer competent to become ectodermal derivitives such as neuroepithelium and epidermis. The *competence* of the responding tissue is a subset of its *potency*, since it comprises all the outcomes achievable by a region of tissue in response to environments present in the embryo at the stage in question.

Morphogen gradient, appositional and permissive inductions

There are two different types of instructive induction which have somewhat different consequences in terms of regional specification (Fig. 2.6). It may be that the signalling centre lies at one end of a cell sheet and is the source of

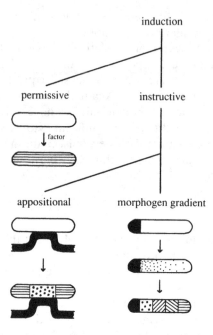

Fig. 2.6. Classification of inductive interactions into permissive; instructive appositional; and instructive morphogen gradient.

a concentration gradient of the signal substance. The competence of the surrounding tissue embodies n different *threshold responses* at different concentrations and hence $n + 1$ territories are formed in response to the gradient. It has become common usage to refer to the signal substance in a case of this sort, where there is more than one positive outcome, as a *morphogen*, and the properties of morphogen gradients will be discussed at greater length in the next Chapter. The morphogen gradient is roughly equivalent to the *positional signal* in the positional information theory of Wolpert (1969). Something like this happens in the *Drosophila* embryo where the protein products of the *bicoid* and *dorsal* maternal effect genes both form gradients.

The other possibility is that the signalling centres lie in one cell sheet and the responding cells in another. When they are brought together the appropriate structures are induced as a result of a single threshold response in those parts of the responding tissue immediately adjacent to the signalling centres. This probably happens, for example, in the induction of nasal, lens and otic placodes from the head epidermis of vertebrate neurulae under the influence of the underlying tissues. We shall refer to this sort of process as *appositional induction*. Typically only one threshold response would be made by the responding tissue and the inducing factor for this reason would not be called a morphogen.

There is a further kind of inductive interaction which is called *permissive*. Here the signal is necessary for the successful self-differentiation of the responding tissue but cannot influence the developmental pathway selected (Fig. 2.6). Permissive interactions are certainly very important in late development, for example, in the formation of the kidney or the pancreas. It is quite possible that permissive inductions occur also in early development, but the interactions which make up much of the subject matter of this book are, as far as is known, instructive in character.

There has been a lot of confusion about instructive and permissive inductions but the essential difference is that instructive inductions lead to a subdivision of the competent tissue while permissive inductions do not. Of course the degree of 'instruction' is limited by the range of choices embodied in the competence of the responding tissue. This is clearly illustrated by experiments involving inductions between tissues from different species where the rule is that the structures formed by the responding tissue are always those of its own species rather than that of the inducer. For example, mouse dermis induces hairs from mouse epidermis but feathers from chick epidermis and scales from lizard epidermis (Sengel, 1976).

Nature of inductive signals

It was suspected for many years that inductive signals were chemical substances but only recently has real progress been made in their identification. There is an obvious problem that most substances are not capable of penetrating plasma membranes and so would have difficulty diffusing through a sheet of cells. Four strategies have been proposed to overcome this, and probably all of them are found in some biological system or another. The first is to use a factor binding to a cell surface receptor which itself provokes an intracellular effect. This appears to be what happens in mesoderm induction in amphibians, where the cytokines, activin and fibroblast growth factor, are implicated (see Chapter 4). The second is to use a lipid soluble molecule which can penetrate the plasma membrane and then bind to a specific intracellular receptor. Retinoic acid, which can enter and leave intact cells with ease, is probably the signal in the vertebrate limb which controls the anteroposterior pattern (briefly reviewed Slack, 1987*a*). The third strategy is to do the signalling during a multinucleate syncytial phase, before the embryo becomes divided up into cells. This happens during the early development of *Drosophila* in which cellularization does not occur until after 13 nuclear divisions and many interactions occur within the syncytial blastoderm (see Chapter 7). The last possibility is to use a low molecular weight substance which is capable of passing through the 'gap' junctions which are widespread in early embryos as well as in adult tissue. There is no decisive evidence in favour of this, although injection of anti-gap junction antibody into *Xenopus* embryos can produce some defects which have been interpreted as indicating an interference with neural induction (Warner, Guthrie & Gilula, 1984).

HOMEOSIS AND THE EPIGENETIC CODING

A *homeotic* gene is one which, when mutated, causes conversion of one part of the body into another. The word was originally introduced by Bateson (1894) and was then restricted to transformations between members of homologous series of structures such as vertebrae or teeth. However, modern usage has tended to ignore this restriction. It is the state of activity of a group of homeotic genes which encodes the developmental commitment of a cell. We believe homeotic genes to be the elements which respond to cytoplasmic determinants or to inductive signals, and the best practical evidence for this at present are the homeotic genes in *Drosophila* whose activity is regulated by early effectors such as the *bicoid*, *torso* and *dorsal* products. Homeotic genes have also been called *selector genes* by Garcia Bellido (Garcia Bellido *et al.*, 1979) because it is their activity which selects a particular developmental pathway for the cell.

Homeotic genes work by regulating the activity of other genes and it is important to note that as much information is encoded by the 'off' state as by the 'on' state. In molecular language the absence of a repressor can be equivalent to the presence of an activator. It is easiest to think about genetic regulation if there are, in fact, just the two functional states 'on' and 'off', although in practice intermediate levels of activity may be important as well. If there are just the two states then they can be represented by binary digits where '1' means 'on' and '0' means 'off'. Suppose that a morphogen gradient turns on three homeotic genes at different concentrations. This will subdivide the field into four territories with the codings 111, 011, 001 and 000. Now suppose that the organism is a mutant and that the second gene cannot be turned on. In this case the codings will be 101, 001, 001 and 000. In other words the second territory has been turned into another copy of the third territory. What happens to the first territory cannot be predicted without knowing more about the logical circuitry since this now has a coding not found in the normal organism. This example illustrates the behaviour of a *loss of function* homeotic mutation. Such mutations are genetically *recessive* since function would be restored by a good copy on either chromosome.

On the other hand, suppose that the mutation in gene 2 caused constitutive activity. Now the sequence of codings would become 111, 011, 011, 010. Here territory three has become a second copy of territory two and territory four has an abnormal coding. This is a *gain of function* mutation. It is genetically *dominant* since inappropriate activation will occur with only one copy of the mutated gene.

The codings introduced here are much the same as the hypothetical *positional values* in Wolpert's theory of positional information (Wolpert, 1969). If two cells or tissue regions have the same coding, eg, 011, then we call them *equivalent*. The translation from early coding to final differentiation may be quite complex and so more than one coding may lead to the same tissue or structure. In such a case the early rudiments are said to be *non-equivalent* (Lewis & Wolpert, 1976).

The investigation of homeotic genes has been the key to the massive explosion of knowledge about regional specification in *Drosophila*. However, it is generally believed that homeotic genes must exist in all animals. The existence of homeotic genes is not only revealed by mutation but also sometimes during the regeneration of missing parts in postnatal life. This is particularly marked in crayfish where amputated appendages often regenerate as appendages normally found elsewhere on the body (Needham, 1965). Also, the best evidence for homeosis in vertebrates comes from the metaplastic transformations between different epithelia in Man which sometimes occur during tissue regeneration (Slack, 1985).

By the end of embryonic development, any particular cell will have taken several developmental decisions in response to cytoplasmic determinants or inductive signals or both. Each decision will have been recorded as a particular combination of states of activity of the selector genes. The whole set of combinations of states is the *epigenetic coding* of the cell, also sometimes known as the *epigenetic address*. It is not known whether the coding is stable enough to persist into postnatal and adult life or whether it is a transient feature confined to embryogenesis. In the case of animals capable of a high degree of regeneration, such as those polychaete worms which can regenerate both heads and tails from an extended region of the body (Slack, 1980), persistence into adult life would seem probable. On the other hand, many of the homeotic genes in *Drosophila* pass through a number of different spatial patterns of activity suggesting that at least part of the coding can be dispensed with at later stages.

GLOSSARY

Many terms have been discussed in this Chapter and a number of sources of ambiguity and confusion have been exposed. While no embryologist is likely to agree with another on the best set of terms, there is a high level of agreement on the actual ideas behind them and the glossary that follows is a brief recapitulation of the definitions as used in this book.

Allocation: Clonal restriction of a cell population irrespective of commitment.
Appositional induction: Instructive induction between cell sheets such that the induced pattern corresponds to the disposition of the signal.
Axis: Line in relation to which the embryo displays some morphological symmetry. Used for morphological description.
Cell differentiation: *See* differentiation.

Cell lineage: 'Family tree' showing the ancestry of every cell in the embryo or some region of the embryo.

Cell state: List of chemical concentrations and/or gene activities sufficient to define the intrinsic character of a cell. May be transient or stable.

Clonal analysis: Examination of the positions and nature of labelled progeny after the labelling of a single embryonic cell.

Clonal restriction: Confinement of a labelled clone within a particular embryo region or cell type.

Coding: *See* 'epigenetic coding'.

Commitment: Aspect of the intrinsic character of a cell or tissue region which causes it to follow a particular pathway of development or fate.

Compartment: (Originally) an allocation territory; (later) an allocation territory due to domain of activity of a homeotic gene.

Competence: The total of all pathways of development of a cell or tissue region which can be achieved by exposure to environments present within the embryo.

Cytoplasmic determinant: A substance located in part of an egg or blastomere which guarantees the assumption of a particular state of commitment by the cells that inherit it during cleavage.

Descriptive embryology: Account of development based on microscopic observation alone.

Determinant: *See* cytoplasmic determinant.

Determination: Irreversible commitment of a cell or tissue region. Heritable on cell division.

Developmental pathway: Same as 'fate'.

Differentiation: The synthesis by a cell of species of protein different to those made at an earlier developmental stage, or different to those made by surrounding cells at the same stage.

Epigenetic coding: Combination of states of homeotic gene activity which make up the commitment of the cell.

Equivalence: Cells are equivalent if they have the same epigenetic coding.

Equivalence group: Group of cells with a common competence.

Experimental embryology: Uncovering the mechanisms of development by experimental alteration of normal development.

Fate: The future experience of a region of the embryo; also called 'developmental pathway'. Does not imply anything about the commitment of the region in question.

Fate map: A diagram which shows, for a particular developmental stage, where each part of the embryo will move to at later stages, and what each part will become.

Field, embryonic: Region within which a single process leads to regional specification.

Gradient: Monotonic change in concentration of a morphogen across an embryonic field.

Growth: Increase in the amount of matter in the embryo. Does not necessarily occur during early development.

Herkunftsgemäss: Development of a graft in accordance with its old position. If behaviour is always herkunftsgemäss then the tissue of the graft is said to be determined.

Homeotic gene: Gene which, when mutated, causes conversion of one part of the body into another.

Induction: Effect on the developmental pathway of one group of cells by a substance displayed by or emitted from another. *See* instructive, permissive.

Instructive induction: Situation in which responding tissue develops along one pathway in the absence of the signal and another in its presence. An instructive induction always increases the number of parts in the embryo.

Label: Visible feature, including genetic modifications, applied by the experimenter to a cell or tissue region in an embryo for the purpose of identification of cellular origins.

Lineage: *See* cell lineage.

Marker: Substance whose synthesis by the embryo characterizes a particular stage, region or cell type. Also used by some to refer to labels.

Maternal: Feature of development due to materials preformed during oogenesis.

Morphogen: An inducing factor which can evoke more than one positive response from the responding tissue.

Morphogenesis: Formation of biological structure by changing the spatial relationships of cells or tissues.

Mosaicism: Identity of specification with normal fate.

Non-equivalence: Used of similar structures or tissues which arose from rudiments with different epigenetic codings.

Normal development: Course of development followed by an embryo free from experimental disturbance.

Organizer: Centre emitting an inductive signal, usually of the morphogen gradient type.

Ortsgemäss: Development of a graft according to its new position.

Pattern formation: Usually taken to include cell movement and sorting processes as well as regional specification.

Permissive induction: Situation where an already committed tissue region needs an external signal in order to complete its maturation. Does not result in any increase in the number of parts.

Polarity: Regional difference in commitment along a particular direction.

Polyclone: Cells making up a compartment.

Positional information: A positional signal is much the same as a morphogen; a positional value is much the same as a set of epigenetic codings which vary with position in the embryo.

Potency: The total of all fates of a cell or tissue region which can be achieved by any environmental manipulation.

Regional specification: Mechanism by which cells in different positions acquire different developmental commitments. What this book is about.

Regulation: Deviation of specification from fate. Comprises the formation of complete patterns following division or fusion of embryos; the formation of complete patterns following partial deletions from or augmentations of embryos; and alterations of fate arising from an alteration of the inductive signals.

Selector gene: Same as homeotic gene.

Specification: Commitment of a cell or tissue region which is manifested on culture in a neutral medium but may still be reversible.

Stem cells: Cells which, individually or as a population, both produce differentiated progeny and reproduce themselves.

Threshold: Sharp, discrete change in cell state with the intensity of a signal.

Zygotic: Features of development due to substances newly synthesized in the embryo.

3

Theoretical embryology

As the science of embryology has matured, certain of its ideas have taken on mathematical form. Among the most important of these are the concepts of *cell state*, *bistability*, *symmetry breaking* and the *morphogen gradient*. Although these ideas still exist mainly in the form of mathematical models, we are entering a time when they are also becoming an empirical reality. A few years ago such models were often dismissed as mere speculation, but now no serious student of the subject can afford to ignore them.

CELL STATES

Most of what we think we know about cell states is derived from the observed properties of the histologically distinguishable *cell types* in the adult animal and it is often assumed that embryonic cells are like adult cells except that they are expressing different sets of genes. However, there are important differences. In the human there are about 2–300 cell types ranging from red blood cells through fibroblasts and secretory epithelia to the different pharmacological classes of neuron. The cell types that we can see are notable in three respects. They are, on the whole, qualitatively distinct from one another without intermediate forms. They persist for long periods of time and they conserve their cell type on division. The last is not true for tissues fed from a population of stem cells, but the stem cells themselves conserve their *organ type* on division although they may produce a spectrum of differentiated types. In contrast, embryonic cells often show smooth gradations of property across a field, either in terms of their general histological appearance or in terms of individual molecular markers. States of commitment are not usually stable over long periods of time, and are not necessarily clonally heritable. Moreover, states of specification can, by definition, be altered by a stimulus as mild as grafting from one position to another. These characteristics of differentiated and embryonic cells suggest that we need to think more carefully about embryonic cell states. In fact a natural explanation of the difference between embryonic and differentiated cell states is provided by the lan-

34

guage of *dynamical systems theory* and so that is what much of this Chapter will be about.

DYNAMICAL SYSTEMS THEORY

If we are thinking at the molecular level then in order to define the state of a cell fully we need to have a complete list of all the chemical substances present with their concentrations, and if there is regional inhomogeneity we need to know the concentrations in each compartment and also the diffusion constants for the movement of each substance between each pair of compartments. Evidently a description at this level does not exist for any cell at the present time and only for bacterial cells is it even a realistic possibility. So in practice we are bound to settle for something less.

The simplified definition which is at present most popular is one which ignores all substances in the cell except the genes and also ignores all compartmentation. The state of the cell is then defined as a list of those genes which are active, and if there are degrees of activity, how active they are. The idea is that the genes are ultimately the source of all biosynthesis and so the other characteristics of the cell are automatic consequences of the state of the genome, no extra information being given by stating them explicitly. This proposition is normally accepted as dogma not requiring any experimental support. In fact, the best evidence relating to it comes from some quite old experiments by Boveri on haploid hybrid merogones in sea urchins (reviewed Morgan, 1927). In sea urchins it is often possible to fertilize cytoplasmic fragments of the egg of one species by the sperm of another. The resulting embryos tend to resemble the maternal species up to about the prism stage and then gradually change so that they come more and more to resemble the paternal species. So in these animals at least the assumption that cell state equals gene activity is probably valid for the later stages of development but seems less likely to hold for the early stages. The reason for this is that in early embryos most protein synthesis is directed by maternal mRNA and there is often also some cytoplasmic localization of substances which specify some early subdivisions of the body plan. While the state of a cell may be defined by a list of gene activities for most cells following the onset of zygotic transcription, we still have some way to go before all the gene products have been identified in even a simple multicellular organism and so for practical purposes a still further simplification is necessary. This is to make the assumption that the substances in the cell belong to a number of metabolic subsystems which interact only weakly with one another. If the decisions involved in regional specification can be located in one subsystem then all the other substances could be ignored, at least to a first approximation.

The molecular biologist must assume the truth of this assumption since if regional specification cannot be ascribed to a reasonably small subset of substances then there will probably never be a molecular solution of the problems. However, the theoretician need not worry. From his point of view it does not matter whether the state is defined in terms of a few substances, just the genes or the whole lot; in any case the state will consist of a list of values $c_1, c_2....c_n$ which represent the chemical concentrations of each of the n components which are to be included. Because the n substances undergo many interactions with one another, the state of the cell is liable to spontaneous change and the direction of this change is governed by the particular laws of motion, or dynamics, appropriate to the system in question.

According to the law of mass action the rate of increase of each chemical concentration will depend on the concentrations of the precursors. For example if $A + 2B \rightarrow C$ then $dc/dt = k_1 ab^2$ where k_1 is a rate constant and a,b are concentrations of A,B respectively. If C is then removed by reaction with E in the process $C + E \rightarrow F$, then the rate law for C becomes $dc/dt = k_1 ab^2 - k_2 ce$ where c,e are the concentrations of C,E. In addition, the rate of formation of C might be regulated by other substances which do not participate in the reaction, for example, the rate might be reduced in inverse proportion to the concentration (g) of a negative regulator G. So the full scheme would be:

$$A + 2B \xrightarrow[\substack{\ominus \\ \uparrow \\ G}]{} C \xrightarrow[\substack{| \\ E}]{} F$$

and the rate law for C might in the simplest case be:

$$\frac{dc}{dt} = \frac{k_1 ab^2}{g} - k_2 ce$$

In certain applications we might need also to consider diffusion of C in and out of the compartment in question and this would necessitate adding a diffusion term to the equation:

$$\frac{dc}{dt} = \frac{k_1 ab^2}{g} - k_2 ce + D\nabla^2 c$$

D is the diffusion constant of C and ∇^2 is the Laplacian operator indicating the second derivative of C with respect to space. In fact, we shall not consider diffusion further for the time being. As far as chemical kinetics are concerned, the situation is that the concentration of each substance will change in time as a function of some or all of the other substances in the cell. The actual differential equation for each substance will depend on the details of the kinetics and so in general we can only say that the rate law will be some function of the other concentrations and that it will usually be non-

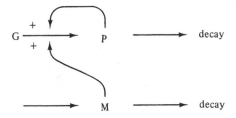

Fig. 3.1. Reaction scheme for a process showing bistable behaviour.

linear. So if there are n substances in the cell the concentrations are governed by laws of the type:

$$\frac{dc_i}{dt} = f_i(c_1, c_2,...,c_n)$$

($c_1,...,c_n$ could of course be regarded either as concentrations or gene activities). If we could measure the concentrations then it would be possible to describe the evolution of the state of the cell by plotting all of them against time. Alternatively the state can be represented at one point in time by a single point in a *state space* whose axes are the concentrations themselves. The evolution of state is then described by movement of this point along a *trajectory* in state space.

The state space

This can most easily be understood by considering a concrete example involving only two concentrations. With two concentrations, the state space can be represented on a plane by plotting the concentration of one substance against the other. The model is adapted from one proposed some years ago to explain discontinuous change in development (Lewis, Slack & Wolpert, 1977, and Fig. 3.1). A gene G produces a product P, and the gene may either be turned on by a regulator M or by its own product P. The differential equations describing the system are in words:

rate of change of product   =   regulator + autoinduced − decay
                                 induced

and:

rate of change of regulator   =   production − decay

or in maths:

$$\frac{dP}{dt} = k_1 M + \frac{k_2 P^2}{k_3 + P^2} - k_4 P$$

$$\frac{dM}{dt} = k_5 - k_6 M$$

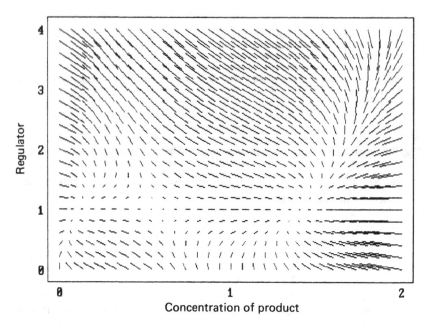

Fig. 3.2. Simulation of reaction scheme in Fig. 3.1. For each point in the concentration space the system will evolve in a direction indicated by the attached line and at a rate proportional to the length of the line. In this and later figures the parameters were given the following values: $k_1 = 0.04$, $k_2 = k_3 = 1$, $k_4 = 0.5$, $k_5 = k_6 = 0.1$.

The state space is shown in Fig. 3.2. Each point represents a particular pair of values of M and P and the line associated with each point represents the rate of change calculated from the equations for these particular concentrations. The longer the line the greater the change of concentration per unit of time. In Fig. 3.3 the same diagram is shown but with the lines joined up to form a set of *trajectories* drawn on the M,P plane. The direction of motion is indicated by the arrowheads. Here for clarity trajectories are only shown originating from the points around the edge, but of course every point in the state space lies on a trajectory. If a cell is suddenly brought into existence with any given values of M and P its state will spontaneously evolve along the trajectory on which it lies and will continue to evolve until it achieves one of the two possible pairs of *stable steady state* values which in this case are $M = 1$, $P = 0.1$ and $M = 1$, $P = 1.4$. There are in fact three pairs of values at which no further change will occur (ie, at which $dM/dt = dP/dt = 0$), but the third one, which lies between the two stable points at $M = 1$, $P = 0.566$, is intrinsically unstable since trajectories lead away from it rather than towards it.

The two stable steady states shown in Fig. 3.3 are stable because in the

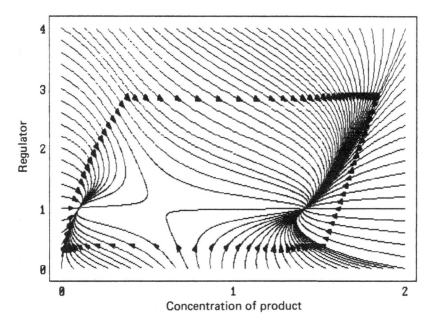

Fig. 3.3. The same system is shown in the form of a set of trajectories starting from each of the points around the edge. The direction of movement is shown by the arrowheads, which also indicate the state of the system after 5 time units.

neighbouring regions of state space all the trajectories converge onto them. This means that once a cell has reached such a state it will persist in it. Any small fluctuations which may occur in *M* and *P* will produce only transient changes of state because the system will spontaneously move back to the stable point. Steady states like these are not the same as states of thermodynamic equilibrium. In a completely isolated system there would be only one steady state and this would be the state of thermodynamic equilibrium. But cells are not isolated. They are always exchanging materials with the environment and the dynamical description includes terms for this, so the steady states are kinetic rather than thermodynamic.

Since in this example there are two stable points and not one it is clear that their stability must be local rather than global. Each stable point is surrounded by a region of state space called a *basin*, consisting of all the other points from which trajectories converge to the stable point. The boundary between the basins is called a *separatrix* and it is the line passing through the unstable point, shown in Fig. 3.4. If the system is in one stable state and is then acted on by an external influence which is sufficient to carry it across the separatrix, it will then settle into the other stable state and remain there even when the force is removed.

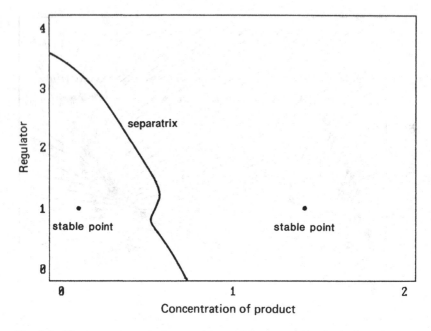

Fig. 3.4. The same system is shown with steady states and basins indicated.

A model for cellular specification

If one is seeking a molecular understanding of development then the language of dynamical systems theory seems a natural one to use. The example we have considered could be regarded as being a model for specification of cells of an animal embryo. The steady states can be identified with the *stable differentiated cell types*, since they are chemical compositions which are stable over time. A region of state space around each steady state is the range of modulation of cellular behaviour during adult life. Only very rarely will changes occur which are large enough to change one cell type into another (metaplasia) since one or more of the system variables must be altered enough to cross the separatrix. A larger region of state space corresponds to *determination*, since this is the group of states which inexorably run down to a given steady state for any environment present in the embryo. The entire region of state space within the basin corresponds to the zone of *specification*. In isolation, cells with these states will run down to one of the steady states but if exposed to inductive signals then M or P may be changed enough to cross the separatrix and run down to the other steady state. The *potency* of a cell population corresponds to all of the steady states present in the state space since all can be reached under some condition or another.

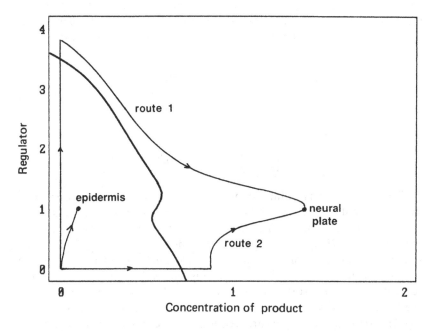

Fig. 3.5. The same system is used as a hypothetical model of neural induction. Route 1 shows cells moving into the neural basin as a result of an increase of M, while route 2 shows them entering the neural basin as a result of an increase of P.

To see how this description might apply to an actual example of cellular decision-making, consider the use of this diagram to provide a hypothetical model for neural induction in a vertebrate embryo. In Fig. 3.5 the state of the ectoderm cells is represented by a point at the origin; in other words the cells start off with $M = 0$, $P = 0$. If left alone, or if isolated from the rest of the embryo, they will run down the trajectory to the epidermal steady state, and so we would say that they are specified to become epidermis. Now suppose that the neural inductor is the substance M. At gastrulation, M is increased in cells receiving the signal to a value of 4. This increase changes their state because they now lie across the separatrix in the zone of neural determination. If left alone or isolated from the embryo they will evolve along their new trajectory and eventually become neural plate (route 1). In this example an increase of P would also serve to drive the cells into the neural basin despite the fact that M and not P is the natural inductor (route 2). This example should warn us that inductive signals need not necessarily be very specific. More than one substance can bring about the same final state, so the degree of specificity will depend on the particular dynamics in question.

Of course, real cells are likely to have thousands of interacting substances rather than two. Our example used two simply in order to be able to represent the state space in two dimensions on a plane. If there are n substances in the cell then the dynamics will be representable as a system of n equations:

$$\frac{dc_1}{dt} = f_1(c_1,...,c_n)$$

$$\cdot$$
$$\cdot$$
$$\cdot$$
$$\cdot$$

$$\frac{dc_n}{dt} = f_n(c_1,...,c_n)$$

The state space has n dimensions, one axis for each chemical concentration, and the state $[c_1,...,c_n]$ and the rate of change vector $[dc_1/dt,...,dc_n/dt]$ each have n components. Although it is not possible to visualize a space with more than three dimensions the principles are not very different. The steady states are still points inscribed in the n dimensional space. The space is still partitioned into basins each having one stable point, although the separatrices are now surfaces of dimension $n-1$. A study has been made of the behaviour of such systems in relation to development by Kauffman (1971) and interested readers are recommended to study it. There are in fact also some other types of *attractor* more complex than the stable point. These include the limit cycle which would manifest itself as a sustained oscillation in two or more of the system variables, and the strange attractors which would be apparent if very similar starting conditions regularly gave rise to a wide variety of final states. Neither of these has so far proved to be of great interest to the embryologist and so they will not be discussed here.

We still do not really know how complex real cells are. The best estimates of mRNA complexity have been obtained from early sea urchin embryos (see discussion in Davidson, 1986, Chapters 2 and 3). They contain perhaps 2×10^4 messages of maternal origin and another 10^4 different ones transcribed from the zygotic genome. Estimates from *Xenopus* and *Drosophila* are of a similar order of magnitude. Although this number of molecules will presumably be catalogued and sequenced in due course, one must hope that the essential dynamical features of the cell's developmental behaviour are manifested by a small subset of these substances in relative isolation from the remainder. This is the way that it presently looks in *Drosophila*, where molecular understanding is most advanced, and has also been the experience in the past with other processes such as oxidative phosphorylation, protein synthesis or electrical excitability.

The point of this section has been to show that if we are serious about attempting to understand the hierarchy of developmental decisions in molecular terms then we do not just need to identify the relevant genes and gene products but also to understand their dynamical behaviour. In the past this has proved to be necessary for understanding such things as the mechanism of nerve conduction or aggregation in slime moulds. In the future it seems probable that it will be through the mathematics of dynamical systems theory that embryological and molecular results can meaningfully be brought together.

GRADIENT MODELS FOR EMBRYONIC INDUCTION AND REGULATION

We believe that the selection of cell states in different parts of the embryo is controlled, at least in part, by extracellular chemical signals called inducing factors. As discussed in the previous chapter, inductions may involve a single threshold response or several, and of particular interest is the situation where several thresholds are triggered in response to a monotonic concentration gradient of the factor (morphogen) emitted from a source region (organizer). 'Monotonic' means that there are no maxima or minima in the concentration profile, and enables the gradient to be counterposed to a *prepattern*, in which there are maxima and minima (see below). To earlier theorists such as Child, gradients in embryos were of 'metabolism', while to later ones such as Crick (1970) they became diffusion gradients of single substances. Morphogen gradients used to be regarded by molecular biologists as a kind of mystical pseudoexplanation and a distraction from the serious business of molecular–genetic analysis, but this prejudice has subsided in recent years as the gradients themselves have begun to be discovered. The purpose of this section is to describe the properties of the most popular gradient models, and discuss what they can and can not explain.

DIFFUSION OF A MORPHOGEN

For ease of computation, models involving diffusion are usually one dimensional, referring either to diffusion along a line of cells or across a sheet of cells in which the sources and sinks are arranged as transverse rows. If two cell populations containing different concentrations of a substance are placed in contact then the material will start to diffuse from the region of high concentration to the region of low concentration and the transport across a unit area is given by:

$$\text{flux, } J = -D(\mathrm{d}c/\mathrm{d}x) \qquad \text{moles s}^{-1}$$

where $\mathrm{d}c/\mathrm{d}x$ is the local concentration gradient and the minus sign indicates that the flux is from high to low concentration: down the gradient. D is the diffusion constant and is measured in units of $\text{cm}^2 \text{ s}^{-1}$. The larger its value the faster the substance spreads, and it depends both on the characteristics of the molecule in question and on the viscosity of the medium. If cell membranes have to be crossed, the diffusion constant will be rather small for all but the most hydrophobic substances. If diffusion across membranes depends on gap junctions or carrier molecules then D can be regarded as a variable which depends on the state of the cells. Diffusion constants have been measured for various substances in cytoplasm, usually by recovery of fluorescence after spot photobleaching of fluorescent conjugates (Slack, 1987b). The values for small molecules are in the range 1×10^{-6} to 5×10^{-6} cm² s^{-1}, about half the corresponding figures for solutions in water. The values for proteins are much lower, at around 0.3×10^{-8} to 1×10^{-8} cm² s^{-1}.

If we consider the change of concentration in a cell due to diffusion, we see that it must be proportional to the difference between the flux in and the flux out; in other words to the rate of change of concentration gradient with position, which is the second derivative of c, hence in words:

rate of change of concentration gradient in time	$= D \times$	rate of change of concentration gradient in space

or in maths:

$$\frac{\partial c}{\partial t} = D \frac{\partial^2 c}{\partial x^2}$$

This is the fundamental equation of diffusion, often known as Fick's second law. If regions containing different concentrations of a substance are placed in contact then eventually diffusion will spread out the substance so that it has a uniform concentration everywhere. However, if the substance is being produced in some places and removed in others then the distribution of material will tend towards a non-uniform steady state in which $dc/dt = 0$ at all positions and there is a continuous flow of material from the sources towards the sinks. For example, suppose that there is a source at one end ($x = 0$) of a line of cells and there is a sink at the other end ($x = l$). At the source the concentration is fixed at a $c = c_0$ and at the sink it is fixed at $c = 0$. In the steady state:

$$\frac{\partial c}{\partial t} = D \frac{\partial^2 c}{\partial x^2} = 0$$

therefore

$$\frac{d^2 c}{dx^2} = 0$$

therefore

$$c = Ax + B$$

At $x = 0$, $c = c_0$ therefore $B = c_0$

At $x = l$, $c = 0$ therefore $A = \frac{-c_0}{l}$

So

$$c = c_0\left(1 - \frac{x}{l}\right)$$

This gives a linear steady-state gradient with the concentration falling from c_0 to 0 across the field. Although this model is of some historical interest as the first quantitative gradient model (Crick, 1970), it has found little application because two specialized regions, the source and the sink, are required to regionalize the tissue in between. Also it is a very slow model: it takes a long time to build up to an approximation to the steady state starting from a uniform concentration across the field, and if a morphogen had a diffusion constant as low as 10^{-8}, as do most proteins, then the gradient could not be established within a realistic time. This led to alternative formulations and the most popular of these has been the 'local source and dispersed sink model'.

The local source-dispersed sink (LSDS) model

This model, referred to below as 'LSDS' for short, puts the source at one end with a constant concentration c_0, and the sink as the entire responding field with each cell destroying the morphogen at a rate proportional to its local concentration. There is no exit from the field so the far end acts as a barrier for the morphogen. Since there can be no flux across this barrier, it follows that the local concentration slope (dc/dx), must be zero. So:

$$\frac{\partial c}{\partial t} = D\frac{\partial^2 c}{\partial x^2} - kc$$

In the steady state

$$D\frac{d^2 c}{dx^2} - kc = 0$$

Let $k/D = a^2$ then

$$\frac{d^2 c}{dx^2} - a^2 c = 0$$

This is a well-known type of differential equation to which the solution is

$$c = Ae^{ax} + Be^{-ax}$$

The boundary conditions are

$$c = c_0 \text{ at } x = 0$$
$$\frac{dc}{dx} = 0 \text{ at } x = l$$

and when these are substituted, we arrive at

$$\frac{c}{c_0} = e^{-ax} + \text{ß} \sinh ax$$

where ß is a constant formed from a and l.

('sin h' indicates a hyperbolic function whose values can be looked up in tables or calculated using your PC.) Some solutions are shown in Fig. 3.6 for three values of a, showing that they are monotonic gradients which are steeper the larger the value of a.

The calculation of transients, which means the form of the gradient before the steady state is reached, is a rather complex exercise if approached by analytical mathematics. However, it is easy by computer simulation and in the era of cheap personal computers such calculations are within the reach of amateurs. A set of transients for the model with $a = 31.6$ is shown in Fig. 3.7. It is clear that over a field of 1 mm length an acceptable approach to the steady state can be obtained in one hour with a diffusion constant of 10^{-7}, starting with no morphogen present. Of course, this assumes that the field length is around 1 mm in length, at larger sizes the times required increase considerably. But as emphasized by Wolpert (1969) embryonic fields *are* small: whether they be whole early embryos or organ rudiments at more advanced stages, there is abundant evidence that regional specification occurs across a territory not exceeding 1 mm and often much less. It seems highly probable that the time requirement for setting up diffusion gradients is one of the main constraints that have preserved this feature throughout the course of animal evolution.

Regulative behaviour of the LSDS model

If this gradient were an inductive signal we can assume that it would turn on biochemical switches in the responding tissue at certain threshold concentrations (Fig. 3.8). We shall not consider the mechanism of thresholds until later, but since it is the positions of the thresholds that determine the anatomy of the organism it is clear that disturbances of the steady-state gradient would produce corresponding disturbances in the anatomy. Let us now consider the effects of certain embryological manipulations of a type which have been used on real embryos and are discussed later in the book.

If the source region is left fixed and other regions of tissue are rearranged, then the gradient will eventually relax back to its original configuration (Fig. 3.9). Similarly if a small region is removed which is fated to form a certain structure, then the structure will form none the less from the tissue on either side. As long as the threshold responses do not take place until the steady state has become re-established, this type of simple gradient model provides a satisfactory explanation of *defect regulation*.

It also explains the special properties of organizer regions by identifying them with sources. If a source region is removed then the steady state concentration will fall to zero and no structures will be formed. If the source is moved from one end of the field to the other then the structures will be formed in the correct sequence but with an overall reverse of polarity. If a second source is grafted to the far end of the field then a *duplication* is

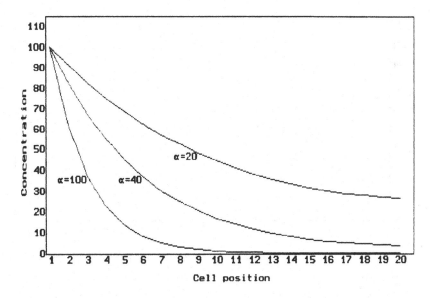

Fig. 3.6. Steady state gradients formed by the LSDS model. $\alpha = \sqrt{k/D}$ and determines the steepness of the gradient. In this case the field length, l is 1 mm.

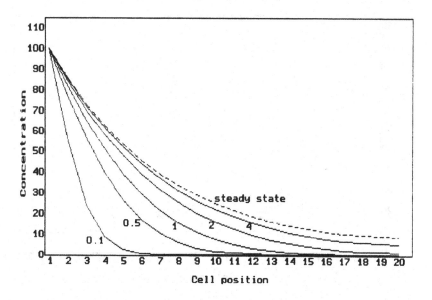

Fig. 3.7. Transient gradients arising from the LSDS model. The starting situation was a concentration of 100 at the source and 0 elsewhere. The number near each curve is the time elapsed in hours. The field length is 1 mm, D is 10^{-7} cm^2 s^{-1}, and k is 10^{-4} concentration units per second, giving an α value of 31.6.

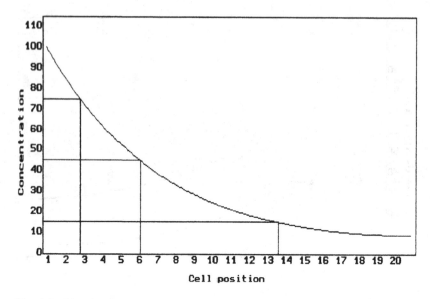

Fig. 3.8. Threshold responses to the morphogen gradient establish four regions with different epigenetic codings.

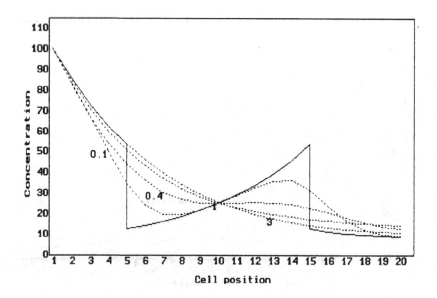

Fig. 3.9. Regulative properties of the LSDS model. A region of tissue is rotated but the gradient relaxes back to its steady state within 3 hours.

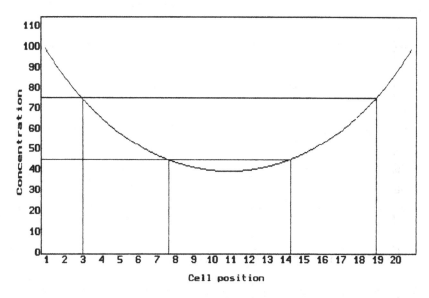

Fig. 3.10. Formation of a mirror symmetrical pattern by grafting a second source region at $x = 0.1$ cm. The new steady state gradient is U shaped and the same threshold responses shown in Fig. 3.8 will produce a mirror symmetrical anatomy.

formed (Fig. 3.10). This consists of two sets of similar structures which are arranged with opposite polarity around a central plane of *mirror symmetry*. This type of mirror duplication will be encountered several times in later Chapters and the gradient model explains the overall symmetry, the fact that normal neighbour relations are maintained between structures and the fact that some structures are missed out because the central minimum of the U-shaped gradient exceeds some of the lower thresholds. It does not, however, explain how the second source arises.

Limitations of the LSDS model

The LSDS model is able to account for many of the facts of regional specification in early development: the essential requirement for the presence of the organizer regions, the coordinated determination of structures at different distances from an organizer, defect regulation, and the formations of mirror symmetrical duplications when a second source is created or introduced. But there are also three phenomena which seem to be incompatible with it: adaptation of proportions to overall size, symmetry breaking and regulation of organizers.

The problem with size regulation is illustrated in Fig. 3.11. If the size is reduced, for example, by insertion of a barrier in the midst of the field, then the morphogen accumulates on one side and disappears on the other. When

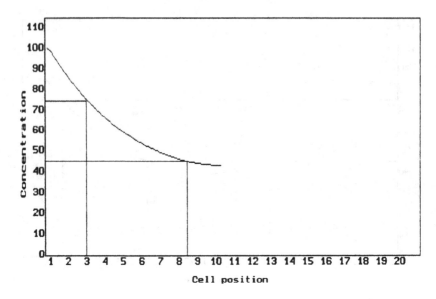

Fig. 3.11. Insertion of a barrier between cell 10 and 11 leads to the loss of *more than* the territories predicted from the fate map. This is because the morphogen piles up at the barrier and so the positions of threshold responses are displaced away from the source. (Compare Fig. 3.8).

the threshold responses are drawn in, it is clear that nothing is formed on the far side and the structures formed on the near side are shifted towards the barrier relative to the positions of their presumptive areas on the normal fate map. In fact, this is similar to what is found in the so-called 'gap phenomenon' of certain insect eggs (see Chapter 7). But the phenomenon is not a general one, since in most types of embryo reduction of size at an early enough stage leads to the formation of a small but normally proportioned larva. In several cases it is also possible to augment the size of embryos and obtain large normally proportioned body plans and the model does not predict this either.

Secondly, there is the matter of symmetry breaking (see below). There are some reasons for thinking that certain regional differences are set up in response to stimuli which are very minor environmental perturbations. Examples of this are the orientation of the dorsoventral axis in the amphibian egg, or of the craniocaudal axis in the isolated anterior half of an avian blastoderm. The LSDS model, or any other model depending on some arrangement of sources and sinks to create the gradient, require some prespecification of the sources and sinks and this seems unlikely to exist in such cases.

Thirdly, there is the regulation of organizers. Although in most cases the removal of a signalling centre causes severe defects in the pattern formed from the rest of the field there are some, such as Hensen's node in the chick, which seem to be able to be reconstituted following their surgical removal.

These objections have caused some workers to produce more elaborate gradient models such as the GM1 model described below. While they may be right it is worth considering whether the case against the LSDS or other simple gradient models really is so substantial. The argument about size regulation is based on the steady-state properties of the diffusion gradient. However, as mentioned above in the context of the timing of events, we do not know whether diffusion gradients in embryos ever reach a steady state. If they do not then there is no problem about size reduction. There can be a transient state in which the gradient spans its normal range of concentrations and if the threshold responses occur at this stage all structures will be represented in the anatomy. This argument is no use in the case of *augmented* size but here it is possible to argue that the double size embryo is in any case an illusion, all that has really happened is that the extraembryonic portion has increased in size. All embryos have an extraembryonic portion even if it is quite small, such as that part of the amphibian endoderm destined to end up within the gut lumen; and we know from the careful studies on the early mouse embryo that the proportions are not necessarily maintained with regard to embryonic and extraembryonic regions. It will take quite a lot more careful research on the proportions of embryos following size changes before we can really be sure that the LSDS model is inadequate. Symmetry breaking is an awkward problem because we can never be sure that it is really happening, rather than being the manifestation of some cryptic pre-existing asymmetry. All animal eggs are polarized in some way when they are laid and though in a few cases they seem to be symmetrical with respect to some axes it is not possible to prove the absence of a significant internal asymmetry which normally sets in train the formation of a source region. As for the matter of the regulation of organizers, in cases where regulation has been claimed we do not know the exact spatial extent of the signalling centre and so it is always possible to argue that part of it remains.

A major boost for the LSDS model has been the understanding of the *bicoid* system in *Drosophila* where the *bicoid* product forms an exponential concentration gradient in the syncytial blastoderm, fed by synthesis from a localized mRNA. Although the action occurs in a syncytium rather than in a sheet of cells the model gives a good qualitative fit to the results. Although it is strictly outside the scope of this book, it is also worth remarking that the LSDS model gives an excellent fit to experiments on the developing chick limb in which gradients of radioactive retinoic acid have been established from implanted carrier beads (Eichele & Thaller, 1987). But as mentioned above the LSDS model does not explain everything in embryology, and signalling systems with different dynamical properties will doubtless be discovered in due course.

SYMMETRY BREAKING PROCESSES

One of the questions addressed at some length by theoretical biologists is how a uniform initial state can spontaneously change into one with regional differentiation. Where the environment of the tissue is asymmetrical there is no problem, but what of an entire embryo suspended in a fluid of uniform composition? Although few, if any, animal eggs are truly uniform when they are laid, there are some cases in early embryogenesis where polarization seems to arise *de novo*, for example, in the position of the blastocoelic cavity in the mouse embryo, or the onset of the dorsoventral polarity in the amphibian embryo. This sort of symmetry breaking process seems to violate the conception of the cell as a deterministic dynamical system presented above in so far as there is no obvious cause for the orientation of structures.

The answer really lies in distinguishing between what Aristotle called the 'formal cause' and the 'efficient cause' of an event. In answer to the question: 'why does the light come on?' we might answer that given the mains voltage and the resistance of the lamp filament, the energy dissipation is such as to make the filament incandescent. We might alternatively answer that it was because someone flipped the switch. The first is the formal cause, roughly the laws of motion of the system; and the second is the efficient cause, or the agent which unleashes a particular process. In the present context we can distinguish a formal and an efficient cause for symmetry breaking in systems made up of chemical reactions. The formal cause lies in the dynamics of the system which must be of a type which has more than one stable steady state, such as the example described above. The state of the system will be represented by a point on the separatrix, the line which demarcates the different basins. This means that a very tiny perturbation of one of the system variables will move the point into one of the basins, following which the normal dynamics will take over. The efficient cause then lies in the molecular nature of matter which means that at a given point in time a well-mixed solution will have very small regional differences in composition arising solely from the random distribution of molecules. The symmetry breaking occurs when one of these microscopic perturbations is amplified by the dynamics into a macroscopic inhomogeneity.

The first symmetry breaking model appropriate to embryology was published by Turing (1952), and the theory was subsequently developed by the school of physical chemists around I. Prigogine in Brussels (see Nicolis & Prigogine, 1977). We are mainly concerned with asymmetries in *space* so the systems involve chemical reactions and diffusion ('reaction–diffusion' models). In general there will be one or more *bifurcation parameters* in the differential equations, for example, a rate constant, or the length of the field. Over a certain range of the bifurcation parameter the system will have a single, homogeneous, steady state. But at a certain critical value the homogeneous state will become unstable and be replaced by (usually) more than one possible spatially inhomogeneous steady states in which some regions of the solution are net sources and other regions are net sinks. An example of how this can occur is given below (GM1 model). So it turns out that there is no fundamental problem about symmetry breaking. Certain types of kinetic network can be guaranteed spontaneously to generate structure and the orientation of the structures can be determined by tiny and imperceptible pre-existing asymmetries.

Double gradient models

The most famous double gradient models are due to Gierer & Meinhart (1972). Because of the difficulties of analytical non-linear mathematics they used the method of computer simulation to investigate a variety of models and apply them to a number of situations in developmental biology. From an explanatory point of view, their models are superior to the LSDS and other simple gradient models in three respects: they have symmetry breaking properties and so can establish a gradient across an initially homogeneous field; they can reconstitute the source region if it is removed; and they have some ability to regulate in the face of changes in size of the field.

Their first model (called GM1 for short) depends on lateral inhibition and involves two morphogenetically active substances called the activator and the inhibitor. The activator stimulates its own production (autocatalysis) and also that of the inhibitor, while the inhibitor represses the formation of the activator. Both substances are removed at a rate proportional to their concentrations. In addition, both substances are diffusible, and the inhibitor diffuses faster than the activator. The reaction scheme is shown in Fig. 3.12 and the kinetic equations for the system in one dimension are:

$$\frac{\partial A}{\partial t} = k_1 \frac{A^2}{I} - k_2 A + D_A \frac{\partial^2 A}{\partial x^2}$$

$$\frac{\partial I}{\partial t} = k_3 A^2 - k_4 I + D_1 \frac{\partial^2 I}{\partial x^2}$$

As in previous examples the final terms represent diffusion and the first and second terms represent production and removal by chemical reactions. In some versions there is a small basal activator production as well. As in other symmetry breaking processes there is a certain parameter range for which the homogeneous state is unstable. It then spontaneously breaks down into an inhomogeneous state in which there are regions of net production of A and I (sources) and regions of net removal (sinks). It is easy to understand how this happens. Suppose we start from the unstable homogeneous state (in which $A = k_1 k_4 / k_2 k_3$ and $I = k_1 k_4^2 / k_2^2 k_3$). If a slight

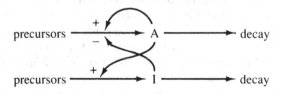

Fig. 3.12. Reaction scheme for the lateral inhibition model of Gierer & Meinhardt (GM1 model).

perturbation should increase the local concentration of A, then because both substances are made at a rate proportional to A^2, there will be a rapid net local increase of both A and I. This will establish a concentration gradient and both substances will diffuse towards the surrounding regions in which their concentrations are lower. But I diffuses faster than A, so A will then predominate over I at the original disturbance centre and I over A in the surroundings. Because of the autocatalysis the inhomogeneity will grow until a steady state is reached in which local activator production in the source region (as it has now become) is balanced by its removal by diffusion. The steady state will show a sharp peak of activator concentration and a rather broader one of inhibitor concentration (Fig. 3.13). The number of activator peaks which may arise depends on the total size of the field relative to the activator diffusion constant but here we consider only the situation where the normal number of peaks is one and the pattern consists of a monotonic gradient of A and I.

With a change of parameters the model is able to reform a new source when the old one has been removed, and the new source will always form at the end of the fragment which was nearest to the old source. This is because it is at this point that there is the greatest predominance of activation over inhibition. In addition, the model can accommodate a certain degree of change in the overall size of the embryo both in terms of retaining the capacity to form a monotonic gradient, and to some extent preserving the proportions of the embryo. In Fig. 3.14 are shown concentration profiles produced on full size, half size and one-sixth size fields. It is here necessary to assume that it is the activator rather than the inhibitor which acts as the morphogen. The concentration span of the activator is quite well preserved in the half size pattern, not so well in the one-sixth size. But half linear size corresponds to one-eighth volume which is the maximum known range of approximate proportion regulation in both amphibian and starfish embryos. If the extreme concentrations determined extraembryonic regions whose proportions were not conserved then the model could

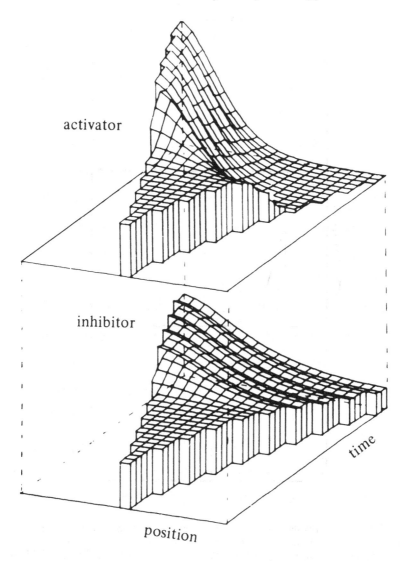

Fig. 3.13. GM1 model. This shows the concentrations of activator and inhibitor at different times in a growing domain. Once a certain critical size has been reached, the homogeneous situation becomes unstable and a peak forms at one end of the field. The peak of the inhibitor is broader than that of the activator because its diffusion constant is greater. (Figure kindly provided by Dr H. Meinhardt.)

explain proportion regulation to a greater degree than is known to exist at present. Furthermore, if there is a saturation of activator production, as, for example, is obtained by substituting $A^2/(1 + KA^2)$ for A^2 in the equations then the proportion regulation becomes even better.

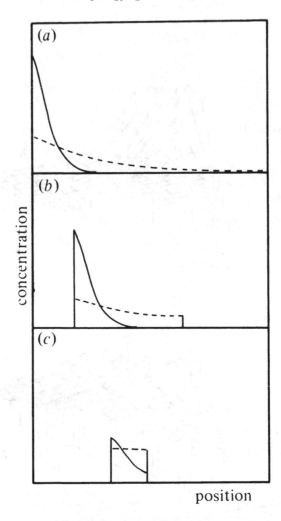

Fig. 3.14. Partial regulation of proportions by the GM1 model. In this version there is a slight intrinsic gradient of activator production which imposes a polarity on the field. (*a*) shows steady state gradients of the activator (continuous line) and inhibitor (dashed line). (*b*) shows the gradients which arise if half of the field is isolated. The height of both peaks has been slightly reduced but the concentrations spanned by the activator are equivalent to most of the original field. (*c*) shows what happens if one third of the half size field is isolated. A gradient is still formed but only spans a fraction of the original concentration range. (After Gierer & Meinhardt, 1972.)

Other double gradient models

The particular kinetic relationships in the GM1 model are not the only ones which show the essential properties of the double gradient models, namely symmetry breaking, defect regulation and proportion regulation. In fact, any pair of chemical reactions will do so as long as they satisfy the following conditions (Gierer, 1981):

1. One component must be autocatalytic
2. The other must inhibit the first
3. The inhibitory effect must be strong enough to prevent an explosion
4. The inhibitory effect must be fast compared to the autocatalysis
5. The range of activation must be smaller than the field size
6. The range of inhibition must be sufficiently large relative to the range of activation (in this context 'range' is the mean distance between production and decay of molecules).

This means that there is an infinity of possible models. It is not possible to falsify the whole class of models by morphological experiments because if a particular one is excluded another with different parameters could always be advanced. Nor is there any guarantee that there is only one activator and one inhibitor. There may be many more substances involved in the patterning mechanism, some of which behave collectively as the activator and others which behave collectively as the inhibitor. The value of the GM1 model is therefore not a precise prediction about the molecular basis of any given phenomenon, but rather to show what can be done, even with quite simple chemical kinetics.

The important thing in the present era when the molecular basis of regional specification is being uncovered is that workers in the field will understand this, and be alert to the likely existence of interesting dynamics in the processes they are studying. The above presentation of the LSDS and GM1 models shows that an explanation for regional specification can be provided by simple chemical kinetics and diffusion. We are not obliged to postulate the involvement of other processes such as vectorial pumping or electrophoretic transport of materials although we cannot exclude them either. The theoretical work can tell us what is possible but it cannot predict what is actually going on in reality. This must be found out by experiment.

THRESHOLDS AND MEMORY

The problem of *thresholds* is the problem of the emergence of the discontinuities between regions of the embryo, that is to say the selection, controlled by a cytoplasmic determinant or an inductive signal, of a particular cell state from a set of possibilities. A threshold is a line which

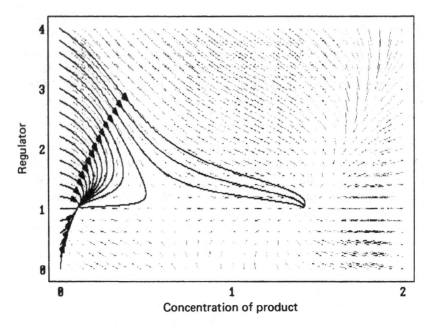

Fig. 3.15. The effect of raising the concentration of M (now regarded as a morphogen) from 0 to 4, using the same model as shown in Fig. 3.3.

separates two qualitatively distinct regions. It is sometimes used to refer to geographical boundaries in the embryo, which are lines separating regions of different states of commitment, and it is also sometimes used to refer to a line, or surface, in state space which separates two basins of attraction, being then the same as a separatrix.

It is sometimes assumed that the cooperative properties of allosteric enzymes are sufficient to explain the existence of thresholds. But this is not enough. However great the enhancement of binding of substrate to enzyme by positive cooperativity there will still be a smooth relationship between state and position and this is not a threshold. Furthermore, any equilibrium process such as this will be reversible. Thresholds on the other hand correspond to changes in state that are not reversible. Once they are formed they persist even in the absence of the interactions which brought about the regionalization in the first place. If a threshold is going to be truly discrete and persistent then it must be produced by some sort of switch mechanism such as the bistable reaction system discussed earlier in this Chapter. In Fig. 3.15 the state space is shown again in such a way as to emphasize the choice of trajectories controlled by the morphogen concentration. If the cells start off in the lower stable state and are then exposed to more of the effector M (which we can now regard as an inducing factor) such that the internal concentration is raised above 3.4, then they will follow a trajectory leading to the upper stable point. If M is now withdrawn and its internal concentration falls, even as far as zero, the cells are still left in the upper basin since there is no way that they can now leave it. The bistable system therefore has a 'memory' for the transient elevation of the effector level. If this mechanism were operating in a field of cells exposed to a gradient of morphogen then all cells with $M > 3.4$ would enter and remain in the upper state while all with $M < 3.4$ would stay unchanged in the lower state. Evidence for a mechanism rather similar to the hypothetical example presented here has been produced for the activation of the oestrogen receptor gene by oestradiol (Barton & Shapiro, 1988). Also the products of several genes in *Drosophila* such as *fushi tarazu*, *even-skipped* and *Deformed* (see Chapter 7), have been shown to activate their own transcription, although it is not yet certain that the auto-activation is sufficiently strong to give bistability.

It is not only inducing factors which need biochemical switches to create sharp boundaries in the tissue; any other mechanism of regional specification also entails the existence of thresholds. For example, cytoplasmic localization of a determinant will produce a boundary because some cells will inherit a concentration above the critical threshold and others will not. A growth zone from which structures emerge in a sequential manner might involve a continuous rise in a morphogen in the growing region so that all the cells passed through a sequence of states while they remained in the zone. As cohorts of cells were displaced from it, they remain in the highest state so far attained. This is a possible mechanism for the 'progress zone' model, proposed to explain limb development but often taken to apply to other growth zones as well (Summerbell, Lewis & Wolpert, 1973). When looked at in this way, it is not surprising that part of the body may appear to be formed by a gradient and the remainder by a growth zone mechanism, as is for example found in insect embryos of intermediate length germ anlage, since at the molecular level the elements of the mechanisms may be the same.

DNA methylation

The requirement that the bistable switches underlying threshold formation should have a memory is not affected in principle by the possibility that one of the chemical substances involved is DNA. There has been much interest in models for the stabilization of developmental decisions which are based on DNA methylation (Razin & Riggs, 1980; Holliday, 1987). The idea is that certain cytosine residues in CpG doublets can be methylated in the 5-position. The original methylation is accomplished by an enzyme which recognizes a non-palindromic sequence so that only one of the DNA strands becomes modified. Another enzyme called 'maintenance methylase' will then methylate the C in the complementary GpC of the other strand, but will not work on unmethylated DNA. The methylation pattern is clonally heritable because every replication of fully methylated DNA produces two molecules consisting of a new unmethylated strand and an old methylated one, both of which will be substrates for the maintenance methylase. That the basic idea is correct has been confirmed experimentally on several occasions although the enzymes involved are still not yet fully characterized. What we have here is another bistable biochemical switch, the main difference being that the state space is not continuous but consists of three discrete points corresponding to the three possible forms of the double-stranded molecule. The definition of stability is not the same for discrete as for continuous systems, but roughly speaking the unmethylated and doubly methylated states are stable and the hemimethylated state is unstable. In fact, there are two possible hemimethylated states if the two strands are distinguishable, but the second one would never be accessed by the mechanism as described.

In at least one case it has been shown that a DNA strand modification, presumably methylation, leads directly to the formation of two cell types. This is not in an embryo but in the fission yeast *Schizosaccharomyces pombe*, which has two mating types distinguished by the presence of alternative DNA sequences at the *mat1* locus. The sequences are not ordinary alleles but are transposed into place from adjacent loci. The transposition occurs only in chromosomes that contain a double stranded break and introduction of this break depends on possession of the modified strand. Hence mating type switching occurs in only one cell of four produced by the grandparent cell in which one strand was modified (Klar, 1987). It seems quite likely that similar mechanisms may exist in embryos, particularly those in which developmental decisions occur at a low cell number (see Chapter 5); however, direct evidence is still lacking.

Another example of a situation where DNA methylation is probably important in early development is the phenomenon of 'imprinting' in mammalian embryos (see Chapter 6). Here the expression of a gene depends on whether it was derived from the maternal or the paternal chromosome set. It has been possible to create transgenic mice in which the added genes show this effect and show that, for example, the same gene can be methylated during oogenesis and unmethylated during spermatogenesis (Swain, Stewart & Leder, 1987).

Homeogenetic induction

Molecular switch mechanisms can also provide a possible explanation for the phenomenon of *homeogenetic induction*. This has been investigated mainly for neural induction in the amphibia where it has been shown that explants from the neural plate will induce more neuroepithelium from gastrula ectoderm (see Chapter 4). If a bistable switch mechanism were operative, this might be because the switch can be turned on not only by the effector M but also by the gene product P. If enough P is added to raise the concentration transiently above the P-threshold (0.67 at M = 0) then the upper steady state will be reached. Now suppose that P can be transmitted from one cell to another, perhaps through gap junctions, or perhaps via a cell surface molecule, then it is quite easy to see that the spread of the induced state could become self perpetuating. As Meinhardt & Gierer (1980) have emphasized, in normal development there must be mechanisms which limit this sort of infectious spread of determined states, for example, an inhibitor might be produced by the induced region which antagonized further spread of the induced state.

REPEATING PATTERNS

The segments of an insect larva, or the somites of a vertebrate embryo, are repeating patterns consisting of series of structures which appear similar or identical to one another. Such structures are said to possess *serial homology* and the implication is that at least part of their epigenetic coding is the same, having been set up by a mechanism that generates a spatial series of identical determined states. In most cases repeated structures are formed sequentially, although in *Drosophila* segmentation is more or less simultaneous at all body levels. It has seemed unlikely to most workers that repeated structures are formed directly in response to a morphogen gradient and alternative mechanisms have been sought. Once again the correctness of this belief has now been established with the discovery, in *Drosophila*, of periodic patterns of expression of the genes responsible for setting up the pattern.

Reaction–diffusion models

Although processes such as the LSDS model involve chemical reactions and diffusion, the term *reaction–diffusion model* has come to be applied specifically to those with symmetry breaking properties such as the GM1 model. In fact, the first model of this type (Turing, 1952) was advanced as an explanation for repeating patterns. To see how the repeating patterns can arise, it is important to recognize that the diffusion constants in a reaction–diffusion model, which have units of $cm^2 s^{-1}$, set an absolute scale of size for the field. We have already seen that there is a critical size below which no prepattern can be set up and above this a size range in which a monotonic gradient can form (Fig. 3.13). Above this, things become more complicated since more than one inhomogeneous state may be possible and the one which is reached may depend on the nature of the initial perturbation. However, in general, the larger the field the more concentration peaks it is possible to fit on it, and for any given length there will be one

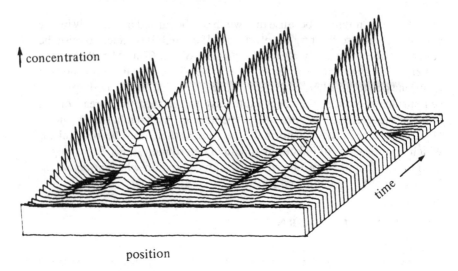

concentration

position

Fig. 3.16. Formation of a periodic pattern in space by the lateral inhibition model of Gierer & Meinhardt. The figure shows the concentration of the inhibitor, which is initially uniform. When subjected to small random fluctuations, a number of peaks begin to grow. Some die away and others persist as a stable pattern. (Figure kindly provided by Dr. H. Meinhardt.)

prepattern which has the fastest growth rate starting from homogeneity. For example, the GM1 model can give the multiple peaks shown in Fig. 3.16. The main difference from the previous figures in which this model produces monotonic gradients is that the inhibitor diffusion constant has been reduced relative to the size of the field.

Another type of reaction–diffusion model which has been advanced to account for repeating patterns depends on two substances whose synthesis is locally mutually exclusive but which activate one another at a distance. This property is possessed by another model produced by Meinhardt and Gierer called by them the 'lateral activation' model and called here the GM2 model (Meinhardt & Gierer, 1980). Here there are two genes G_1 and G_2 which produce products g_1 and g_2. There is a common repressor whose synthesis depends on the concentration of both products and which inhibits both genes. The two products decay into metabolites s_1 and s_2 which are more highly diffusible and each of these activates the other gene of the pair. The net result is that each gene activates itself at short range but activates the other at longer range. The main difference between this and the GM1 model is apparent in two dimensions (Fig. 3.17). The GM1 model will generate isolated peaks of activation surrounded by zones of inhibition. By contrast, the GM2 model will lead to the formation of long stripes because the two types of zone are each dependent on the proximity of the other.

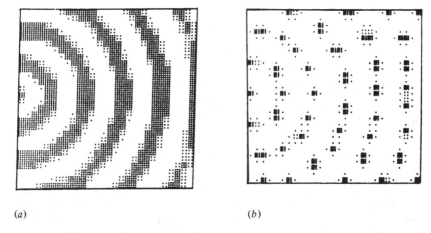

(a)   (b)

Fig. 3.17. Formation of stripes and spots in a two-dimensional space. (a) is the GM2 model and (b) the GM1 model. The development of the patterns was initiated by a disturbance at centre left which becomes the focus of the stripes or spots. (Figure kindly provided by Dr H. Meinhardt.)

If there were a single threshold response made by the cells, so that peak and trough regions became differently determined, reaction–diffusion models could produce sets of identical structures in either a sequential or a simultaneous manner. However, a property of such models is that the total number of peaks will be proportional to the total size of the field and this has caused much controversy because embryos with a segmented body plan can often show proportion regulation, for example, a *Xenopus* embryo developing from one of the first two blastomeres will be half normal size at the time of somite formation but will have a normal number of somites (Cooke, 1988). It is important to note that although a monotonic gradient arising from a reaction–diffusion system may regulate its proportions reasonably well, a multiple peak pattern cannot. This is because the natural wavelength of the system is a function of the kinetic and the diffusion constants. Roughly speaking, a given pattern will remain in existence over a change in linear dimensions of up to one wavelength which is proportionately a large range for a monotonic gradient but a small range for a multiple peak pattern.

The GM1 and GM2 models have been applied in detail to insect segmentation but as the molecular data has come in from *Drosophila* they have not been supported. Although there are gene products which are distributed in periodic patterns, these patterns do not seem to be set up by a reaction–diffusion process, but rather by independent regulation of each stripe of expression by a different combination of gene products expressed earlier in development (see Chapter 7). The models may of course be more successful elsewhere, for example in the determinationn of polyps in colonial hydroids where there is abundant evidence for long-range inhibition.

Mechanical instabilities

Although we have presented the onset of spatial pattern as a bifurcation effect in systems of chemical reactions, rather similar dynamical properties can also shown by mechanical systems. In the embryological context this has been studied in some detail by Odell, Oster and coworkers (Odell *et al.*, 1981; Oster, Murray & Harris 1983), with the object of explaining the characteristic types of cell and tissue movement. They show that a repeating pattern of cell condensations can be set up from homogeneous starting conditions by a mesenchymal–extracellular matrix interaction. This is a complex model with no fewer than nine groups of dimensionless parameters, so will not be described in full. The basic idea is that cell movement is partly random, rather like the diffusion of substances considered above, and partly directed by the extracellular matrix (ECM). In particular, cells will tend to align themselves along strain lines in the matrix and will tend to migrate up adhesive gradients. The cells also remodel the matrix by exerting tractive forces on it. These forces can stretch the matrix in particular directions and can increase or decrease the density and hence the adhesiveness of the matrix at different positions.

Earlier we considered the effect of increasing the size of a field in which a reaction–diffusion process was operating. Now consider what happens to a uniform field of fibroblasts scattered on an elastic extracellular matrix as the traction exerted by the cells is increased. When the traction is low, the remodelling of the matrix may occur but it is insufficient to resist the dispersive forces, so the cell density remains uniform. But when the traction reaches a certain threshold value, any small chance aggregation of cells will become self-amplifying. There are two reasons for this: firstly, because the increase in matrix density brought about by the aggregate will attract in further cells by its greater adhesiveness, and secondly because a radial strain guidance field is set up. So cells become denser in the regions where cell density is already high and become sparser in the regions where cell density is already low. This autocatalytic effect is ultimately limited because the ECM is also an elastic medium and the elastic resisting force limits both the maximum distortion attainable and also its geographical extent. The end result consists of a series of peaks in one dimension or of isolated peaks in two dimensions. This is rather like the GM1 model except that the spatial pattern is of cell density rather than the concentration of a morphogen.

Since the mathematical formalism of the mechanical models is quite similar to that of the reaction–diffusion models, it is not surprising that they have some of the same problems. Although no detailed studies have yet been published, the spacing of peaks clearly depends on the size of the field and so some special pleading is required to accommodate proportion regulation. The model is capable of producing repeated structures in a sequential manner if there is, for example, a gradient of some critical parameter value across the field. However, for somite formation, in particular, it is well known that the visible segmentation process can 'jump across' surgical gaps, suggesting that some cryptic repeating pattern has already been laid down. For a chemical model the essential events will be invisible at the light microscope level and so a cryptic prepattern is to be expected, but for a mechanical model the proposed cell movements would be visible and so the existence of cryptic prepatterns poses a problem.

A clock model

It was with the object of accounting both for proportion regulation and for sequential formation of vertebrate somites that the 'clock and wavefront' model was advanced (Cooke & Zeeman, 1976). It should really have been called the 'clock and gradient' model to indicate more clearly its two principal components, since the term 'wavefront' has been almost universally misunderstood.

The gradient is set up early in development and specifies the *rate* at which cells progress towards the act of segmentation. When they reach a critical time, they change their adhesive properties in some appropriate way. The clock is a biochemical oscillator (*see* eg, Winfree, 1980) which operates synchronously throughout the tissue. It generates an oscillatory concentration which is 'added' to the value specified by the gradient and thus causes groups of cells to change their adhesiveness together rather than in a smooth progression from one end of the field to the other. Each group then forms one segment. The so-called 'wavefront' refers to the visible process of segmentation which sweeps from one end of the field to the other. It was called a wavefront rather than a wave or a front because at this stage there is no communication between one cell and the next.

The model accounts for proportion regulation by assuming that the original gradient is able to regulate. It accounts for the sequential character of segmentation also by reference to the original gradient. It also accounts for the fact that the segmentation process appears to jump gaps where a hole is made in advance of the latest formed segment which is a well known feature of somite formation not easily explained by mechanical instability models. Although the clock and wavefront model is of some interest in that it invokes time as an essential element in the generation of a spatial pattern, the main problem is that it has never really been spelled out in detail and so has never really been properly tested. Readers are invited to consider whether the model of Stern *et al.*, for somitogenesis (Chapter 6) does or does not fall into the clock and wavefront category.

As they read the experimental parts of this book, readers are asked to keep in their minds the ideas of Chapters 2 and 3. The concepts and categories enumerated in Chapter 2 really are a useful aid to clear thinking and enable a particular type of event or process to be recognized from the data. The models of Chapter 3 represent those dynamical behaviours of chemical systems which underly regional specification. In the modern world of molecular detail it is worth asking about a particular gene: is there a sharp threshold around its domain of expression? If so, then is this maintained by autoactivation? Is any observed autoactivation sufficiently strong to cause bistability? If a molecule has a graded distribution then where are the sources and sinks? How do they work? Does this substance evoke thresholds in the synthesis of other substances? How many thresholds? Answers to such questions will both put molecular flesh onto the models and bring dynamical understanding to the molecular biology.

4

Hierarchies of developmental decisions

XENOPUS AND OTHER AMPHIBIANS

Because of the many technical advantages which they offer, amphibian embryos have been a favourite material for the experimentalist for over 100 years. The eggs are large, usually 1–2 mm in diameter, and develop from egg to tadpole outside the mother so remaining accessible to experimentation at all stages. This makes them particularly suitable for micro-operative procedures. In addition, as we shall see below, fragments of early embryos and even isolated cells are able to continue development if incubated in simple salt solutions. This is possible because every cell in the embryo contains a supply of yolk platelets which serves as its nutrient supply until the larval blood circulation becomes established. Recent years have seen intense activity involving the combination of the longstanding microsurgical techniques with cell lineage labelling and the techniques of molecular biology.

Although many different species have been used for experimental work in the past, the African clawed toad *Xenopus laevis* is now the world standard amphibian for the study of early development because of its ease of maintenance, ease of induced spawning and robustness of the embryos. Other laboratory species such as *Rana pipiens*, the American leopard frog, and *Ambystoma mexicanum*, the axolotl, have become much less important. This concentration on a single species has also been apparent in mammalian and insect embryology where the mouse and *Drosophila* respectively have assumed hegemonic positions. There are obvious advantages in standardization and the rapid transfer of results between laboratories, but there are also problems, principally that different species *are* slightly different and concentration on a single one can lead to unwarranted generalizations.

The two principal taxonomic groups of amphibian are the anura (the frogs and toads) and the urodeles (newts and salamanders). *Xenopus* is not unlike the other anurans formerly used for experimental work except that its eggs are relatively small (1.4 mm diameter) and development relatively rapid (1 day to general body plan stage). Urodeles generally have larger

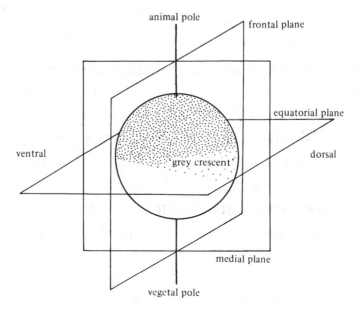

Fig. 4.1. Axes of the amphibian egg after fertilization.

eggs, develop more slowly, and have significant differences in their gastrulation movements. The majority of classical studies in the 1920s and 1930s used newt embryos gathered in the field, and although many of the results are true also for *Xenopus* a certain index of suspicion needs to be maintained about transferring results between species.

The experimental production of *Xenopus* embryos is very simple. The male and female are both injected with human chorionic gonadotrophin, are put together overnight and the next morning there are embryos. Nowadays it has become more common to perform *in vitro* fertilizations which generates smaller numbers of embryos, but ones whose time of fertilization is known and whose development shows a high degree of synchrony.

NORMAL DEVELOPMENT

This account refers to *Xenopus* unless otherwise stated. When laid the egg has a dark pigmented *animal hemisphere* and a light-coloured *vegetal hemisphere*. It lies within a transparent vitelline membrane inside the jelly coat. After fertilization the membrane lifts from the egg surface and the egg rotates under the influence of gravity so that the less dense animal hemisphere lies uppermost. As in other types of embryo the two polar bodies appear at the animal pole and contain the unused chromosome sets from the first and second meiotic divisions. The first polar body is produced

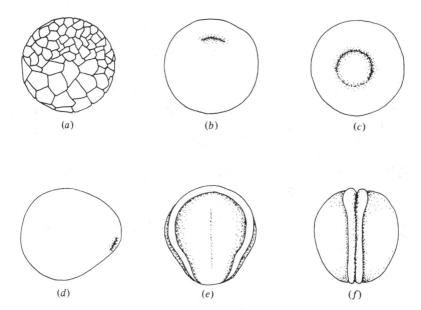

Fig. 4.2. Stages of amphibian development. (*a*) blastula, (*b*) early gastrula, (*c*) late gastrula, (*d*) slit blastopore, (*e*) early neurula, (*f*) closed neural tube. (*a*) is a view from the side, (*b*) and (*c*) are views from the vegetal pole, (*d*) is a view from the side, and (*e*) and (*f*) are dorsal views.

after maturation and the second after fertilization. The axes of a generalized amphibian egg are shown in Fig. 4.1 and stages of early development in Fig. 4.2. The sperm enters the animal hemisphere and, as we shall see, the side of sperm entry later becomes the ventral side of the embryo. In *Xenopus* and other anurans the first sperm to enter imposes a block to any others. Urodele eggs are usually fertilized by more than one sperm although only one sperm pronucleus actually fuses with that of the egg. Shortly after fertilization there is a rotation of the egg cortex relative to the interior which is associated with the transient appearance of an oriented array of microtubules in the vegetal hemisphere. This leads to a reduction in the pigmentation of the animal hemisphere on the prospective dorsal side, opposite the sperm entry point. In some batches of eggs from some other species a similar pigmentation change appears as the famous '*grey crescent*'.

The first cleavage is vertical and usually bisects the egg meridionally, separating it into right and left halves. The second cleavage is at right angles to this and usually separates prospective dorsal from ventral halves. The third cleavage is equatorial, separating animal from vegetal halves. As in other embryos, the large cells resulting from early cleavage divisions are called *blastomeres*. As cleavage takes place a cavity called the *blastocoel* forms in the centre of the animal hemisphere and the embryo is referred to

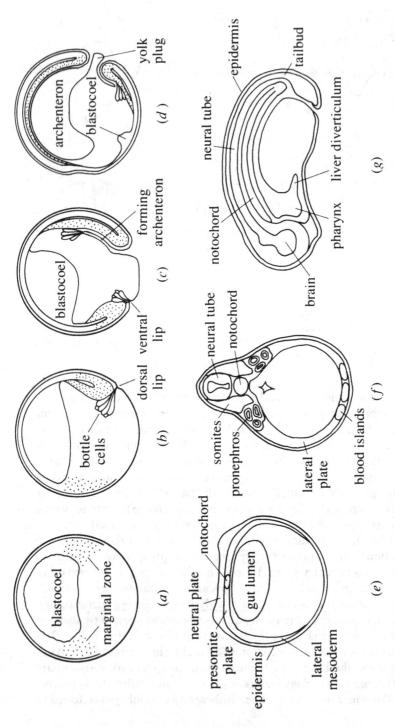

Fig. 4.3. Gastrulation and body plan formation in *Xenopus*. (*a*) blastula, (*b*) early gastrula, (*c*) mid-gastrula, (*d*) late gastrula. (*a*) to (*d*) are shown in medial section with the deep marginal zone shown stippled. (*e*) transverse section through neurula, (*f*) transverse section through tailbud stage, (*g*) medial section through tailbud stage.

as a *blastula*. The outer surface of the blastula consists of the original oocyte plasma membrane which is physically more rigid and binds a number of lectins and antibodies more avidly than the newly formed cleavage membranes. A complete network of close junctions around the exterior cell margins seals the blastocoel from the exterior and renders the penetration of almost all substances highly inefficient. This is why radiochemicals and other substances need to be introduced into the embryo by microinjection. Internal cells are connected by junctions less 'close' than those at the periphery and readily disrupted by removal of calcium ions from the medium. Desmosomes are not found until the neurula stages but all the cells of early cleavage stages are connected by gap junctions which allow limited exchange of molecules below about 1000 molecular weight. Cleavage continues rapidly for 12 divisions after which synchrony is lost, the division rate slows, and zygotic gene transcription commences. This time is known as the *mid-blastula transition* or **MBT** (Newport & Kirschner, 1982*a,b*). There is also a distinct compaction event at which intercellular adhesion increases and the blastula appears as a smooth instead of a knobbly ball.

The next phase of development is one of extensive morphogenetic movements called *gastrulation* (Fig. 4.3 (*a*)–(*d*)). The motor for gastrulation is mainly provided by a belt of tissue around the equator of the embryo called the *marginal zone*, of which the surface layer ends up as the lining of the *archenteron* cavity while the deep part is the prospective mesoderm. Although some autonomous expansion of the animal hemisphere and some dorsal convergence of the marginal zone have already occurred, the conventional start of gastrulation is marked by the appearance of a pigmented depression in the dorsal vegetal quadrant. This is the *dorsal lip of the blastopore*. The blastopore becomes elongated laterally and soon becomes a complete circle. When it is circular, the part referred to as the dorsal lip is the dorsal segment of the complete circle and the part referred to as the ventral lip is the ventral part of the complete circle. The blastopore is associated with the formation of bottle cells and is the locus of invagination of the superficial marginal zone which results in the creation of the archenteron. The invagination occurs all round the blastopore but is much more extensive on the dorsal side where it proceeds until the leading edge of tissue is well past the animal pole, with the archenteron roof being closely apposed to the overlying ectoderm. In the lateral and ventral parts of the blastopore there is only a small invagination. By the end of gastrulation the entire archenteron roof and floor have been formed from the dorsal invagination except for a small crescent of archenteron floor at the posterior end which is formed from the lateroventral invagination. Simultaneously with the formation of the archenteron, the deep marginal zone 'involutes'. The deep marginal zone can be imagined as a cylinder around the equator of the embryo and involution consists of the cylinder turning inward at the lower edge and then extending towards the animal pole. The dorsal involution is most extensive while the ventral involution is slight and the process is accompanied by a general shift of tissue in a dorsal direction. Simultaneously with the movements of the marginal zone, the animal cap tissue spreads vegetally so that the blastopore lip becomes a smaller and smaller ring around the vegetal pole and eventually narrows to a small slit. Once the vegetal pole has been internalized it occupies a position on the archenteron floor in front of which the floor has been derived from the dorsal invagination and behind which it was formed from the ventral invagination.

Extensive studies have been made by the group of Keller who consider that it is not the bottle cells but the movements of the deep marginal zone cells that provide the motive force for the invagination and involution processes. These can be considered as made up of two components: *convergence* and *extension*. Convergence represents the packing together of cells to shrink the total marginal zone circumference and occurs equally around the marginal zone. Extension refers to an active process of cell intercalation leading to elongation in the animal–vegetal direction, and this occurs only in the dorsal region (Keller *et al.*, 1985; Keller & Danilchik, 1988).

In urodeles, the gastrulation movements are somewhat different. The archenteron is entirely formed by the dorsal invagination while in the lateral and ventral regions of the blastopore lip there is a separation of marginal zone from vegetal tissue and the future mesoderm extends toward the animal pole as a cell sheet with a free edge. In *Xenopus* the archenteron roof is at all times lined with endoderm while in urodeles the forming notochord is the archenteron roof and is the part which is common to the dorsal lip-derived invagination and the whole blastopore-derived involution of mesoderm. There is no clear distinction between surface and deep marginal zone, both contributing substantially to the mesoderm. In the newt *Pleurodeles* the injection of antibodies to fibronectin can inhibit gastrulation so the tissue movements are thought to depend on adhesion between the cells at the leading edge and fibronectin deposited on the blastocoelic surface of the animal hemisphere (Boucaut *et al.*, 1984). But this is probably not true of *Xenopus*.

By the end of gastrulation the former animal cap ectoderm has covered the whole external surface of the embryo, the yolky vegetal tissues have become a mass of endoderm in the interior, and the former marginal zone has become a layer of mesoderm extending from the slit-shaped blastopore and reaches the anterior end on the dorsal side but only a short distance on the ventral side. In other words, the three classical germ layers, ectoderm, mesoderm and endoderm, have achieved their final trilaminar arrangement. The archenteron has become the principal cavity at the expense of the blastocoel, and the embryo has rotated so that the dorsal side is uppermost. It now has a true anteroposterior (or craniocaudal) axis which runs from the leading edge of the mesoderm to the blastopore.

The next stage of development is called the *neurula* (Fig. 4.3(*e*)), in which the ectoderm on the dorsal side becomes the central nervous system. The *neural plate* becomes visible as a keyhole-shaped region delimited by raised neural folds and covering much of the dorsal surface of the embryo. Quite rapidly the folds rise and move together to form the *neural tube* which becomes covered by the ectoderm from beyond the folds, now known as epidermis. Tissue from the folds which comes to lie dorsal to the neural tube is the *neural crest*. In the mesodermal layer the posterior part of the dorsal midline segregates as a distinct *notochord*, and the mesoderm on either side begins to become segmented in anteroposterior sequence to form paired *somites*. In *Xenopus* and other anurans the entire archenteron is lined by endoderm from its formation but in urodeles the archenteron roof initially consists of the notochord and presomite plate and the endoderm later sends up folds which meet beneath the notochord. The end of neurulation is a convenient place for us to stop since by this stage the rudiments for all the major structures of the body are in their definitive positions (Fig. 4.3 (*f*),(*g*)). The central nervous system is formed from the neuroepithelium of the neural tube. The mesoderm of the trunk region develops into several structures in dorsoventral sequence: notochord, myotomes, pronephros and mesonephros, and ventral blood islands. Anteriorly the prechordal mesoderm forms part of the jaw muscles and branchial arches. The heart, limbs and future haemopoietic system also exist as rudiments in the mesodermal mantle. The endoderm becomes the epithelial components of the pharynx, stomach, liver, lungs (if any), intestine and rectum, and the tail arises from the tailbud.

Although by this stage the cells still appear to be yolk-filled bags with little visible histological differentiation most of the major tissues, at least in *Xenopus*, are now known to commence their terminal differentiation at the end of gastrulation. It should also be noted that, although an amphibian embryo does become a little bigger over the period covered, this is entirely due to the uptake of water. The embryo has no supply of nutrient from an extraembryonic yolk mass or a placenta and therefore is actually losing matter throughout development by metabolic activity. All early cell divisions are cleavage divisions in which daughter cells are half the size of the mother cell. These things are true of all holoblastic eggs which develop independently of the mother, but are often forgotten by mammalian or chick embryologists whose organisms undergo growth and development simultaneously.

The stages of *Xenopus* development are described by Nieuwkoop & Faber (1967). According to this series, stage 8 is the midblastula, stage 10 the early gastrula, stage 13 the early neurula and stage 20 the end of neurulation. Axolotl stages are very similar to this but the series for *Pleurodeles* and *Rana pipiens* are different and so care is again needed when making interspecies comparisons.

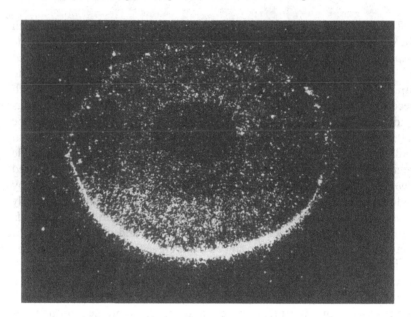

Fig. 4.4. Vertical section through a late oocyte showing the location of the *vg-1* mRNA by *in situ* hybridization. (Figure kindly provided by Dr D.A. Melton.)

Molecular markers

Traditional embryological experiments have relied on the histological differentiation of cells or on characteristically shaped multicellular arrangements to identify structures. In recent years a search has been made for early molecular and immunological markers which can be used to identify cells at a time nearer to that of the determinative events, or even be the molecular basis of the determinative events themselves. These have become far too numerous to catalogue here and so we shall mention only those which crop up later in the Chapter in the course of embryological experiments.

In *Xenopus*, as in many animal embryos, there is little or no transcription during early embryonic stages and protein synthesis is directed by maternal mRNA accumulated during oogenesis. Although rather few messages have turned out to be localized, a few have now been found by differential cDNA screening. The most important of these to date is *vg-1* (Weeks & Melton, 1987), a molecule which is a member of the TGF beta superfamily of cytokines, raising the possibility that it has some role in mesoderm induction (see below). *vg-1* mRNA was shown to become localized in the oocyte at the onset of vitellogenesis, after which time it is present as a thin shell close to the plasma membrane of the vegetal hemisphere (Fig. 4.4). Synthesis of the protein continues during the early embryonic stages and is confined to the vegetal hemisphere.

Zygotic gene expression commences at the midblastula transition (Newport & Kirschner, 1982*b*) and number of the newly expressed genes have been identified using differential cDNA cloning. Genes turned on at this stage include two interesting ones containing homeobox sequences: *mix-1* which is expressed throughout the endoderm (Rosa, 1989) and *xhox-3* which is expressed in the ventroposterior mesoderm (Ruiz i Altaba & Melton, 1989*a*). The homeobox is a coding sequence for a DNA binding polypeptide found as a part of many homeotic genes in *Drosophila*. The muscle specific α-actin gene is expressed in the somites from the end of gastrulation (Mohun *et al.*, 1984) and has been much used as a marker in studies of mesoderm induction. It is preceded by *Myo-D* which was previously identified as an important regulatory element in myogenesis from studies on mammalian cells (Hopwood *et al.*, 1989*a*). The non-muscle parts of the mesoderm and the cephalic neural crest start to express a gene called *twist* during gastrulation (Hopwood *et al.*, 1989*b*). This gene was first identified as necessary for the formation of the mesoderm in *Drosophila* and is likely to be used as a marker in the future. In the neural plate expression of the neural cell adhesion molecule N-CAM is greatly increased compared to the remainder of the embryo (Kintner & Melton, 1987), and certain genes for intermediate filament proteins have been found whose expression is restricted to the anterior or the posterior regions, thus being potentially useful markers for the study of regional specificity in neural induction (Sharpe, 1988).

Apart from localized maternal sequences and early-expressed zygotic sequences, the other classes of gene which have seemed reasonable candidates for some role of developmental importance are those known as oncogenes in mammals or those containing DNA binding motifs such as homeobox or zinc fingers. The homeobox was first discovered in homeotic genes of *Drosophila* but homeobox-containing genes are now thought to exist in all animals, and the products of some of them are suspected of being coding factors for anteroposterior body levels (see also Chapters 6 and 7). Several homeobox genes have been cloned from *Xenopus* and their expression patterns studied, the majority are expressed preferentially in the posterior mesoderm and neural tube although there are also some restricted to an anterior position. Overexpression studies have been conducted with some by injection of synthetic mRNAs into zygotes or early developmental stages and these will be mentioned in the appropriate context below. At the time of writing it was still not possible reliably to ablate the activity of a specific gene although the ability to do this is only a matter of time and will go some way towards remedying the fact that amphibians are rather poor animals for experimental genetic work.

Studies with mono- and polyclonal antibodies and with lectins have not revealed any regional expression prior to the early neurula with the exception of the outer, oocyte-derived, membrane which binds a number of reagents not bound by the inner, cleavage derived, membrane. By contrast, a complete panel of reagents is available for the principal tissues which

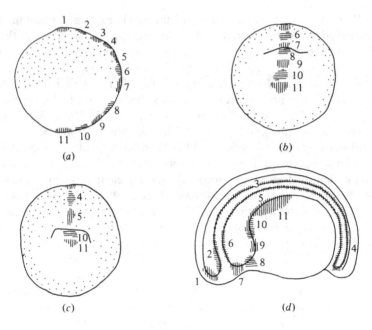

Fig. 4.5. Movement of vital stain marks applied to the external surface of early urodele embryos. (*a*) The marks immediately after application to the blastula. (*b*) and (*c*) the invagination of the marks through the dorsal lip. (*d*) the positions of the marks in the neurula. (After Vogt, 1929.)

become visible as cell clumps during neurulation. These have often been used to add apparent rigour to experiments in recent years but usually add little to what can be seen by conventional histology. An exception to this is the very useful anti-muscle antibody made by Kintner & Brockes (1984), which can be expressed in cells which do not yet look like muscle morphologically.

Fate maps

Like many animal embryos, amphibians have a continuous topographic projection from the fertilized egg to later stages. In other words it is possible at any stage to construct a fate map of the embryo which shows what each volume element will become and where it will move in the subsequent stages. The map cannot be exact, however, because there is some short-range mixing of cells during gastrulation and this means that the prospective regions at early stages are somewhat larger than the later structures. As discussed in Chapter 2 the prospective region includes the entire volume from which some cells are contributed to the structure in question and will therefore overlap prospective regions for other structures.

Classical fate maps for amphibians were constructed by staining parts of the embryo with vital dyes and locating the stained patch at later stages

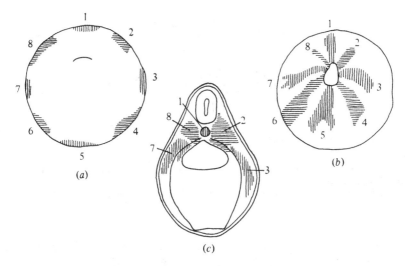

Fig. 4.6. Movement of vital stain marks applied to the marginal zone of an anuran gastrula. The marks are originally equally spaced around the equator and label both superficial and deep tissues. They invaginate through the blastopore and end up as longitudinal stripes concentrated towards the dorsal side of the mesodermal mantle. (After Vogt, 1929.)

(Vogt, 1929; Pasteels, 1942). Although they have recently been refined in detail, the overall picture obtained from these experiments has stood the test of time well. In Fig. 4.5 are shown the trajectories of a number of marks applied to the dorsal meridian of a urodele. Those near the blastopore move in sequentially through the dorsal lip and eventually end up disposed in a continuous ring around the long axis of the archenteron. The marks near the animal pole never reach the blastopore and so end up inside the neural tube. In Fig. 4.6 a similar experiment is depicted in which equatorial marks are used. These become greatly elongated as the marginal zone enters through the ring-shaped blastopore but dorsoventral continuity is maintained in the mesodermal mantle as is shown in the transverse section of the tailbud stage (the ventral marks are present only posterior to the section shown).

Despite their honourable history, vital dyes do have limitations. It is difficult accurately to mark regions within the embryo, and the dyes have a tendency to spread and fade so cannot be regarded as cell autonomous labels. For these reasons high molecular weight lineage labels are now preferred for fate mapping and for cell labelling as an adjunct to other experiments. Two of the most popular are horseradish peroxidase (HRP) (Jacobson & Hirose, 1978) and fluorescein dextran amine (FDA) (Gimlich & Braun, 1985). The labels can be used for fate mapping in two ways: either by injection into a single identified blastomere of an early cleavage stage, or

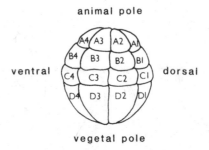

Fig. 4.7. Nomenclature of blastomeres at the 32-cell stage, following the convention of Nakamura.

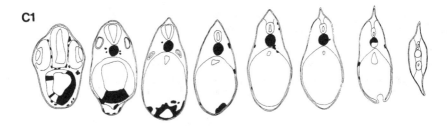

Fig. 4.8. Three-dimensional reconstruction of a typical specimen following labelling of a C1 blastomere at the 32-cell stage.

by injection of a fertilized egg and the use of the resulting uniformly labelled embryo as a donor for grafts. In Fig. 4.7 is shown the most popular nomenclature for blastomeres at the 32-cell stage and in Fig. 4.8 is shown a reconstruction of a typical injection into a C1 blastomere. A fate map which shows the origins of the principal tissues is shown in Fig. 4.9 (Dale & Slack, 1987*a*), and these results are given again in a traditional but less precise form in Fig. 4.13. A similar lineage label study on cleavage stage embryos has been published by Moody (1987) and extensive data for the CNS by Jacobson (Jacobson & Hirose, 1981; Jacobson, 1983). All the lineage label studies have shown that there is more local cell mixing than was previously suspected, particularly in the lateroventral region. The prospective regions are not therefore sharply delineated as in classical fate maps but rather they are fuzzy patches which overlap with their neighbours to a greater or lesser extent. A vital stain fate map for the early gastrula stage of *Xenopus* was published by Keller (1975, 1976) and has also been widely used, complementing the more recent lineage label studies.

Although the fate map has a statistical character, it remains an essential baseline for interpretation of embryological experiments. Unless we know what a group of cells will do in normal development, we cannot very well

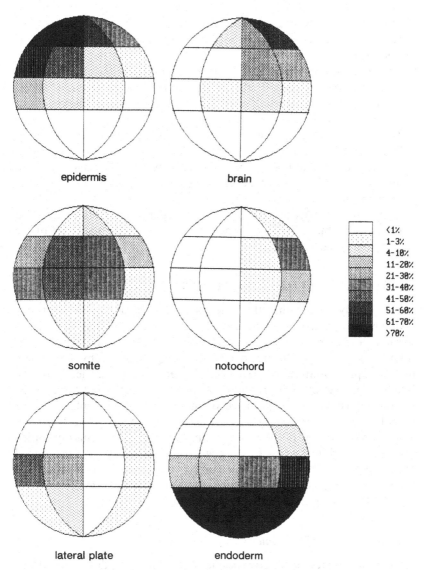

Fig. 4.9. Fate maps for the 32-cell stage *Xenopus* embryo (animal pole up; dorsal side on right). The stipples show the proportion of the volume of each blastomere entering various tissues in the course of normal development.

claim to have changed their behaviour by experimental treatment. The features which need to be borne in mind are as follows:

1. The epidermis and neural plate arise from the animal hemisphere, the neural plate from the dorsal and the epidermis from the ventral region.

2. The mesoderm arises from a belt around the equator but, in *Xenopus*, not from the surface layer of cells carrying the oocyte derived membrane.
3. The dorsal side of the blastula contributes mainly to the anterior of the later body and the ventral side of the blastula to the posterior; however, the dorsal marginal zone also populates the dorsal midline axial structures along the entire body length.
4. There is a rough mapping of mesodermal tissue types in the fate map from dorsal to ventral in the order: notochord, heart, myotomes, pronephros, lateral plate, blood islands. However, the region forming muscle is rather extensive, and about 60% of muscle normally derives from the ventral half of the blastula.
5. The yolk mass part of the endoderm comes from the yolk rich vegetal hemisphere, but the lining of the pharynx from the dorsal part of the marginal zone.

REGIONALIZATION WITHIN THE EGG

It is generally agreed that the unfertilized egg is radially symmetrical, having no prospective dorsal or ventral side. Along the animal–vegetal axis (egg axis) there are various inhomogeneities which become established by unknown mechanisms during oogenesis. Among the more obvious of these are a deep pigmentation of the superficial layer of the animal hemisphere; a difference in size of yolk granules (small in the animal and large in the vegetal hemisphere); the presence of more non-yolk cytoplasm in the animal hemisphere including the remains of the oocyte nucleus (germinal vesicle). Whether any of these visible inhomogeneities are causally related to the later regionalization of the egg axis is not known. As mentioned above there are a few mRNAs which are known to become localized to the animal or the vegetal hemisphere during oogenesis and of these at least *vg-1* is generally thought to be of some significance.

The dorsoventral polarization

The establishment of distinctly specified zones along the dorsoventral axis occurs shortly after fertilization. Although in *Xenopus* the normal development of the light pigmented area and later the dorsal blastopore lip appear opposite the point of sperm entry, it is possible to alter the prospective dorsoventral axis after fertilization by tipping the egg through 90°. This can readily be done by immersing the eggs in a solution of Ficoll, a polysaccharide which can remove the water osmotically from the space below the vitelline membrane so that the egg can no longer rotate within the membrane. Under these circumstances a tipped egg will stay put and internal cytoplasmic rearrangements take place in response to the gravitational field. If an egg is tipped through 90° up to about 75 minutes

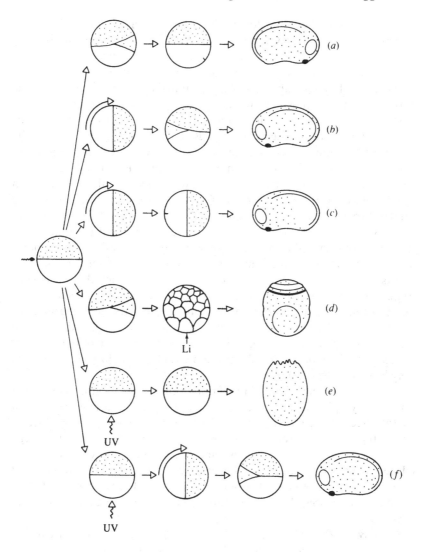

Fig. 4.10. Experiments on determination of dorsoventral polarity in the early amphibian embryo. (*a*) normal development in which the dorsal side develops opposite the point of sperm entry. (*b*) inversion of the axis by a 90 degree rotation after fertilization. (*c*) alteration of the animal–vegetal axis of the embryo cortex by sustained rotation through 90 degrees. (*d*) production of hyperdorsal embryo by lithium treatment of the early blastula. (*e*) production of extreme ventral type embryo by irradiation with ultraviolet (UV) light shortly after fertilization. (*f*) 'Rescue' of UV-irradiated embryo by subsequent 90 degree rotation.

postfertilization, then the dorsal side appears on the face which was uppermost (Fig. 4.10(*b*); Gerhart *et al.*, 1981). If this conflicts with the side opposite the point of sperm entry then the rotation overrides the sperm.

The motor of normal dorsoventral polarization appears to involve an array of parallel microtubules which appear in the vegetal hemisphere at the appropriate time (Elinson & Rowning, 1988). In an undisturbed egg, the outer cortex rotates by about 30° relative to the inner cytoplasm. The axis of rotation is the future mediolateral axis, and the direction is such that the cortex of the animal hemisphere moves toward the sperm entry point (Vincent & Gerhart, 1986). These movements are inhibited by drugs such as colchicine, or by other treatments which depolymerize microtubules such as low temperature. After tipping of the whole egg it is likely that some similar displacements would be initiated by the slippage of the vegetal yolk mass. Since tipping will repolarize up to but not beyond the time of the normal cytoplasmic shifts it is probable that the gravitational movements are sufficient to break the symmetry of the newly fertilized egg, and the effects of the sperm can be overridden because its perturbation is even weaker. If eggs are maintained in a 90° rotated position throughout early development then the dorsal blastopore lip appears near the original vegetal pole and the parts of the embryo are drastically misplaced relative to the cortical pigmentation (Fig. 4.10(*c*)), although because of the coherent movement of the inner cytoplasm the internal regionalization of the egg may be similar to the normal.

It seems clear that the definitive dorsoventral specification is brought about by cytoplasmic movements inherent to the structure of the egg, not by the effect of the sperm or of gravity. In other words, we have here a symmetry breaking process where the dynamical properties of the egg cytoskeleton are the formal cause and the sperm or gravity merely the efficient cause of dorsoventral polarization. A possible mechanism based on mutual reinforcement of tubule alignment and shear is suggested by Gerhart *et al.* (1989). For more about symmetry breaking and causes, see Chapter 3.

Black & Gerhart (1986) showed that centrifugation of one- or two-cell stages could result in *twinning* and although the mechanism of this is not understood it does seem to be the gravity driven cytoplasmic re-arrangement after the spin that is responsible. The resulting embryos resemble the partial twins that may be produced by a variety of other methods such as partial constriction with a hair loop at the two-cell stage, or organizer grafts at the early gastrula stage (see later). They are characterized by the presence of two separate sets of dorsal axial structures over at least part of the anteroposterior axis with a plane of mirror symmetry in between them.

It was formerly thought that the dorsoventral polarity could be altered at later stages by such treatments as imposing a temperature or an oxygen gradient favouring the former ventral side. However, these experiments

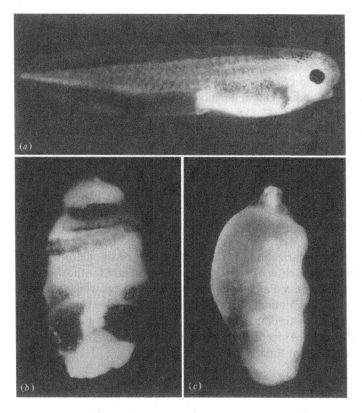

Fig. 4.11. External appearance of the extreme lithium and UV embryos. (*a*) normal, (*b*) lithium treated, (*c*) UV irradiated, ((*b*) and (*c*) kindly supplied by Dr K. Kao.)

have not proved to be repeatable. It seems that after the two-cell stage the dorsoventral polarity, at least in *Xenopus*, is fixed and unalterable, although the level of dorsoventral commitment of each region may still be altered either by grafting or by treatment with lithium (see below).

UV irradiation

A stimulus which is known to disrupt the normal cytoplasmic shifts is ultraviolet (UV) irradiation (Fig. 4.10(*e*) and 4.11(*c*); Malacinski, Benford & Chung, 1975). If fertilized eggs are irradiated from the vegetal side, the resulting embryos develop with a deficiency in the structures normally formed from the dorsal side of the blastula. As we have seen above, the dorsal part of the blastula contributes most cells to the anterior part of the postgastrulation body, and also populates the dorsal midline structures along the entire body length. Hence a mild UV effect involves a reduction of the head, a more severe one the absence of the head, and a most severe one, absence of the head and absence of the dorsal midline structures in the trunk. The effects of UV are dose dependent, and the limit form resulting from the highest doses is a radially symmetrical embryo which contains a normal proportion of mesoderm, but mainly in the form of blood islands, normally the tissue characteristic of the ventroposterior part of the mesodermal mantle (Cooke & Smith, 1987). Since the radiation only penetrates a short distance, it has relatively little effect on cellular viability and the limit forms can survive for many days.

Originally the UV effect was thought to be due to the destruction of a dorsal or axial determinant, but it was later found by Scharf & Gerhart (1980) that embryos which had been subjected to UV irradiation just after fertilization could still form a normal pattern if they were tipped before the normal time of subcortical rotation (Fig. 4.10(f)). So the present majority view is that the UV effect involves damage to a component required for the dorsoventral polarization, perhaps the subcortical microtubule array, rather than destruction of a cytoplasmic determinant for the dorsal region. However, Elinson & Pasceri (1989) showed that eggs which had been UV irradiated as *oocytes* underwent the normal subcortical rotation and could not be rescued by tipping. The pattern defects from irradiation of oocytes are similar to those from irradiation of eggs and this result has given the dorsal determinant theory a new lease of life.

Lithium

Kao, Masui & Elinson (1986) showed that treatment of cleavage stages with high concentrations of LiCl could have drastic effects on pattern (Figs. 4.10(d), 4.11(b)). In principle, the change is the converse of the UV effect, embryos behaving as though they contained a higher than normal proportion of dorsal tissue. The limit form here is a radially symmetrical hyperdorsal (or hyperanterior) form in which a continuous band of pigmented retina and of cement gland extends around the body and there is a central core of notochord with little muscle (Cooke & Smith, 1988 and Fig. 4.11). The maximum sensitivity of whole embryos is found at the 32–64 cell stage, but this may simply reflect the period of maximum exposure of new cleavage membrane at the surface and hence penetrability of the embryos. Although external concentrations of the order of 0.3 M are used, the intracellular concentrations measured have been a few mM. It has also been shown that tissue explants, which have new cleavage membrane exposed to the exterior, can be dorsalized at much later stages (see below). Embryos which have been ventralized by early UV irradiation can later be rescued back towards a normal pattern by Li treatment, and this is true not only of cases irradiated as eggs, but also of cases irradiated as oocytes, which as we have seen above are refractory to rescue by tipping. This suggests that Li is mimicking some event which occurs at a multicellular stage and is normally a consequence of the subcortical rotation.

If LiCl is injected into ventrovegetal blastomeres of normal embryos at the 32-cell stage, it is possible to induce twinning, implying that a second centre of dorsal character has been formed on the ventral side. In a dramatic experiment it was shown that this effect could be inhibited by coinjection of *myo*-inositol (Busa & Gimlich, 1989). To understand the significance of this, it is necessary to know that many hormones, neurotransmitters and cytokines exert their action by causing the hydrolysis of phosphatidyl inositol bisphosphate (PIP$_2$) into two biologically active components: inositol triphosphate (IP$_3$) and diacyl glycerol (DAG). One of the biochemical effects of Li is to block the re-synthesis of the PIP$_2$ by inhibition of the enzyme inositol monophosphatase, normally responsible for regenerating the free inositol from the IP$_3$ and the inhibition of the cycle eventually reduces the levels of IP$_3$ and DAG. The proposition then is that the ventral side normally has an active cycle and the dorsal side a depressed cycle, but confirmation of this will have to await the relevant metabolic studies.

Curiously enough, Li had been known to have pattern altering effects on amphibian embryos for decades preceding the recent burst of activity. But it had always been applied to much later stages, usually gastrulae, and the results were somewhat less interesting consisting of reductions of the head and in extreme cases suppression of the trunk notochord as well (Lehmann, 1937; Bäckstrom, 1954). At least the head reductions are probably secondary to an inhibition of gastrulation movements which mean that the leading edge of the dorsal invagination reaches less far than usual, and the induction of anterior neural structures cannot occur. The suppression of the notochord is less easy to explain as an indirect effect. It occurs without cell death, the dorsal midline cells becoming muscle rather than notochord and does therefore seem to indicate a ventralization of the dorsal midline occuring during gastrulation. This is more or less the opposite effect to that shown on the cleavage stages and according to Yamaguchi & Shinagawa (1989) the switchover in effects occurs at the MBT. So if Li really is affecting the inositol phosphate cycle, the effects of the cycle on regional specification must be different at different stages.

A hyperdorsal phenotype can also be provoked by treating eggs with heavy water (D$_2$O) during the first third of the first cell cycle (Scharf *et al.*, 1989). This is less extreme than the Li effect and seems to depend on a quite different mechanism. D$_2$O stabilizes microtubules, it makes eggs more rigid, it causes the microtubule array to form precociously and some subcortical rotation is still necessary for the formation of dorsal centres. D$_2$O treated eggs can be rescued towards the normal pattern by a subsequent cold shock, which depolymerizes the tubules again. If D$_2$O treated eggs are subsequently irradiated with UV then they tend to become completely ventralized.

REGIONAL ORGANIZATION DURING CLEAVAGE AND BLASTULA STAGES

States of commitment during cleavage

In *Xenopus* and other anurans the first cleavage plane is usually medial, the second is frontal and the third equatorial. A deterministic sequence of cleavages inevitably means that particular regions of zygote cytoplasm will be included in particular blastomeres and inhomogeneities will become differences between blastomeres. It is, however, known from the many embryos which depart from the normal cleavage pattern, and from compression experiments in which the normal sequence of cleavages is altered, that the positioning of plasma membrane relative to cytoplasmic regions is not critical for the operation of subsequent developmental decisions.

Comprehensive blastomere ablation studies have been carried out on *Xenopus* by Kageura & Yamana (1983, 1984) which are much more complete than the various studies done in former years. Separation of the first two blastomeres or medial division of the eight cell stage yields two half size but complete larvae. Although no structures are missing in these twins, there is a slight departure from normality in that slightly more tissue than usual is devoted to the dorso-anterior parts (Cooke & Webber, 1985). However, separation of four cell embryos in the frontal plane to produce dorsal and ventral halves produces dissimilar embryos: the dorsal member has dorsoanterior parts strongly overrepresented while the ventral half is severely deficient in axial structures and may achieve the radially symmetrical appearance of the UV limit form mentioned above. At the eight-cell stage, removal of either of the two dorsovegetal cells also produces UV type effects. All this is consistent with the formation of a dorsal signalling centre in the dorsovegetal region as a consequence of the subcortical rotation. The mechanism is still a matter of speculation but the apposition of regions of membrane and inner cytoplasm in the area of maximum displacement might lead to the synthesis of some localized determinant later involved in the signalling process. Blastomere transplantations point to the dorsal cells as being dominant over the ventral. Replacement of one ventrovegetal or two ventro-animal cells with dorsal ones will generate double dorsal duplications, while the converse transplantations have no effect (Kageura & Yamana, 1986).

An alternative way of assaying the dorsalizing capacity of blastomeres is to implant them into the corresponding position of an embryo which was UV irradiated as an egg with a dose sufficient to produce the UV limit form (Gimlich, 1986). The dorsal blastomeres of the third and fourth tiers (C1 and D1: Fig. 4.7) at the 32-cell stage are capable of rescuing such embryos back towards a normal pattern. The usual fate of the C1 blastomere includes much of the dorsal midline, but the D1 cell typically contributes its progeny to yolky endoderm vegetal to the prospective axial region. Because of these experiments and others to be described below under the heading of mesoderm induction, the concept has arisen that the ultimate dorsalizing centre, called the *organizer* in amphibian embryos, arises in the region of the B1 and C1 progeny cells because of an early induction from the C1/D1 region.

Another aspect of regionalization in the fertilized egg is the possible existence of a germinal determinant in *Xenopus* and other anurans (Smith, 1966; Whitington & Dixon, 1975). Anuran eggs contain a region of special cytoplasm in the vegetal cortical region which is later found in the primordial germ cells. In *Rana pipiens* irradiation of the vegetal hemisphere with low doses of UV light (lower than those which produce dorsoanterior defects) can produce larvae lacking germ cells but otherwise apparently normal. The action spectrum for this effect shows a peak at 260 nm, suggesting a nucleic acid target. Eggs can be 'rescued' by subsequent injection of vegetal but not animal cytoplasm, although this is not quite sufficient proof for the existence of a determinant because it is not known whether the germ cells in the rescued hosts actually inherit the donor cytoplasm.

States of commitment in the blastula

Commitment of nuclei

The nuclei of blastula cells do not appear to be restricted in potency since one nucleus can support the development of an entire embryo. This was first shown by Spemann in 1928 (reviewed by Spemann, 1938). A hair loop was used to constrict a fertilized egg of *Triturus* into a dumbbell shape in which one half contained the zygote nucleus and the other half did not. Cleavage takes place only in the nucleated half but if the loop is not too tight a nucleus may pass through the constriction up to the 64-cell stage and initiate development in the previously enucleated half. These halves developed into complete embryos despite the fact that their nucleus had already undergone several rounds of DNA replication and mitosis. A similar result was obtained by Briggs & King (1952) who injected blastula nuclei into enucleated host zygotes. This procedure has subsequently been used to clone amphibians since a number of embryos can be grown using nuclei from a single blastula (Gurdon, 1974; McKinnell, 1978). Nuclei from later stages become progressively less able to support the development of enucleated eggs, although the few successes with nuclei from fully differentiated cells have achieved wide publicity and are usually described in the first chapter of developmental biology textbooks since they are the best evidence that no genetic material is lost from the somatic genome during development (Gurdon, 1974; DiBeradino *et al.*, 1986).

Commitment of single cells

As described above, the use of cell lineage labels has revealed a fair amount of short-range cell mixing during gastrulation and this has enabled clonal analysis studies to be carried out on *Xenopus*. The strategy of clonal analysis is fully discussed in Chapter 2, but briefly the main point is that a cell is not yet determined if its progeny can later populate two structures or tissues.

In situ clonal analyses have been carried out by Jacobson (1983) who has argued that the dorso-animal quadrant of the embryo becomes divided into seven compartments between the 256- and 512-cell stages. Up to the 256-cell stage there is no clonal restriction and from the 512-cell stage onward the progeny of a cell spread out within each compartment but not between compartments. It should be noted that the restrictions are regional but not histospecific; all cells at the 512-cell stage may contribute descendants to more than one tissue type. Jacobson's interpretation has not gained universal support from other workers although it does seem probable that there are some regional differences in cell adhesivity before gastrulation which could bring about effects of this kind.

A different type of clonal analysis has been performed by Heasman and colleagues (Snape *et al.*, 1987; Wylie *et al.*, 1987) who have studied the behaviour of single labelled cells inserted into the blastocoel of a host embryo. The transfers are made into late blastulae and during gastrulation the labelled progeny spread out across the embryo and are thus exposed to a variety of different environments. By this method, vegetal pole cells, which form endoderm in normal development, are shown, as a population, to become progressively determined to form endoderm during the blastula stages and by the onset of gastrulation nearly all vegetal cells are so determined. Animal pole cells, which form epidermis and neural tube in normal development, become determined at a somewhat later stage, during gastrulation. At earlier stages they enter mainly endo- or mesodermal tissues presumably because of exposure to mesoderm inducing factors (see below). These commitments, which correspond to the classical definition of determination, seem rather late in the context of other events and are probably preceded by more labile forms of commitment which can be reversed by cell dissociation or changes in the extracellular environment, for example, Jacobson & Xu (1989) grafted single cells between dorsal and lateral marginal zones at the 512-cell stage and found completely position specific (ortsgemäss) behaviour.

Methods have recently been devised for the *in vitro* culture of isolated cells from middle blastulae. Since all early embryo cells are packed with yolk granules, they can divide and differentiate in a simple medium containing salts and a few defined protein additives (Godsave & Slack, 1989). If single cells are explanted into this medium from anywhere in the animal hemisphere then they self-differentiate into neurons, while groups of animal cells tend to form epidermis. If cells are taken from the marginal zone they may also form these types, or may form muscle or mesothelium. This shows that despite the equivalence of blastula nuclei, there are none the less regional differences at this stage in the state of specification of cells. Single clones are usually of a single differentiated type but sometimes mixed clones are formed, which suggests that cells can change their commitment in a neutral medium. In other words there are forms of commitment even more labile than the (operationally defined) specification measured by *in vitro* culture.

Commitment of tissue regions

It is an experimental fact that the commitment of a single cell and the tissue region from which it came may be different. The reasons for this presumably lie in the fact that a chunk of tissue carries with it a certain extracellular environment which is not the same as the bulk culture medium in which it sits. Also, in tissue explants but not in single cells, the normal types and arrangement of cell contact are preserved. The explantation of small regions followed by culture in buffered salt solutions is a method introduced by Holtfreter in the 1930s (Holtfreter, 1938 *a, b*). As with the isolated cell culture, the cells of the explant are sustained by their own internal yolk supply and so will survive, divide and differentiate in a very simple medium. The situation is, however, intrinsically more complex than for single cell culture. Explants often do not differentiate well unless they become surrounded by epidermis, either produced from the explant itself or supplied by the experimenter. This means that most of the cells are sitting in a microenvironment within the epidermal vesicle and this is to a large extent an unknown quantity. Also, almost all explant cultures give rise to more than one histological cell type, suggesting that regional differences exist within the explant, either inherited from the intact embryo or generated internally after isolation. Fortunately, several cell types arise in characteristic locations in the vertebrate body plan and so the regional identity acquired by an explant can often be known from the cell types which it contains or from the arrangement of structures which form within it.

Many isolation studies have been performed over the years using a variety of species. This account will concentrate on the more recent work and on that using *Xenopus* (eg, Dale & Slack, 1987*b*). In the early blastula, explants from most parts of the animal hemisphere will develop into solid balls consisting entirely of epidermis. The outer layer is somewhat different from the rest in that the cells are more regularly arranged and certain epidermal markers normally only expressed on the outer cell layer are found. However, in favourable cases it can be shown that no cells die, no cells are lost, and every cell in the explant expresses epidermal markers. On

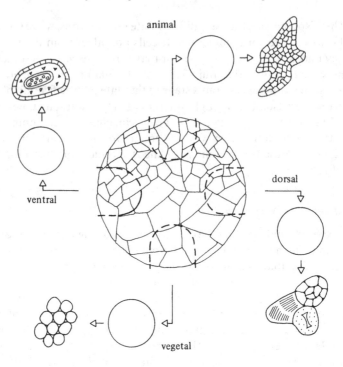

Fig. 4.12. Self-differentiation of isolates from different regions of the amphibian blastula. Isolates from the animal cap produce only epidermis, isolates from the vegetal region remain as large yolky cells, isolates from the dorsal marginal zone produce notochord, muscle and neuroepithelium, while isolates from the ventral marginal zone produce epidermis, mesenchyme and erythrocytes.

the dorsal side, many animal hemisphere explants from near the pigment boundary will also form some neural tissue, especially at the later stages (MBT-gastrula), and most explants from this position taken at any stage will contain cement glands. Explants from the vegetal hemisphere coming from the regions nearer the pigment boundary tend to form dense staining yolky cells which express endodermal markers, while those from the vegetal pole region remain as clumps of large light staining yolky cells.

Explants from the equatorial region (the marginal zone) will usually produce some mesodermal tissues together with endoderm, all surrounded by a well-organized two-cell layer epidermis. There is a sharp difference between explants from the dorsal marginal zone (DMZ) which produce clumps of notochord, muscle and neuroepithelium and those from ventro-lateral marginal zone (V + LMZ) which usually produce mesenchyme and erythrocytes (Fig. 4.12). This difference is also apparent in the shape changes shown by these explants at the time that intact control embryos are gastrulating. The DMZ explants will elongate in an attempt to reproduce

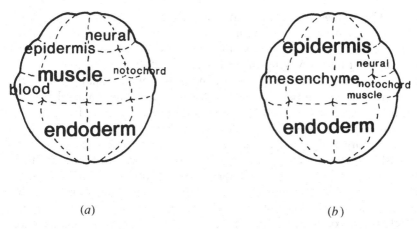

(a) (b)

Fig. 4.13. Comparison of the fate map (a) with the specification map (b) for cleavage stage *Xenopus* embryos. (a) is a simplification of the data shown in Fig. 4.9, (b) is drawn from isolates from 128-cell stage but projected onto the cell outlines of the 32-cell stage for comparison.

the convergent extension they undergo *in situ*, while the VMZs remain fairly spherical. The DMZ region is the famous 'organizer' of Spemann, and although textbook accounts of its properties over the years have tended to be misleading and inaccurate, there is now general agreement that its primary effect is as a dorsalizing centre (see more below). In the early blastula, little mesoderm is obtained from explants above the pigment boundary, while by the late blastula a belt about 30° wide will produce mesoderm.

If this type of information is put together in one diagram, we get a specification map as shown in Fig. 4.13. We know from the single cell culture experiments that there are regions containing mixtures of differently specified cells so it should not be thought that at this stage the embryo is necessarily a mosaic of regions between which sharp lines can be drawn. As to the origin of the regional differences these are certainly partly derived by passive partition of the egg cytoplasm. As discussed above we have reasons for thinking that the oocyte becomes partitioned at least into two zones along the egg axis, as witnessed, for example, by the localization of the *vg-1* mRNA. After fertilization the subcortical rotation is responsible in some way for generating a small dorsovegetal and a large ventrovegetal region with axial and non-axial mesoderm inducing properties respectively. At multicellular stages the marginal zone appears to consist of a small organizer zone and an extended ventral mesodermal zone. The animal hemisphere is mainly specified to become epidermis, with a rather small neural/cement gland zone on the dorsal side.

A number of extirpation and grafting experiments have been carried out at the 32-cell stage which also shed some light on the regional specification of the marginal zone. If the B1 and C1 cells (see Fig. 4.7 for nomenclature) are destroyed then the resulting embryos show dorsoanterior deficiencies rather like a mild UV syndrome (Takasaki, 1987). This experiment and many others like it shows that the organizer cannot be regulated if enough tissue is removed, an issue which was discussed in Chapter 3 in relation to the differences between the LSDS and GM1 models. If the B1C1 cells are grafted to the ventral side then a double dorsal body plan is produced, just as in the organizer graft in the early gastrula (see below) (Takasaki & Konishi, 1989). When rescue experiments are performed by grafting blastomeres into UV irradiated embryos, it is found at the 32-cell stage that the maximum rescue activity is found in the D1 cells, which do not themselves become incorporated into the axis. At later stages, after the MBT, the maximum activity is found in grafts from a dorso-equatorial position, which are the direct precursors of the axis and are themselves derived mainly from the B1C1 region of the 32-cell stage (Gimlich, 1986). These results show that the DMZ region of the blastula has dorsalizing activity, but that the acquisition of this activity depends on the D1 cells at the 32-cell stage.

Regulation of proportions

The last section dealt with the cryptic regional specification of the blastula region by region. It is clear that to obtain a complete pattern it is necessary to include at least some animal material, some vegetal material and some dorsal material. We shall now look at the pattern as a whole and ask what controls the different proportions of the embryo volume devoted to each specification.

One way of changing the size of an embryo is by removing parts of the fertilized egg. Kobayakawa & Kubota (1981) bisected *Xenopus* eggs into a nucleated and enucleated half by slow constriction with a glass rod. This procedure can be repeated to produce embryos having only a quarter of the normal volume. The enucleated fragments do not develop, but the nucleated ones do and produce approximately normally proportioned embryos. As already mentioned, it is also possible in *Xenopus* and other species to separate the first two blastomeres, each of which develops into a half size larva (Kageura & Yamana, 1983; Cooke & Webber, 1985). These are not quite normally proportioned: they contain all of the usual body parts but the size of the dorsoanterior region is a little greater relative to the ventroposterior than in normal embryos. Cooke (brief review, 1988) has also made quantitative studies of somite number and of the number of cells in the subdivisions of the mesoderm and has shown that reduction of blastula volume by about 50% does not affect the proportions significantly.

It is also possible to create double-sized embryos by the aggregation of two 2-cell stages (Mangold & Seidel, 1927). This procedure frequently yields embryos with multiple axes, but a single embryo can be formed when one embryo with medial cleavage is fused with one with frontal cleavage in such a way as to align the dorsoventral axes. Such embryos are twice the normal size and are at least approximately normally proportioned. Taking these experiments together we see that it is possible to vary the volume of an amphibian embryo by a factor of up to 8, and hence vary each linear dimension by a factor of the cube root of 8, which is 2. Over this range the size of each part of the body is scaled to the size of the whole and each part is composed of a different number of normal-sized cells from the usual.

It is also possible to vary the cell size but keep the total embryo volume constant. Application of a pressure shock at the time that the second polar body is expelled can cause its resorption and, often, its incorporation into the zygote nucleus along with male and female pronuclei. Such embryos are then triploid, having three chromosome sets. Triploid cells are 50% larger than diploid ones but are otherwise normal and able to carry out mitosis. Pressure shocks, and also hot and cold shocks, can on occasion give rise to embryos of other ploidy ranging from haploid to heptaploid (Fankhauser, 1945). These animals have been studied exhaustively by Fankhauser & Humphrey and it seems clear that, although the embryos are of normal size, each structure within them is composed of a different number of cells from usual. In haploid embryos there are twice as many cells as usual and in polyploids there are fewer than usual in inverse proportion to the increase in cell volume.

So the size of parts does not depend on the size of individual cells, nor does it depend on some absolute dimension specified in the developmental program; rather it depends on the size of the whole embryo. This suggests that the specification of territories depends on signals which extend across the whole extent of the embryo and can scale each territory to the amount of tissue available.

EARLY INDUCTIVE INTERACTIONS

It is clear from the fate map shown in Figs 4.9 and 4.13 that much of the mesoderm comes from the animal hemisphere. However, we have seen that cells and explants from the animal hemisphere of the early blastula develop only into epidermis, neural tissue or cement gland. This suggests that an interaction with vegetal tissue is necessary for the formation of mesoderm and indeed such an interaction was discovered by Nieuwkoop (1969) in the axolotl and shortly afterwards, by the groups of Nieuwkoop and Naka-mura, in *Xenopus*. Further studies using cell lineage labels, antibodies and molecular probes have abundantly confirmed the original findings (Dale, Smith & Slack, 1985; Gurdon *et al.*, 1985; review: Smith, 1989). The basic experiment is shown in Fig. 4.14(*a*). A combination is made from an animal

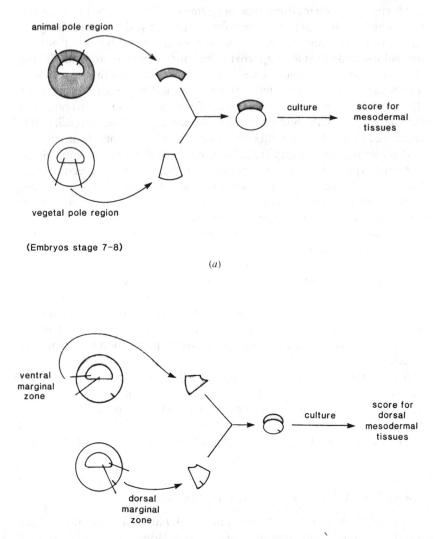

Fig. 4.14. The standard combination experiments showing (*a*) mesoderm induction; (*b*) dorsalization.

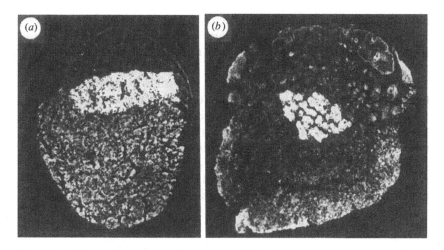

Fig. 4.15. Mesoderm induction in animal–vegetal combinations. In both cases the animal part is at the top and contains a clump of muscle cells stained with a fluorescent antibody. In (b), cell divisions and movements have been suppressed with cytochalasin. (Photographs kindly provided by Dr J.B.Gurdon.)

pole explant, preferably lineage labelled, and a vegetal pole explant. On its own, or in normal development, the animal pole explant would form only epidermis but in the combination it forms abundant mesodermal tissues as well. Studies of the times of signalling and competence using heterochronic combinations suggest that mesoderm induction occurs during the morula and blastula stages and is over by the early gastrula (Jones & Woodland, 1987). Interestingly, the time at which events occur in the responding tissue, such as the gastrulation movements or expression of the muscle specific gene α-actin, do not occur at a fixed time after the combinations are assembled but rather when the responding tissue has reached the appropriate age. In other words, not only the whole embryo but also explants from it contain some kind of clock which controls when things happen (Gurdon, 1987). Gurdon (1989) has exploited the insensitivity of the process to cytochalasin B and been able to study the spatial arrangement of the induced tissues after all cell division and cell movement has been blocked. This showed that the effective range for the induction of muscle was about 80 microns or four cell diameters at stage 8 (Fig. 4.15). The calculations presented in Chapter 3 show that it would be possible to establish a concentration gradient of a small protein over such a distance within a few hours.

Although many experiments have concentrated on the formation of muscle, a variety of tissues conventionally classified as mesodermal are formed. Dorsal vegetal tissue induces 'dorsal type' structures containing notochord and large muscle masses together with abundant neural tissue, presumably formed by secondary neural induction within the explant. Ventral vegetal tissue induces mainly 'ventral type' structures containing a concentric arrangement of mesenchyme, mesothelium and blood cells, or sometimes 'intermediate type' structures containing smaller muscle masses and mesenchyme. At the 32-cell stage it is possible to isolate single cells from the vegetal octet (D octet) and test their inductive activity. Only the D1 cells induce dorsal structures (Dale & Slack, 1987b) and only the D1 cells can 'rescue' UV irradiated embryos (Gimlich, 1986) so it seems likely that these activities are one and the same thing: a dorsal mesoderm inducing signal. The D2, 3 and 4 blastomeres produce small ventral or intermediate type inductions and show no rescue activity so these appear to be the source of a distinct ventral mesoderm inducing signal.

Dorsalization

Since there are several mesodermal tissues, and these arise in explants from different dorsoventral levels of the marginal zone of the early gastrula, it is evident that there must be some mechanism of regional specification along the dorsoventral axis, either within the mesodermal annulus or across the whole embryo and affecting the mesodermal annulus as well as animal and vegetal regions. We have seen that the mesoderm inducing signal itself has at least two qualities, suggesting some dorsoventral organization of the signalling region presumably derived from the postfertilization cytoplasmic movements. However, this is not the whole story. In the fate map, about 60% of the somite comes from the ventral half of the embryo. But if blastulae are divided in half, ventral halves typically form extreme ventral structures containing little or no muscle, reminiscent of UV embryos (Dale & Slack, 1987b). This suggests that some signal from the dorsal half is required for formation of somite in the ventral half and we have called this process 'dorsalization' (Smith & Slack, 1983; Dale & Slack, 1987b).

Dorsalization can be demonstrated directly by combining a dorsal marginal zone explant with a labelled ventral explant or ventral half embryo and showing the formation of abundant muscle in the latter (Figs 4.14(b), 4.16). The area with dorsalizing activity occupies not more than 90° of the marginal zone circumference, being wider than the prospective notochord but smaller than the prospective somite. It may represent the response to the original dorsal mesoderm induction and at the early gastrula stage is identical to the classical 'organizer' region of the embryo around the dorsal blastopore lip. An independent measurement of the size of the organizer was made by Gerhart's group (Gerhart *et al.*, 1989) by combining vertical halves of normal and extreme UV embryos at the blastula stage. If a lateral half of an unirradiated embryo is used then the resulting combination forms a fairly normal larva, in which part of the axis has been formed from the UV irradiated half by dorsalization. But if the

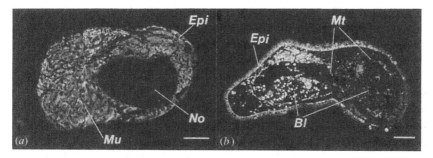

Fig. 4.16. Dorsalization in dorsal–ventral combinations. In both pictures the fluorescent part of the specimen is derived from a labelled explant of ventral marginal zone. In (*a*) it is combined with dorsal marginal tissue and has formed a large muscle mass, in (*b*) it is combined with another ventral piece and has formed blood and mesenchyme. (Photographs from Dale & Slack, 1987*b*.)

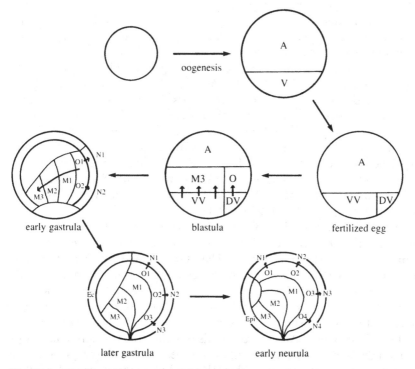

Fig. 4.17. The three signal model. The three territories A, VV and DV are set up in the egg as a result of oogenesis and the postfertilization cytoplasmic movements. In the blastula stages the region DV emits a signal which induces the organizer, and the region VV emits a signal which induces ventral type mesoderm. During gastrulation the organizer emits a third, graded, signal which regionalizes the mesoderm into zones forming somite, lateral plate and blood islands. The diagram also shows neural induction by the archenteron roof, occurring progressively during gastrulation.

unirradiated half is cut more than 30° away from the dorsal meridian then there is no dorsalization and the combination resembles a UV limit form. This would mean that the organizer occupies a sector of about 60° around the dorsal meridian.

The events of mesoderm induction and dorsalization have been summarized as the 'three signal model' shown in Fig. 4.17. This is undoubtedly grossly oversimplified, but it does explain the bulk of the micromanipulation data and has so far stood up rather well in the recent era of inducing factor research.

Mechanism of mesoderm induction

Mesoderm induction will work when one or both of the explants comes from a zygote that was injected with antibody to gap junction protein and in which the junctions are blocked (Warner & Gurdon, 1987), so this shows that the signal does not need to pass through gap junctions. The interaction also works if the explants are separated by a nucleopore filter of sufficiently small pore size to exclude cytoplasmic processes (Grunz & Tacke, 1986). These two experiments provide proof that the agent of induction consists of one or more *extracellular diffusible substances.* Mesoderm inducing factors (MIFs) had been obtained for many years from heterologous sources (reviewed Tiedemann, 1978) but they were insoluble in physiological solutions and of questionable purity, so did not attract much attention. In recent years it has been shown that the active factors are actually known cytokines (often called growth factors) and this has led to a revival of interest in the problem. The recent results have been heavily reviewed and the relevant references will be found in the reviews by Smith (1989), Slack *et al.,* (1989) or Melton & Whitman (1989).

The active factors fall into two groups: those belonging to the transforming growth factor beta (TGF β) superfamily and those belonging to the family of fibroblast growth factors. The TGF β superfamily comprises the TGF β themselves, which are closely related by primary sequence and which have similar biological activities often involving a stimulation of extracellular matrix synthesis (Massagué, 1987), together with various other factors more distantly related by primary sequence and with distinct biological activities. All members exist as dimeric proteins linked by disulphide bridges and are initially synthesized as longer precursors. Members of the superfamily with MIF activity are the activins, TFG β 2 and 3, and certain of the bone morphogenetic proteins (BMPs). Activins are otherwise known as factors from the ovary which can stimulate FSH release from the pituitary, and as erythroid differentiation factors (Ling *et al.,* 1988). The XTC–MIF purified by Smith from a *Xenopus* cell line is activin A (Smith *et al.,* 1990) as are factors purified from various other sources. Bone morphogenetic proteins were isolated as factors causing

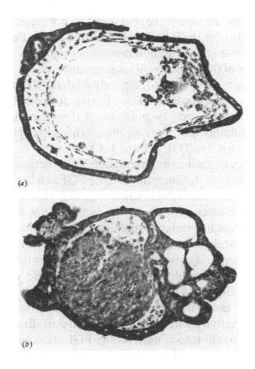

Fig. 4.18. Mesoderm induction provoked by *Xenopus* bFGF. (*a*) A low dose produces a ventral induction, (*b*) a higher dose produces a muscle mass (author's photograph).

osteogenesis from dermal fibroblasts (Wozney, 1990), and are closely related in primary sequence to the *Drosophila* gene *decapentaplegic* (see Chapter 7). The TGF β superfamily also includes *vg-1* which has been mentioned above as an mRNA localized in the oocyte, but this is thought probably not to have MIF activity. The fibroblast growth factors were first isolated as mitogens, being particularly active on capillary endothelial cells. They are single polypeptide chains and most of the family bind tightly to heparin (Gospodarowitz *et al.*, 1987). At least aFGF, bFGF, kFGF and FGF-5 have been shown to be active as MIFs, and the product of the oncogene *int-2* also has limited activity.

The various MIFs have much in common in their action on animal hemisphere explants. Only a brief treatment of the explant is necessary during its period of competence, and this causes elongation into a sausage shape at the time that intact embryos are gastrulating. Over the next 1–2 days the induced explants differentiate and a significant proportion of the cells form mesodermal tissues (Fig. 4.18). As with mesoderm induction in tissue combinations, the time of the response does not depend on the time of treatment but rather on the internal clock of the tissue. For example, the elongation will occur at gastrulation time rather than at a fixed time after treatment. Studies of the induction of single animal cells *in vitro* have shown that all inner cells are competent to respond and that different responses do occur at different concentrations. The effective dose of MIF required to achieve a given response may be much lower if certain other factors are administered simultaneously. Among these synergistic agents, which are inactive on their own, are TGF β-1 and LiCl.

There are, however, various important differences between the biological effects of the activins on the one hand and the FGFs on the other (Green *et al.*, 1990). Firstly, activins induce both dorsal and ventral mesoderm, the former at higher concentration and the latter at lower concentration. Dorsal inductions are like isolates from the dorsal marginal zone, containing notochord, neural tissue and muscle blocks, while ventral inductions include mesothelium and mesenchyme, but, unlike ventral explants from embryos, do not contain blood cells. The FGFs also produce ventral inductions at low dose and yield more and more muscle with increasing concentration but rarely induce notochord. Secondly, activin inductions acquire organizer activity, meaning that they will dorsalize mesoderm with which they are brought into contact, while FGF inductions probably do not (Cooke, 1989*b*, but see also Ruiz i Altaba & Melton, 1989*b*). Thirdly, the period of competence for activins is longer than for FGF, extending into the mid gastrula stage, and hence rather resembling the period of competence for the natural signal which was inferred from heterochronic combinations. Fourthly activin induces early expression of a homeobox containing gene, *mix-1*, which is normally expressed in the endoderm, while FGF does not (Rosa, 1989). This rather supports the view of workers such as Nieuwkoop and Tiedemann that parts of the endoderm as well as the mesoderm is induced in the process called 'mesoderm induction'. Finally, FGF is rather more effective than activin at inducing expression of *xhox-3*, which is normally expressed in posteroventral mesoderm (Ruiz i Altaba & Melton, 1989*b*).

Activins are obvious candidates for the putative dorsal inducing signal of Fig. 4.17, and, at the time of writing, activin B mRNA was known to be expressed after MBT, and activin A mRNA somewhat later. It is possible that the ventral inducing signal also consists of activin at a lower dosage, or alternatively it might be qualitatively different, for example one of the FGFs. bFGF mRNA and protein are known to be present in *Xenopus* oocytes and embryos and there is also a receptor for FGFs whose concentration at the cell surface rises and falls during the blastula stages and thus correlates well with the window of competence for FGF induction. The receptor is present all over the embryo at about the same level which suggest that marginal zone cells, which become induced in normal development, have a similar competence to the animal pole cells which are usually used for experiments (Gillespie *et al.*, 1989). There is thus quite good circumstantial evidence for the involvement of bFGF in mesoderm induction although there is also a problem because this molecule is devoid of a signal peptide and is known to be secreted from cells very inefficiently. So perhaps it is not an inducing factor but has some intracellular role such as amplification of the response to the extracellular signal.

The intracellular signal transduction pathway of FGF is known to involve a tyrosine kinase activity of the receptor, so it is of interest that the polyoma middle T protein, which activates an endogenous tyrosine kinase called pp60c-src, can also induce mesoderm if introduced into embryos in the form of a synthetic mRNA (Whitman & Melton, 1989). In such embryos isolates from the animal hemisphere will 'auto-induce', since they have inherited the mRNA and the activity of the protein product presumably bypasses the need for the extracellular signal.

Some further results which have a bearing on the mechanism of mesoderm induction are those showing that *homeogenetic* induction is possible. This means that newly induced mesoderm can itself induce more mesoderm from competent ectoderm. Several studies have suggested that this can occur, for example one by Represa & Slack (1989) in which presomite plate from early neurulae was shown to provoke further mesoderm formation in labelled blastula ectoderm with which it was brought into contact. Homeogenetic induction suggests some sort of autocatalytic spread of the induced state, and logically requires some countervailing force to prevent the entire embryo becoming induced. Gurdon (1989) has proposed that the spatial range of the induction is controlled simply by the loss of competence of the ectoderm at the beginning of gastrulation, but a quantitative study by Cooke (1989a) showed quite a high degree of proportion regulation in induced animal hemispheres of different sizes, and this would argue in favour of a proportioning mechanism based, for example, on the release of an inhibitor by the already induced region.

The organizer

Long historical shadows lie over experimental embryology and the biggest of them all derives from the famous organizer experiment published in 1924 (Spemann & Mangold, 1924). The original organizer graft was performed on newt embryos at the early gastrula stage and is a graft of tissue from the dorsal lip region into the ventral midline of the marginal zone. Although the original paper contained few cases, the experiment has been repeated on many species, and a study on *Xenopus* using lineage labelled grafts was published by Smith & Slack (1983). The graft continues its invagination to form a second set of dorsal mesodermal structures, mainly notochord and prechordal plate, on the ventral side of the host. It also dorsalizes a certain

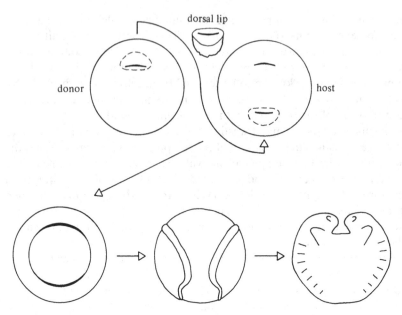

Fig. 4.19. The organizer graft. A piece of tissue from the dorsal marginal zone is grafted into the ventral marginal zone and induces the surrounding tissue to participate in the formation of a double-dorsal embryo.

region of the surrounding host mesoderm so that this becomes the outer part of the secondary mesodermal axis (somites, kidney, lateral plate) (Figs 4.19 and 4.20). The subsequent events of neural and endodermal regionalization follow from this, so that the end result is the formation of a mirror symmetrical double dorsal embryo. Except for the fact that the induced axis is often smaller than the host axis, these are rather similar to the double dorsal embryos arising from centrifugation, hair loop constriction of the zygote, injection of Li into ventral blastomeres, or transplantation of extra D1 blastomeres to the ventral side (see above). Using explants cultured *in vitro*, Dale & Slack (1987*b*) showed directly that the ventral marginal zone could be dorsalized by co-culture with organizer tissue. Ventral explants which normally produce blood islands formed large masses of muscle after a certain period in contact with the dorsal lip region (Fig 4.16). If the signal is weakened by using as inducer an explant from one side of the dorsal midline, or by reducing the area of contact, then kidney tubules may predominate in the responding tissue, showing that dorsalization is not just a matter of muscle differentiation but of shifting from a ventral to an intermediate epigenetic coding.

An alternative method of performing the organizer graft, introduced in the 1920s by O. Mangold, involves the insertion of the graft into the blastocoel through a hole made in the ectoderm. This method is easier than

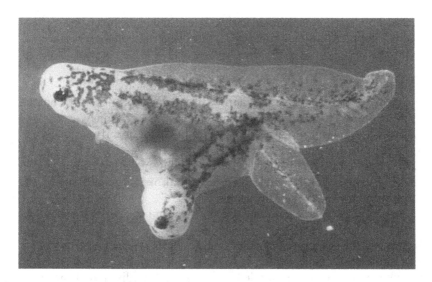

Fig. 4.20. A double embryo produced by an organizer graft on an axolotl gastrula (author's photograph).

implantation into the ventral marginal zone and much of the classical work on the organizer made use of it. Unfortunately it is difficult to control exactly where the graft ends up and so the result is a product of several processes which operate to different extents depending on the details of the experiment; namely mesoderm induction of ectoderm in contact with the graft, dorsalization of this induced mesoderm or of host mesoderm, self-differentiation of the graft, and, last but not least, direct neural induction of ectoderm. It is also not a graft that can easily be performed on *Xenopus* whose blastocoelic cavity is smaller than that of newts and where such procedures may cause splitting of the axis by mechanical means, thus further confusing the outcome. Magnificent though the achievements of Spemann and his school were, they did not recognize that they were dealing with a two- or even three-dimensional problem of regional specification. They saw the whole embryo as uncommitted except for the organizer itself, out of which somehow flowed the information to construct the whole body plan. When the 'goldrush' for the chemical nature of the organizer took place in the 1930s attention became focused on one aspect of the response which was easy to measure, namely induction of neural plate from ectoderm. This misconception has regrettably been propagated by text-book writers ever since, even though precursors to our present way of thinking did exist in the 1930s, in particular, the far-sighted two-gradient model of Dalcq & Pasteels (1937). So for most of its history the organizer has been seen principally as a simple agent of neural induction rather than an agent of mesodermal dorsalization.

Based on the results reviewed in this section and in the preceding sections of this Chapter, we can summarize the properties of the organizer as follows:

1. Organizer mesoderm will self-differentiate into notochord.
2. It will dorsalize ventral mesoderm with which it is brought into contact. A stronger signal will induce somites, a weaker one kidney.
3. In the normal embryo the organizer is found in the dorsal marginal zone.
4. In extreme UV-treated embryos there is no organizer.
5. In extreme Li treated embryos the entire marginal zone is organizer.
6. An organizer can be induced from animal or marginal tissue by the blastomeres D1 at the 32-cell stage.
7. An organizer can be induced from animal hemisphere tissue by activin.

In terms of the models discussed in Chapter 3 this evidence is all consistent with the idea that the organizer is the source region for a dorsoventral morphogen gradient responsible for the regional subdivision of the mesoderm. Although the organizer itself arises in the early blastula, its dorsalizing effect is probably active during gastrulation and there is some evidence in urodeles that dorsoventral interconversions can continue to occur in the mesodermal mantle of the neurula (eg, Forman & Slack, 1980).

Specification of the anteroposterior pattern

During gastrulation the dorsal marginal zone undergoes a process of involution and convergent extension so as to produce an anteroventral clump of pharyngeal endoderm together with an extremely elongated dorsal midline for the mesoderm comprising the prechordal plate, the notochord, and the tailbud. In *Xenopus* there is little or no topographic projection within the DMZ from the pre- to the postgastrulation situation, meaning that cells from anywhere in the DMZ can end up anywhere in the series of structures listed. In urodeles there probably is some projection, with the region that moves furthest originally lying nearest to the dorsal lip (Okada & Hama, 1945). Because UV irradiation of the egg causes dose-dependent reductions of pattern from the anterior end, certain authors have been tempted to assume that there is some specification of anteroposterior codings prior to gastrulation. But it is really not necessary to assume this since the dorsal part of the blastula contributes quantitatively more to the anterior than the posterior, hence a reduction of the size of the dorsal territory will reduce the number of cells available for formation of the whole axis but will affect the most anterior structures disproportionately.

In urodeles where there is some topographic projection from the preinvagination to the postinvagination dorsal midline, the self-differentiation behaviour of small explants from the DMZ at early gastrula stages is rather different from the fate map. The region closest to the blastopore which forms pharyngeal endoderm *in situ* tends to form notochord and neural tissue in isolation, and the region further away which forms trunk axis *in situ*, tends to form epidermis and pigment cells in isolation. The prospective trunk mesoderm only acquires a self-differentiation behaviour of notochord/neural type by the mid-gastrula when it is just about to invaginate (Kaneda & Hama, 1979). A head to tail specification of the different anteroposterior levels does not, therefore, seem to exist as early as the blastula stage in urodeles either.

It seems reasonable to postulate that the specification of anteroposterior levels occurs coincidently with the involution of the successive cohorts of cells during gastrulation. Unfortunately, there is a curious dearth of recent experimentation on this problem; most of what we have was done some time ago and on urodeles. The effects of removing the dorsal lip region at different stages of gastrulation are consistent with the idea of successive specification. Removal of the dorsal lip from early gastrulae leads to deficiencies in the head region, removal of the dorsal lips from mid-gastrulae leads to deficiences in the trunk region, and removal of dorsal lips from late gastrulae leads to deficiencies in the tail (Shen, 1937).

The same is true of ventralization of the dorsal midline caused by short term treatment with LiCl. Gastrulae treated with lithium chloride tend to develop into embryos that have axial defects (Lehmann, 1937; Bäckstrom, 1954). If early gastrulae are treated, the defects are concentrated in the head while if later gastrulae are treated defects still occur in the head and also more posteriorly. The primary effects are certainly on the mesoderm but gaps arise in the central nervous system as a consequence of the failure of neural induction. Hence cyclopia and anophthalmia are common defects. Note that these effects of Li are more or less the opposite of the dorsalizing effect shown on the early stages. Rather similar effects can also be obtained by injecting polyanions such as suramin into the blastocoelic cavity (Gerhart *et al.*, 1989).

When grafted to the ventral marginal zone, an early dorsal lip can, as we have seen, form a complete axis with head, trunk and tail. But a dorsal lip from a mid-gastrula grafted to the ventral marginal zone of an early gastrula will form only trunk and tail, and a dorsal lip from a late gastrula only tail (Spemann, 1931). Again these results are consistent with the idea that the anteroposterior level is specified at the time of invagination.

All these three sets of data are consistent with successive specification, but none of them provide proof and none really places any constraint on the possible mechanisms. All we know is that before gastrulation the state of specification of the DMZ is still governed by the signals in the blastula, while after involution, the dorsal mesoderm is specified to become head, trunk or tail, and, as we shall see below, to induce corresponding neural structures from the overlying ectoderm. Future research on the mechanism of anteroposterior specification will probably focus on the changing sequence of microenvironments to which the invaginating cells are exposed, using as markers mRNAs for genes such as those homeobox genes which show some anteroposterior restriction in their expression pattern.

NEURAL INDUCTION

The essential facts about neural induction were established in the 1920s and 1930s and have stood the test of time well, although it has regrettably often been called 'primary embryonic induction' which has served to obscure the sequence of regional specification events occuring in the fertilized egg and the blastula. In the original organizer graft of Spemann & Mangold (1924) the graft induced a second nervous system from that part of the host ectoderm which came to lie above it at the end of gastrulation. In the absence of the graft, this tissue would have become ventral epidermis. Since it was also known that prospective epidermis and neural plate (dorsal and ventral ectoderm) would each develop according to their new position if exchanged in the early gastrula, Spemann (1938) argued that the nervous system was formed in *normal development* as a result of an inductive signal from the dorsal mesoderm. This interpretation was strengthened by the discovery of Holtfreter (1933) that the entire ectoderm could develop into epidermis if isolated from the mesoderm. This occurs in 'exogastrulae' which arise if embryos are cultured in stronger salt solution than usual; the morphogenetic movements work to exclude the mesoderm and endoderm from the ectoderm so that the ectoderm becomes isolated as an empty sac.

In 1932 it was shown that *killed* dorsal lips would induce some neural tissue following implantation into the blastocoel, suggesting that a chemical substance was responsible (Bautzmann *et al.*, 1932). This led to the famous 'gold rush' for the biochemical basis of what was then called the organizer but what we would now call the neural inducing factor. It was found that many tissues from adult animals or indeed many purified chemical substances would induce neural tubes or patches of neuroepithelium from gastrula ectoderm. After a period of confusion, it became clear that the ectoderm was delicately balanced between epidermal and neural pathways and that a variety of stimuli could tip the balance one way or the other.

In recent years it has been confirmed using lineage labels and molecular markers that neural tubes can be induced from the ventral ectoderm, although the dorsal ectoderm may be more susceptible to induction (Jacobson & Rutishauser, 1986; Sharpe *et al.*, 1987 and Fig. 4.21) The required contact time is about 2 hours and the signal can pass across a small pore nuclepore filter, showing that it consists of one or more extracellular diffusible substances (Tarin, Toivonen & Saxen, 1973). Although in urodeles autoneuralization of animal explants is common and led to the failure of the original gold rush, this turns out not to be the case for

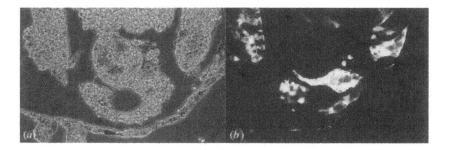

Fig. 4.21. An induced neural tube resulting from an organizer graft (*a*) phase contrast (*b*) lineage label. Both blastomeres C4 of the host were labelled so it may be seen that the induced neural tube and somites are host derived (author's photograph).

Xenopus. We have seen above that *isolated* ectoderm cells may show neuronal differentiation on suitable protein substrates, but ectoderm *explants* always become balls of epidermis. No pure proteins have been described at the time of writing that will elicit neuralization in *Xenopus* ectoderm explants although it has from time to time been reported that crude gastrula extracts will do so. Neural induction is provoked to a limited degree by the tumour promoting agent TPA, and more effectively by TPA together with dibutyryl cyclic AMP (Otte *et al.*, 1989). TPA activates the enzyme protein kinase C, which is a serine–threonine kinase normally activated by diacylglycerol generated from phosphatidyl inositol or other membrane phospholipids in response to extracellular effectors. The TPA effect suggests that the activation of protein kinase C might be an essential step in neural induction and the enzyme is in fact found to be activated during normal neuralization. Cyclic AMP is often found as a second messenger synthesized in response to a number of hormones and neuro-transmitters, and it works by activating another serine threonine kinase: protein kinase A. Dibutyryl cAMP is simply an analogue of cAMP capable of penetrating cell membranes. Otte *et al.* showed that activatable adenyl cyclase as well as protein kinase C activity became elevated during normal neuralization, and suggested that both second message pathways need to be stimulated for effective neural induction. This probably means that more than one extracellular neural inducing factor is involved. We have already encountered the phosphatidyl inositol pathway in connection with the lithium effects, presumed to be due to an inhibition of the cycle, but cAMP has no known effects on mesoderm induction or dorsalization.

Regional specificity of neural induction

Neural induction is often discussed as though it is a simple process with one signal and one response. Actually things are more complex because neural induction, just like mesoderm induction, shows regional specificity: different parts of the dorsal mesoderm inducing different parts of the central nervous system. This was first shown by Mangold (1933) who implanted explants from different craniocaudal regions of archenteron roof into the blastocoel of host *Triturus* embryos, and similar results were obtained by Horst (1948) in combinations of archenteron roof explants with gastrula ectoderm in isolation. The most anterior inductions, with forebrain structures and eyes, are called *archencephalic*. Hindbrain type inductions are called *deuterencephalic* and trunk–tail inductions are called *spinocaudal*. This implies that more than one signal is involved in neural induction, and it is interesting that there is a regional difference in the neural inducing capability acquired by ectoderm explants treated with different mesoderm inducing factors. Ectoderm treated with activin will yield archencephalic inductions while that treated with FGF gives mainly spinocaudal ones (Ruiz i Altaba & Melton, 1989*b*). Another indication of regional specificity of the signals from the dorsal mesoderm is the cement gland inducing activity which is, paradoxically, maximal in the middle part of the axis although the cement gland itself forms at the anterior margin of the neural plate (Sive *et al.*, 1989).

There is also evidence that there can be interactions within the neural plate itself. It has been known for a long time that, at least in urodeles, explants of neural plate are themselves neural inductors, although it now seems that this may not be so for *Xenopus* (Jones & Woodland, 1989). This *homeogenetic induction* was also shown to have some degree of regional specificity, in that prospective brain induced brain and prospective tail induced tail (Mangold, 1933). Nieuwkoop (1952*a,b*) showed that, if small folds of gastrula ectoderm are implanted into a neural plate, the basal parts of the folds develop into similar structures as are formed by their surroundings while the apical parts of the folds form more anterior structures that tend to be arranged in the normal sequence. So anterior neural plate could induce only forebrain, but posterior neural plate could induce a complete sequence of structures. A possible clue about the mechanism of this process comes from the finding that retinoic acid can provoke a posterior transformation within the neural tissue, either in whole embryos or in explants (Durston *et al.*, 1989). Retinoic acid is well known for its striking effects on the digit pattern of developing and regenerating limbs. It does occur in the early *Xenopus* embryo although at present its relationship to the rest of cellular metabolism is obscure. The simplest interpretation of the effect might be that neuroepithelium is induced by stimulation of protein kinases A and C, but the anteroposterior coding was dependent on a posterior to anterior gradient of retinoic acid. This is a modern rephrasing of the 'activation-transformation theory' of Nieuwkoop (1952*c*).

The existence of homeogenetic induction means that there must exist some mechanism to limit the spread of the induction, otherwise the entire ectoderm would eventually be induced. According to a study by Albers

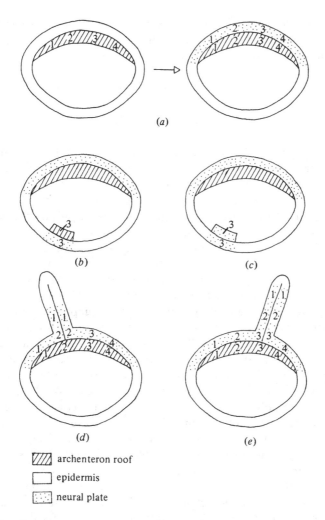

Fig. 4.22. Neural induction. (*a*) In normal development the neural plate is induced by the dorsal mesoderm. (*b*) The dorsal mesoderm can induce neural tissue from prospective epidermis. (*c*) Neural plate can induce further neural tissue of the same regional specificity from prospective epidermis (homeogenetic induction). (*d*) and (*e*) serially ordered structures are induced in ectodermal folds which were implanted into the neural plate.

(1987) on the mediolateral spread of neuralization in the axolotl, the spread continues simply until the normal time for the loss of competence. These indications that interactions can occur within the neural plate, although perhaps only in abnormal situations, suggest that the signals from the dorsal mesoderm may be representable as some sort of serial hierarchy as implied in Fig. 4.22, where they are shown as a series of numbers. The alternative model would be that the CNS is divided into a number of territories each induced by a different substance, but this would be difficult to reconcile with Nieuwkoop's experiments.

LATER EVENTS

Somitogenesis

During neurulation, the mesoderm on either side of the neural tube becomes segmented to form a row of somites. These later form the myotomes, the vertebrae, and part of the dermis of the skin. The cell movements which lead to segmentation differ somewhat from species to species but in all cases the segmentation occurs in an anteroposterior sequence and is temporally correlated with events occurring elsewhere in the body. It is unlikely that the size of somites depends on a cell counting process since haploid embryos have been shown to contain normal-sized somites each containing twice the normal number of cells (Hamilton, 1969). When embryos are transected at a level posterior to the last-formed somite, both halves continue segmentation at the normal tempo and end up with the number of somites which those parts would have developed in the intact animal (Deuchar & Burgess, 1967). To this extent the segmentation pattern behaves as a mosaic from the end of gastrulation. However, although it is not possible to alter the fate map it is possible to disturb the arrangement of cells by exposing the embryos to brief temperature shocks at 37°C. During the neurula and tailbud stages the somites which become deranged are those which are formed a few hours after the temperature shock, the exact time depending on the species (Pearson & Elsdale, 1979). Therefore there seems to be an anteroposterior progression of temperature sensitivity which runs in advance of visible segmentation. The process is also unaffected by transection of the embryo and thus presumably does not depend on any long-range cellular interactions. The authors call the temperature-sensitive event 'somite determination' and argue that the cells must be able to recover from the temperature shock after a few hours since only a short length of the somite file is deranged and the somites posterior to the lesion are normal. Temperature shocks given to late blastulae or early gastrulae produce more serious defects in segmentation which may involve the whole somite file or any part of it. This is interpreted as being the stage at which the general anteroposterior pattern is established in the mesodermal mantle. Shocks given to the late gastrula are without any effect at all, and this refractory period between two periods of sensitivity indicates that somite determination involves two distinct processes: one early and the other just preceding visible segmentation. These results are compatible with, but do not prove the correctness of, the 'clock and wavefront' model for the determination of repeating structures which was discussed in

Chapter 3. It should be noted that these temperature experiments were performed on *Rana*. Interestingly, rather similar disruptions of somite segmentation in *Xenopus* occur in embryos injected with synthetic mRNA for the homeobox containing gene *xhox-1A*, a gene normally expressed only in the somites (Harvey & Melton, 1988).

The endoderm

The regionalization of the endoderm has not been the most popular of research topics but the results we have suggest that the territories become determined by the middle neurula under the influence of the adjacent mesoderm. A fate map of the early neurula endoderm was produced by Balinsky (1947) and used to examine the results of interchanging prospective regions from stomach and liver (Balinsky, 1948). Normal embryos arose from these grafts, indicating that determination had not yet occurred. The question was later examined by Okada using explants (Okada, 1953, 1957, 1960). He found that gastrula endoderm alone would not differentiate *in vitro*. Prospective pharyngeal endoderm would form pharynx in combination with archenteron roof tissue, and intestine in combination with lateral plate mesoderm. Prospective gastric and intestinal endoderm would form branchial or oesophageal structures in the presence of head mesoderm, and intestinal structures in the presence of lateral plate mesoderm.

SUMMARY OF REGIONAL SPECIFICATION IN THE EARLY AMPHIBIAN EMBRYO

The results which have been described here may perhaps be brought together and summarized as follows (Fig. 4.17). Animal–vegetal polarity is established during oogenesis and dorsoventral polarity is established after fertilization. Both of these apparently involve cytoplasmic localization, the animal–vegetal perhaps depending on the differential localization of the *vg-1* and other mRNAs, and the dorsoventral in some way depending on the subcortical rotation. The first inductive interactions occur along the egg axis and result in the formation of a ring of mesoderm around the equator. As soon as the mesoderm is formed the dorsal and ventral regions seems to be differently specified. The dorsal region, which may perhaps be induced by an activin-like molecule, becomes the organizer and later emits a morphogen gradient which regionalizes the rest of mesoderm along the dorsoventral axis. The lateroventral region may possibly be induced by an FGF-like molecule and initially has a ventral specification, although it becomes dorsalized during gastrulation to a degree depending on its distance from the organizer and hence contributes to all mesodermal structures except the notochord. In the gastrula and neurula stages the neural plate is induced by the action of the mesoderm on the overlying ectoderm by some mechanism probably involving the activation of adenyl cyclase and protein kinase C. The neural plate is formed as an array of territories committed to become different parts of the nervous system. The endoderm is similarly regionalized under the influence of the mesoderm to form different parts of the viscera.

Up to the time at which the basic body plan has become determined we therefore have at least two decisions involving cytoplasmic localization and at least four inductive signals, some of which have multiple outcomes and may be complex signals involving several biochemical substances. Although there is obviously a vast amount of work to be done, and many unsolved problems remain, we do at least have an outline theory of the events which take us from egg to body plan. The work on amphibians also enables us to make the following generalizations which, as we shall see in the remainder of this book, probably hold for all other animal embryos as well.

1. Both cytoplasmic localization and induction are important.
2. The body plan is formed by a *sequence* of events, not all at once.
3. Inductive interactions show regional specificity and this is a property of the *signals* not of the responding tissue.
4. The forms of commitment preceding terminal gene expression are labile and reversible.

SEA URCHINS

Sea urchins share with amphibians the status of 'senior citizens' in embryological research for they have been subjects of experimental investigation since the late nineteenth century. They are familiar marine animals and make up the class *Echinoida* of the phylum *echinodermata*. The true sea urchins are radially symmetrical and the related sand dollars and heart urchins are bilaterally symmetrical with shorter spines.

NORMAL DEVELOPMENT

A variety of species have been used for experimental work and some of their individual peculiarities are mentioned by Hörstadius (1973). In most cases the eggs are 75–150 microns in diameter, they mature in the ovary before being shed and are surrounded by jelly coats. After fertilization a membrane is formed by combination of the vitelline membrane with material released from cortical granules. A surface coat known as the hyaline layer is secreted shortly afterwards.

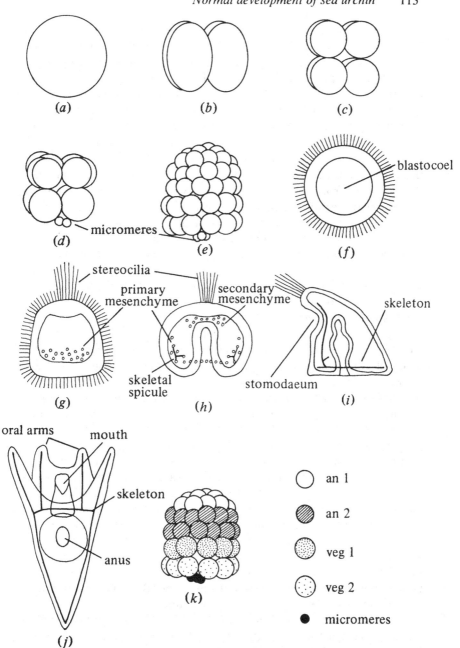

Fig. 4.23. Normal development of the sea urchin *Paracentrotus lividus*. (*a*)–(*e*) Cleavage and segregation of micromeres. (*f*) blastula. (*g*) early gastrula. (*h*) late gastrula. (*i*) prism stage. (*j*) pluteus larva. (*k*) 64-cell stage shaded to show the nomenclature of the different cell layers. (After Hörstadius, 1935.)

The first two cleavages are vertical and the third is equatorial (Fig. 4.23) and from the beginning the cells are polarized with microvilli on the outer surface, which is in contact with the hyaline layer. In the fourth cleavage the four cells at the animal pole cleave equally to form eight *mesomeres* while the four at the vegetal pole cleave unequally to form four *macromeres* and four *micromeres*. Each micromere subsequently divides to form two *small micromeres* and eight *primary mesenchyme* cells. After about six cleavages the divisions become asynchronous and a blastocoelic cavity develops in the interior with a basement membrane on the blastocoelic surface. From the seventh division, septate desmosomes form between the cells and isolate the blastocoel from the external medium. At the seventh–eighth cleavage, one cilium appears on each cell and shortly afterwards the blastula hatches from its fertilization membrane. By this stage there are in *Strongylocentrotus purpuratus* a total of 424 cells; 384 produced by five cleavages of the meso- and macromeres, together with 32 primary mesenchyme cells and eight small micromeres. In contrast to amphibian embryos, zygotic transcription starts at fertilization and continues throughout the early developmental stages, so by the beginning of gastrulation most of the maternal mRNA has been replaced.

After hatching, only about one further cell division takes place before the feeding stage. An apical organ develops at the animal pole consisting of an apical plate bearing long immotile cilia (*stereocilia*). The blastula wall also thickens to form a plate at the vegetal pole and from here the 32 primary mesenchyme cells invaginate into the blastocoel and arrange themselves roughly in a ring. During gastrulation, these commence formation of the calcareous spicules which later become the larval skeleton. Gastrulation proper consists of an invagination of the remaining vegetal plate tissue. It occurs in two principal phases, an initial buckling to form an *archenteron* (in the sea urchin 'archenteron' means the invaginating tissue itself and not just the cavity within it), at the tip of which are a number of *secondary mesenchyme* cells; and a second stage in which fine pseudopodia from the secondary mesenchyme cells pull the archenteron tip up to the animal pole. Then the future oral side flattens and forms a stomodeal rudiment which fuses with the archenteron tip to give the mouth. At the other end of what is now the alimentary canal the blastopore becomes the anus. This is the *prism* stage and is the first stage at which the oral–aboral (also sometimes called ventral–dorsal) polarity becomes clearly visible.

After this the embryo elongates at all extremes to form the two long oral arms, two shorter anterolateral arms, and the apex, and these are reinforced by the skeletal rods laid down by the primary mesenchyme. The secondary mesenchyme cells form two lateral masses which become coelomic vesicles, muscle and pigment cells. Also in the coelomic sacs are found the eight small micromeres which are destined to contribute to the echinus rudiment. By now, the embryo has developed into the *pluteus* larva which is the usual end-point for embryological experiments since it is reached in a few days without feeding. The adult sea urchin is produced one or two months later by a complex metamorphosis from the *echinus rudiment* which is a structure formed by fusion of the left body wall with an outgrowth from the coelomic cavity. The tempo of development depends on the species and the temperature, but for the popular Californian sea urchin, *Strongylocentrotus purpuratus*, the blastula would hatch at about 21 hours, gastrulation occur about 35–52 hours, and the pluteus be reached by 72 hours after fertilization.

This account, based on Hörstadius (1973) and Okazaki (1975), is valid for most sea urchin species used for experimental purposes. However, some variations are worth mentioning. As discussed by Raff (1987), a number of species with large and yolky eggs undergo direct development to the adult with no feeding larval stage. An intermediate case is provided by some sand dollars which have a reduced pluteus and can get through metamorphosis within a few days without feeding. In such species the primary mesenchyme cells continue to divide after invagination. Other classes of echinoderm such as starfish and brittle stars, which have been used occasionally by experimentalists, have a comparable early development to sea urchins, but tend to have an equal 16-cell stage with no micromeres.

Molecular markers

Various monoclonal antibodies have been described which promise to be useful for the study of regional specification (Wessel & McClay, 1985; Livingston & Wilt, 1990; Fig. 4.24). A 380 kD protein known as meso-1 appears on the primary mesenchyme from the time when the cells first enter the blastocoel. This is one of quite a large number of primary mesenchyme markers discovered in various laboratories, most of which come up at about the time of ingression. A 320 kD protein known as endo-1 appears on the vegetal plate of the blastula, is retained by the posterior two-thirds of the invaginating archenteron and is later found in the mid and hind gut of the pluteus. A third antigen, called ecto-V, appears at the mesenchyme blastula stage with a uniform distribution. By the mid-gastrula stage it has become restricted to the ectoderm on the prospective oral side. It then appears on the anterior archenteron and may be involved in the formation of the stomodeum. Its localization enables an earlier identification of the oral side than is possible by direct observation.

Among the markers discovered by nucleic acid methods are the *Spec* gene family whose mRNAs are expressed on the aboral side from the early blastula stage onward (Carpenter *et al.*, 1984). These code for calcium binding proteins of unknown function. There is a whole crop of genes expressed only in the primary mesenchyme, including the mRNA for a 50 kD spicule matrix protein called SM50 (Benson *et al.*, 1987). Considerable regional specificity is also shown by the actin gene family (Cox *et al.*, 1986).

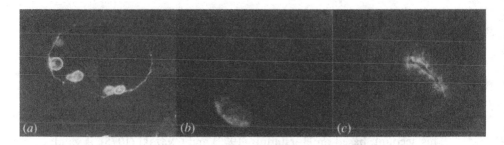

Fig. 4.24. Monoclonal antibodies against regional markers in the early sea urchin embryo. (a) primary mesenchyme specific, (b) endo-1 staining of vegetal plate, (c) endo-1 staining of archenteron lining. (Photographs kindly provided by Dr D.R. McClay.)

The cytoskeletal form CyIIIa mRNA is turned on in the aboral ectoderm from the early blastula, and is later joined by CyIIIb mRNA. CyIIa mRNA appears in the primary mesenchyme during ingression and then disappears. It also appears for a while in the secondary mesenchyme during invagination of the archenteron. CyIa and b mRNA are initially present all over the embryo but become elevated in the vegetal plate, and remain on in the forming archenteron and later in the stomach and intestine of the pluteus while simultaneously falling in the aboral ectoderm. The muscle specific M-actin mRNA appears first in the archenteron wall of the midgastrula and in the pluteus is found in muscle cells of the coelomic sacs and connecting the oesophagus to the ectodermal wall.

Although the embryologist is interested in finding markers which represent *de novo* zygotic synthesis characteristic of particular regions, this does not necessarily mean that all or indeed even most of the gene expression in the early embryo will be of this form. In the 1960s and 1970s there was extensive investigation of the quantitative aspects of transcription in the sea urchin (reviewed Davidson, 1986). It was found that a high proportion of the mRNAs in the early embryos was preformed in the oocyte ('maternal mRNA') and that the complexity of mRNAs actually declined during development even though the number of cell types was increasing. The initial number of different mRNAs is around 11 000 and by gastrulation this has fallen to about 8500 of which about 10% are newly activated zygotic genes.

Early regional differences are not confined to macromolecules. Sea urchins have the distinction of being the material in which the first biochemical gradients were demonstrated which correlated with morphogenetic behaviour, beating gradients in *Drosophila* by half a century. The so-called *reduction gradient* can be visualized by staining the embryos with a dye which is decolourized by reduction and then depriving them of air (Child, 1936; Hörstadius, 1952). In cleavage stage embryos there is a gradient of reducing rate with its maximum at the animal pole, and in the mid- to late blastula the gradient becomes reversed with its high point at the vegetal pole. Although not presently fashionable it is perhaps worth taking these early metabolic regional differences seriously since they have a greater potential for displaying the lability so characteristic of early embryos, and they may also relate to signal transduction pathways which connect inductive signals to new patterns of gene expression.

Fate map

The first sea urchin fate map was constructed for *Paracentrotus lividus* by Hörstadius (1935, 1939) using vital stain marks applied to single blastomeres, and also using the subequatorial pigment ring which forms after maturation in the eggs of this species. He showed that the primary mesenchyme arises wholly from the micromeres and that archenteron and secondary mesenchyme come from the ring of macromeres abutting them (called veg 2 in the 64-cell stage: Fig. 4.23(*k*)).

Recently the method of lineage label injection has been applied by Cameron *et al.* (1987, 1989) who injected blastomeres of *Strongylocentrotus purpuratus* with fluorescein-dextran at the two- to eight-cell stage (Fig. 4.25). At the two-cell stage there are only two complementary labelling patterns, suggesting that the first cleavage plane has a fixed relationship of 45° to the future oral–aboral axis. This relationship means that at the four- and eight-cell stages there are prospective dorsal, ventral, right and left blastomeres. At the eight-cell stage they saw six discrete labelling patterns in the ectoderm. The four animal blastomeres contribute only to the appropriate quadrants of the pluteus ectoderm. The four vegetal cells all contribute to the gut and mesoderm and also to different regions of the ectoderm. VO contributes ectoderm from mouth to anus, VA the posterior aboral ectoderm, and VL the anal plate. There are coherent labelling patterns in the ectoderm suggesting an absence of cell mixing, whereas in the gut and mesoderm there is extensive mixing of labelled and unlabelled cells. The results of Cameron *et al.* have been obtained on a single species and may not apply to all sea urchins. For example, in the classical vital staining studies of Hörstadius & Wolsky (1936) on *Paracentrotus lividus* and also in a recent study on *Hemicentrotus pulcherrimus* using Lucifer Yellow injection (Komamini, 1988) it was found that the first cleavage plane bore no particular relation to the future dorsoventral polarity.

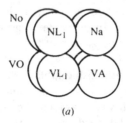

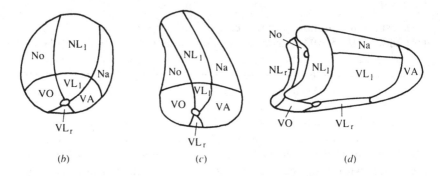

Fig. 4.25. Fates of the blastomeres at the 8-cell stage in *Strongylocentrotus purpuratus*. (*a*) shows nomenclature of blastomeres; (*b*)–(*d*) contributions of each blastomere to the ectoderm of gastrula, prism and pluteus stages. (After Cameron *et al.*, 1987.)

REGIONAL ORGANIZATION

Unfertilized egg

Evidence of some regional specification along the egg axis has been obtained by Maruyama *et al.* (1985) who bisected and then fertilized eggs of *Hemicentrotus pulcherrimus*. Either haploid or diploid embryos are capable of some development so both fragments can develop even though only one contains the egg nucleus. The animal halves formed ciliated balls while the vegetal or lateral halves formed approximately normally proportioned half-size larvae (Fig. 4.26). This suggests that some component necessary for complete pattern formation along the egg axis is localized to the vegetal hemisphere before fertilization.

Low speed centrifugation can stratify the cytoplasm of sea urchin eggs into zones in which yolk granules, mitochondria or clear cytoplasm predominates, the order of zones depending on the species (Brandriff &

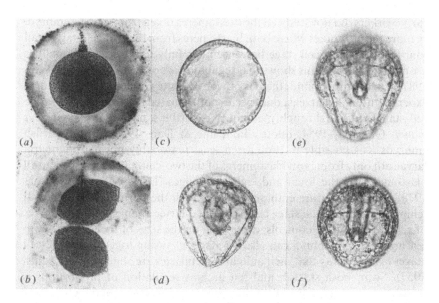

Fig. 4.26. Results of bisection of unfertilized eggs of *Hemicentrotus pulcherrimus*. (*a*) unfertilized egg, animal pole marked by jelly canal. (*b*) Animal–vegetal bisection. (*c*) isolated animal half after 2 days. (*d*) isolated vegetal half from same egg as (*c*), after 2 days. (*e*),(*f*) a pair of isolated lateral halves after 2 days. (Photographs kindly provided by Dr Y.K. Maruyama.)

Hinegardner, 1975). Following fertilization, stratified eggs develop quite normally suggesting that whatever is localized in the vegetal region is not free to move and is therefore presumably associated with the cell cortex. It is found that the oral side nearly always arises on the centrifugal side of the egg, so perhaps the oral–aboral polarity can be influenced by rearranging the cytoplasm, as in amphibians. More severe centrifugation can break the eggs into two halves of different composition. With *Arbacia punctulata* the centrifugal 'red' half is rich in yolk and pigment and the centripetal 'white' half contains mitochondria, clear cytoplasm, lipid droplets and the nucleus. Both halves are capable of forming small plutei after fertilization although the yield is very low and the centripetal halves also sometimes develop into ciliated balls. These experiments do not then rule out the existence of a determinant associated with the vegetal cortex, although it seems unlikely that there is much regional specification in the egg beyond this.

Blastomere isolation and fusion

Following work by Driesch, Boveri and Herbst around the turn of the century it was believed that both of the first two blastomeres and all of the first four blastomeres can develop to form miniature plutei. However, the early workers did not study all the blastomeres taken from a single embryo. In more recent studies where both blastomeres from the two-cell stage or all four from the four-cell stage have been carefully followed, abnormalities are often found and this shows that regulation is not perfect (reviewed Wilt, 1987). On the other hand, there is obviously quite good regulation in these experiments and in at least one species both isolated blastomeres of the two cell stage can get through metamorphosis (the sand dollar *Astriclypeus manni*: Okazaki, 1954, quoted in Kumé & Dan, 1957). From starfish embryos it is possible to produce at least reasonably normal bipinnaria larvae not only from both blastomeres of the two-cell stage but also from all blastomeres of the four- and eight-cell stages (Dan-Sohkawa & Satoh, 1978). These miniature embryos develop with the normal developmental tempo but at any given stage have a half, a quarter or an eighth the number of cells possessed by controls, depending on the sizes of the isolates. With sea urchins, giant plutei can also be made by fusing together two 32-cell stages in such a way that their animal–vegetal axes are aligned (Hörstadius, 1957). So in both starfish and sea urchins regulation of proportions is possible over at least an eight-fold volume range, corresponding to a two-fold linear range, and possibly even more. Taken together these results suggest that the classical view about regulation in sea urchins is basically correct, but that there may be variations in developmental potency depending on the species and on the manner of blastomere separation. The implication for regional specification is that not much happens during the early cleavages and so long as a blastomere inherits some vegetal cortical material it can develop into a reasonably complete pattern.

The same conclusion may be drawn from the ability of early stages to tolerate unnatural cleavage planes. As long ago as 1892 Driesch compressed embryos under a coverslip which causes all the cleavages to occur in one plane and to yield a flat plate of cells. After release of the pressure the embryos rapidly recover a fairly normal form by cell movements. Although the cytoplasm cannot become parcelled up in quite the normal way during compression, it is probable that the blastomeres inheriting the critical vegetal material will be adjacent to each other so this experiment does not represent a serious disturbance to the egg axis.

Morula

Any account of sea urchin development is dominated by the experiments performed by Hörstadius (1935; reviewed 1939,1973) in which rings of cells from different levels of the 32–64 cell stage were isolated and recombined. As has been emphasized by Wilt (1987) it is important to remember that, unlike the classical grafts of amphibian embryology, many of these experiments have not yet been repeated in recent years using lineage labels and molecular markers. This means that we can not be sure that all the

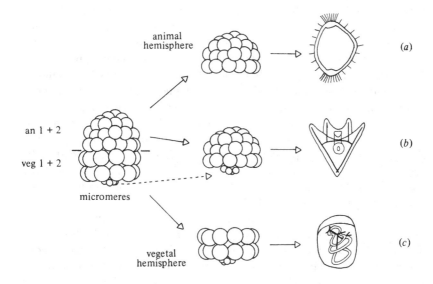

Fig. 4.27. Evidence for an inductive interaction along the egg axis of the sea urchin embryo. The isolated animal hemisphere forms a ciliated ball (*a*), the isolated vegetal hemisphere forms an 'ovoid' (*c*), but the animal hemisphere in combination with the micromeres forms a reasonably well proportioned pluteus (*b*).

results are really due to respecification, as claimed by Hörstadius, rather than to other factors such as cellular selection, alterations in cell movements or developmental delay. Bearing this caveat in mind, the principal results are as follows: isolated animal halves tend to produce ciliated balls, sometimes containing a stomodaeum; while vegetal halves tend to produce ovoids containing skeletal spicules, a gut and pigment cells, but lacking stereocilia, oral arms and stomodaeum (Fig. 4.27). However, the embryo is able to form a complete pattern following minor defects in the egg axis: it is possible to remove the an_1 ring of cells (see Fig. 4.23(*k*)) or the micromeres, or to combine an animal half with two or more micromeres (ie, remove veg1 and veg2). This suggests that the regional pattern along the egg axis depends on an interaction which occurs at or after the 32-cell stage. Although there does not seem to be any specific level of tissue which is essential for establishment of the pattern, it may be significant that it is possible to remove more tissue form the middle than from the extreme positions. It has often been suggested that the micromeres are an organizing centre since they can restore a near normal pattern when combined with a mesomere cap and they can induce a secondary archenteron if implanted at unusual sites in the vegetal hemisphere (Hörstadius, 1935). However, since micromeres can be removed from an embryo without greatly affecting its pattern it is probable that the organizing centre extends into the veg2 layer.

Some of Hörstadius's recombination experiments have recently been repeated using modern methods. Livingston & Wilt (1990) confirmed his conclusions on the behaviour of isolated animal and vegetal halves using the markers ecto-V (oral ectoderm) and endo-1 and alkaline phosphatase (gut). They found, using combinations of rhodamine-labelled micromeres with unlabelled mesomeres, that gut and pigment cells could be induced from the mesomeres. On the other hand, a similar study on the same species (*L. pictus*) by Henry *et al.* (1989) showed that, while intact mesomere caps would not do so, some isolated mesomeres could spontaneously produce gut, pigment cells or spicules. This suggests that, as in amphibians, the specification of individual cells is not necessarily the same as that of the tissues from which they came. Evidently, the issue of induction along the sea urchin egg axis is still not fully resolved.

Recently it has become possible to examine the expression of individual genes in dissociated cells and this is of interest in the present context if they are normally expressed in a regionally specific way. *CyIIIa* and *Spec-1*, normally expressed in the aboral ectoderm, and SM50, normally expressed in the primary mesenchyme, all turn on at about the right time in dissociated cells. SM50 is expressed at about half the normal level, and in the expected fraction of cells. The aboral ectoderm markers are expressed at much lower levels than usual suggesting that they may require some activation by inductive signals from elsewhere in the embryo. (Stephens *et al.*, 1989; Hurley *et al.*, 1989). The control of expression of these genes has also been examined using reporter constructs injected into eggs. For example, the *CyIIIa* gene has about 15 protein binding sites 5' to the coding sequence. When the entire regulatory region is attached to a chloramphenicol acetyl transferase (CAT) reporter gene and injected into eggs it becomes integrated in the nuclei of a proportion of the cells and the expression pattern of the CAT is in the aboral ectoderm only, corresponding to the normal pattern for *CyIIIa*. Some of the regulatory sites are positive and some negative, as may be shown by coinjecting the isolated sites. An excess of a positive site will compete for factors and depress expression of the CAT, while an excess of a negative site will mop up repressors and cause ectopic expression of CAT in gut or mesenchyme (review: Davidson, 1989). Interesting though this work is, there is still a gap in our knowledge between the presumed extracellular inducers and the intracellular genetic regulatory proteins.

Hörstadius presented his results in terms of a 'double gradient' model. This is *not* a double gradient in the sense used in Chapter 3 but rather two simple gradients, one emanating from the animal pole and the other from the vegetal pole. Like other gradient models of the *BC* (Before Crick, 1970) era, it is not sufficiently defined to enable clear predictions to be made, and so, for example, the detailed criticisms made in the review of Davidson (1989) may not strike their target. Davidson himself prefers a hierarchy of

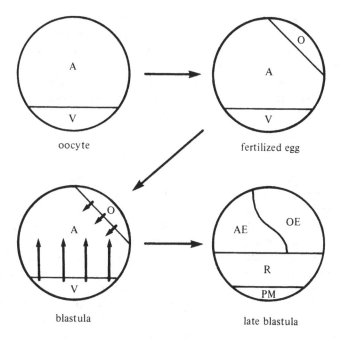

Fig. 4.28. Minimum model for regional specification in the early sea urchin, greatly simplified from Davidson (1989). V represents a vegetal signalling centre and O an oral centre. The territories formed by the late blastula are oral ectoderm (OE), aboral ectoderm (AE), archenteron (R) and primary mesenchyme (PM).

inducing signals travelling up the embryo with each cell layer displaying receptors on its lower surface and exporting the appropriate inducers from its upper surface. This seems unnecessarily complicated given the present data and there seems no particularly good reason why the LSDS model should not apply to the egg axis with the source located in the micromeres- + veg 2 and the sink across the remainder of the embryo (Fig. 4.28). This would explain why animal halves form extreme animal structures (deprivation of signal), why regulation can occur following removal of slices of tissue (re-establishment of gradient over remainder) and why vegetal halves do not regulate. Of course, a separate system is necessary to specify thresholds along the oral–aboral axis and this is explicitly recognized in the model of Davidson.

When embryos are disaggregated in calcium-free sea water the polarity of the cells is retained and when the calcium is restored they can reaggregate with the microvillous apical side out and the basal side in (Nelson & McClay, 1988). It has been claimed by some authors that cells from as late as blastula stages can sort out after reaggregation to give normal larvae (eg, Giudice, 1986 using *Paracentrotus lividis*). However, a recent study by

Freeman on *Hemicentrotus pulcherrimus*, using a rather strict criterion of normality, found that normal larvae could not be formed from reaggregates after the 16 cell stage and then only at low yield (Freeman, 1988). Again this suggests a rather low degree of spatial organization by the 16-cell stage since the more different kinds of cell there are, the less likely is it that a reaggregate will be able to sort out correctly. The signals which cause specification along the egg axis presumably operate later on.

Micromeres

A number of studies have also been made of the formation of the primary mesenchyme under conditions where the micromeres have been interfered with. Using the sand dollar *Dendraster excentricus*, Langelan & Whiteley (1985) showed that the fourth or fifth or both cleavages could be equalized by brief treatment with the detergent SDS. If the 32-cell stage is made to consist of equally sized cells then in about half the cases the primary mesenchyme is prevented from forming. However, the skeletogenic tissue is later regenerated by cells migrating out of the secondary mesenchyme. Not only are the cleavages equalized by SDS treatment but also the arrangement of cells at the 16- or 32-cell stage is deranged. Despite this, the majority become normal plutei. If the primary mesenchyme is removed microsurgically then skeletogenic tissue can again arise from the secondary mesenchyme (Ettensohn & McClay, 1988).

Vegetalization

If whole embryos are treated with lithium, they develop rather like isolated vegetal halves, with an enlarged archenteron and suppression of the animal tuft. Lithium can also vegetalize isolated explants from the animal hemisphere which undergo a shift toward the normal pattern instead of forming the usual ciliated balls. For example, treatment of isolated mesomeres of *Strongocentrotus purpuratus* with 20 mM Li for 6–8 hours yielded 50% expression of SM50 and alkaline phosphatase, associated with spicules and gut respectively (Livingston & Wilt, 1989). As discussed in the amphibian section, the best understood biochemical effect of lithium is the inhibition of inositol monophosphatase, so this may implicate the inositol lipid cycle as part of the interpretation mechanism of a vegetal–animal morphogen gradient. We would then predict that the cycle would be active in the animal hemisphere and inhibited in the vegetal hemisphere of the normal embryo.

Certain substances were supposed to cause a reverse transformation of 'animalization'. But animalizing agents are under suspicion at present since a reinvestigation of the animalizing action of zinc ion (Nemer, 1986) concluded that there was no genuine respecification but that the zinc simply inhibited differentiation. Some protein fractions have been extracted from sea urchin embryos which have animalizing and vegetalizing activity (Runnström, 1975) but there is obviously a long way to go before our knowledge of the inducing factors reaches even the level now obtained for amphibians.

Oral–aboral axis

The fate mapping data mentioned above suggest that, at least in *Strongylocentrotus purpuratus*, there is some asymmetry already in the fertilized egg, since the first cleavage occurs at a constant 45° to the future oral–aboral axis. However, the fact that centrifuged eggs frequently form their ventral structures on the centrifugal side suggests that an early specification can also be reimposed from without by a sufficient stimulus. Experiments on this subject by Hörstadius & Wolsky (1936), using *Paracentrotus lividus*, also indicate that any early specification is reversible. In one of these, 16-cell stages were divided into meridional halves and the cut surface was vitally stained. Often the resulting plutei had complementary halves stained, ie, the right half of one and the left half of the other, but sometimes they both had the *aboral* half stained suggesting that the polarity had become inverted in one member. In another experiment, meridional halves of 16-cell stages were joined at the cut surface, and in almost all cases yielded normal plutei. Since the orientation of the cut is not known, it must be the case that many of these composite embryos consisted of mismatched halves but they were none the less able to regulate.

The first molecular markers to show asymmetric expression patterns are the aboral specific actin Cy IIIa and the Ca binding protein *Spec-1*, both of whose mRNA comes up at 128–256 cells. But these molecular events may be preceded by differences at a metabolic level such as the difference in cytochrome oxidase activity apparently found to be greater on the oral side as early as the 16-cell stage (Czihak, 1963).

SEA URCHINS AND AMPHIBIANS

Although the descriptive embryology of these two 'senior citizen' groups seems very different, it is possible to identify a number of formal similarities which may point towards common mechanisms for regional specification. Both types of egg are formed during oogenesis with a definite animal-vegetal polarity and in both the vegetal region is necessary for normal pattern formation in isolated egg or embryo fragments. It seems clear that in both the vegetal region emits an inductive signal at a morula–blastula stage which is responsible for specification of structures along the egg axis. Whether this can properly be called a gradient, really depends on whether there is more than one annulus of induced structures formed in response to different morphogen concentrations. Even in the amphibia where we have a tentative understanding of the nature of the signals and the initial responses it is not really clear whether this is so. In any case, there seems no more reason in the sea urchin than in the amphibian for invoking a two gradient model for the egg axis. In the LSDS model the fact that the entire responding field is a sink for the morphogen accounts for the fact that isolated fragments will develop into the 'ground state' tissue: epidermis in amphibians and ectoderm with stereocilia in sea urchins. A few years ago the effects of lithium seemed similar on the two groups, in both cases producing a vegetalization. However, in *Xenopus* we now know that the early effect of lithium is a dorsalization, so at present all we can really say is that the lithium experiments offer a tantalizing suggestion of the involvement of the inositol phosphate signal transduction mechanism in early regional specification.

In both groups the dorsoventral (or oral–aboral in sea urchins) polarity is labile at early stages and can be experimentally reversed. However, it probably becomes determined earlier in the amphibian since only the first two blastomeres will form reasonably normal twins whereas in sea urchins it is possible for the first four blastomeres or later meridional halves to do so. In the amphibian the organizer region behaves very much like the source of a morphogen gradient which is responsible for the dorsoventral pattern. In the sea urchin there is no compelling evidence at present for any inductive signal in a dorsoventral direction although it does seem that at least three levels of different specification must arise, as shown by the markers *endo V*, *CyIIIa* and the space in-between.

Both amphibians and sea urchins used to be cited by those who claimed that embryo fragments could regulate with perfect retention of proportions over a wide size range. Careful studies of the products of such experiments now seem to show that there is not perfect regulation of proportions and so there may be no need to invoke models with a high degree of mathematical complexity to account for it. None the less isolated blastomeres do regulate well enough to exclude all but the most limited degree of cytoplasmic localization: in both cases probably just a vegetal signalling centre laid down during oogenesis.

GENERAL REFERENCES

Czihak, G. ed. (1975). *The Sea Urchin Embryo. Biochemistry and Morphogenesis.* Berlin and Heidelberg: Springer-Verlag.

Hamburger, V. (1988). *The Heritage of Experimental Embryology. Hans Spemann and the Organizer.* New York: Oxford University Press.

Hörstadius, S. (1973). *Experimental Embryology of Echinoderms.* Oxford: Clarendon Press.

Nakamura, O. and Toivonen, S. eds. (1978). *Organizer: A Milestone of a Half Century from Spemann*, North-Holland, Amsterdam: Elsevier.

Slack, J.M.W. ed. (1985). Early Amphibian Development. BSDB Symposium. *J. Embryol. Exp. Morph. 89 supplement.*

5

Development with a small cell number

In this Chapter we shall examine the experimental embryology of some diverse types of animal: molluscs, annelids, ascidians, and nematodes. These groups can be collectively contrasted to the embryos considered in the previous Chapter in so far as the key decisions of early development seem to be made at a very early stage when there are only a few cells in the embryo. This may mean that each individual cell has a unique identity in terms of its biochemical properties and corresponds to a zone of tens or hundreds of similarly committed cells in a vertebrate, insect, or sea urchin embryo. The small cell number means that all individual embryos of a given species are identical or nearly identical and makes it possible in principle to construct fate maps of very high precision by direct observation of the cell lineage. A number of studies of this type were carried out around the turn of the century and some, such as Wilson's (1892) study of *Nereis*, Wolterek's (1904) study of *Polygordius* and Conklin's (1905*a*) study of *Styela*, are masterly works still referred to today. More recently the introduction of Nomarski interference contrast microscopy has made it possible to extend this approach to the nematode *Caenorhabditis* and to carry it to the ultimate limit of a complete cell lineage from egg to adult (Sulston *et al.*, 1983).

The case for using these animal species as experimental models is partly that they are felt to be 'simple' in some sense, but also that they offer favourable material for the study of cytoplasmic localization. These are the archetypal 'mosaic' embryos. Each embryo fragment or isolated blastomere is supposed to develop into the part which it would have formed in the whole embryo. How true this really is we shall consider below, but there certainly is a good body of evidence for some sort of significant cytoplasmic localization. We must then carefully consider four things for each species concerned:

1. When the localization occurs: in the oocyte, in the fertilized egg, or in a blastomere during cleavage.
2. Whether it qualifies as a cytoplasmic determinant, ie a substance causing cells that inherit it to acquire a particular commitment.
3. If so whether it is a determinant for a terminal cell type, an earlier body

128

subdivision, an organizing centre, or something else.
4. Finally, of course, what its molecular nature might be: messenger RNA, genetic regulatory factor, cytoskeletal specialization or whatever.

This Chapter involves three sets of anatomical description which may prove rather heavy going for the uninformed reader. On a first read through readers are advised only to try and remember the essentials: for the molluscs and annelids remember the *polar lobe*, which is the cytoplasmic protrusion confering organizer properties on cells which inherit it. For ascidians remember the *yellow crescent*, a putative determinant for muscle cells. For nematodes remember the stem cell like early divisions of the P blastomere which leads to the initial regional subdivision.

MOLLUSCS AND OTHER SPIRALLY CLEAVING FORMS

The molluscs make up a phylum of animals containing the familiar spiral gastropods, the bivalves and cephalopods, and the less familiar chitons and scaphopods. The species of embryological interest are mainly marine or fresh water gastropods (*Ilyanassa, Patella, Lymnaea, Bithynia, Crepidula*) together with the scaphopod *Dentalium*. The characteristic *spiral cleavage* of the early stages is shared by many annelid worms and also by some other minor phyla.

NORMAL DEVELOPMENT AND FATE MAP

Much work has been performed on the marine mud snail, *Ilyanassa obsoleta*, and it is the development of this species which is described initially. The egg is about 170 microns in diameter and develops quite slowly, taking about 7 days to reach the *veliger larva* stage at which the general molluscan body plan is visibly differentiated. The veliger has a shell on the dorsal side secreted by a shell gland, a foot on the ventral side, and a complex locomotary organ called the velum at the front end (see Fig. 5.1(*j*)).

Fertilization occurs before oocyte maturation, and at the first and second meiotic divisions a cytoplasmic protrusion called the *polar lobe* appears at the vegetal end of the egg. For this and subsequent stages see Fig. 5.1. The polar lobe is not to be confused with the polar bodies which are extruded near the animal pole as usual. The lobe reappears at the first and second cleavages and is a particularly prominent feature of early development. During the first cleavage, the lobe attachment becomes highly constricted and since the lobe is as large as both the forming blastomeres the embryo appears superficially to consist of three cells and this stage is accordingly called the *trefoil* (Figs 5.1(*b*) and 5.3(*a*)). The blastomere which includes the

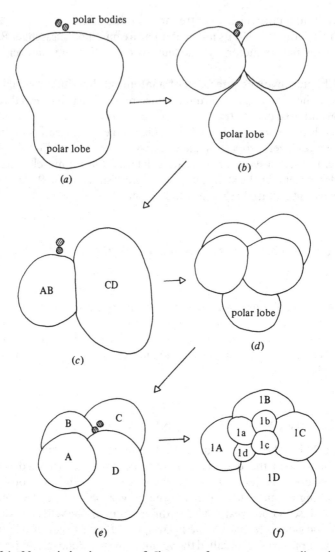

Fig. 5.1. Normal development of *Ilyanassa* from zygote to veliger larva. (*a*) fertilized egg, (*b*) trefoil, (*c*) 2-cell stage, (*d*) second cleavage, (*e*) 4-cell stage, (*f*) first micromere quartet formed by dexiotropic cleavages, (*g*) and (*h*) continuation of spiral cleavage, (*i*) median division of 4d to give two mesentoblasts, (*j*) veliger larva. (After Clement, 1952.)

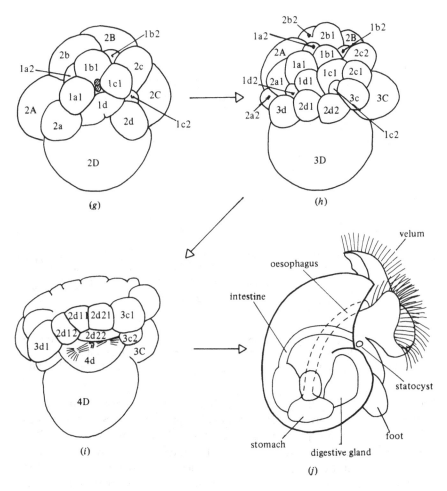

(g)

(h)

(i)

(j)

lobe material is designated CD while the other, smaller, blastomere is called AB. The lobe appears again at the second cleavage in which the four blastomeres A,B,C and D are formed and passes into the D blastomere which thereby becomes larger than the other three.

The plane of cleavage is altered to the equatorial for the third division. Each of the four blastomeres A,B,C,D, now called *macromeres*, divides to form a small cell called a *micromere* at its animal end. The nomenclature of blastomeres for spirally cleaving embryos was introduced by E.B. Wilson at the end of the nineteenth century. It is rather tiresome to learn, but is still universally used by workers in the field. After the third cleavage the macromeres are relabelled 1A, 1B, 1C, 1D and the micromeres are called, respectively, 1a, 1b, 1c and 1d. This type of division is repeated several times, each ring of micromeres being called a *quartet*. So the second quartet comprise 2a, 2b, 2c and 2d and the macromeres then become 2A, 2B, 2C

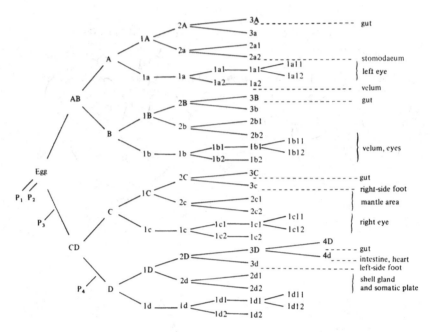

Fig. 5.2. Cell lineage of *Ilyanassa* up to the 29-cell stage. P represents the polar lobe which appears four times altogether. The micromere 2d is the first somatoblast and 4d is the mesentoblast. Tentative fate map assignments are given on the right-hand side.

and 2D. The cleavage pattern is called *spiral* because each quartet of micromeres is somewhat rotated relative to the macromeres. When looked at from the animal pole the first quartet is formed by right-handed or *dexiotropic* cleavages, the second by left handed or *laevotropic* cleavages and the orientation of cleavages continues to alternate in this zig-zag fashion for subsequent quartets. Each quartet of micromeres also divides at a time somewhat after the corresponding macromere division. The off-spring are designated by numbers after the letter with 1 indicating the animal daughter and 2 the vegetal daughter. For example 1a becomes 1a1 and 1a2, and 2c becomes 2c1 and 2c2. The same nomenclature is followed for subsequent divisions, so 1a1 becomes 1a11 and 1a12, and 2c2 becomes 2c21 and 2c22. The eight cells 1a11, 1a12, 1b11, 1b12, 1c11, 1c12, 1d11, 1d12 arise in a characteristic pattern called the *molluscan cross* at the animal pole. The complete cell lineage for these early stages is shown in Fig. 5.2.

Readers who have trouble remembering all this need only retain the names of two cells. The micromere 2d is called the *first somatoblast* and forms much of the veliger body, particularly the shell gland. The micromere 4d is called the *second somatoblast* or *mesentoblast*. In *Ilyanassa* this arises somewhat before the other cells of the fourth quartet, and is much larger than they are. After formation it divides in the plane of bilateral symmetry of the embryo to form two mesentoblasts, one on the right and one on the left side (Fig. 5.1(*i*)). Each of these then cuts off three more micromeres, two of which are enteroblasts which contribute to the gut, and the remaining large cell is called a *mesoblastic teloblast*. These are stem cells that divide repeatedly to give rise to the internal mesodermal bands, which subsequently form the larval muscles and mesenchyme. However, not all tissues classically regarded as mesodermal arise from the bands; some are also formed from the micromeres of the second and third quartets.

Gastrulation occurs by a spreading of the micromere cap to cover the whole embryo and after this the embryo is referred to as a *trochophore*. It has a ciliated apical end bearing paired eyes derived from the first micromere quartet, and ciliated bands around the equator called the *prototroch* derived from the first and second micromere quartets. In later development the remnants of the blastopore become shifted anteriorly and a stomodeal invagination arises at its site to become the mouth. The dorsal ectoderm forms the *shell gland* and the ventral ectoderm the foot. On the two sides of the base of the foot are the *statocysts* (Fig. 5.1(*j*)). The stomach and digestive gland are formed from the macromeres and the intestine from the mesentoblast. The prototroch grows to form the bilobed locomotary organ called the *velum*. The veliger larva represents the definitive body plan of the organism. It hatches after 7–8 days and in later development the bilateral asymmetry becomes more pronounced and the shell assumes its characteristic right handed spiral form.

The fate maps of molluscs are largely based on descriptive studies of cell lineage which cannot be carried beyond the trochophore stage. The assignments made on Fig. 5.2 are drawn from four sources: Conklin's description of the cell lineage of *Crepidula*, (Conklin, 1897), Van Dongen and Geilenkirchen's (1974) description of *Dentalium*, defect experiments by Clement (Clement, 1967) and marking experiments using carbon particles (Cather, 1967). It is regrettable that there still does not exist a more accurate fate map for the veliger structures based on the use of lineage labels, and, if the results from leeches, ascidians and nematodes are anything to go by, the contributions of individual early blastomeres will probably turn out to be more complex than indicated here. Unfortunately, a recent attempt to study the lineage of *Patella* with microinjected HRP was unsuccessful, probably because of exocytosis leading to spread of the enzyme beyond the originally labelled clone (Kühtreiber, Serras & van den Biggelaar, 1987).

Other species and annelids

The description given for *Ilyanassa* is broadly applicable to other molluscs used in experimental embryology except for cephalopods. For species mentioned in this Chapter it should be noted that the scaphopod *Dentalium* follows a very similar course of early development. The other species are,

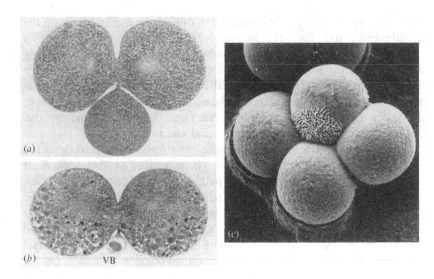

Fig. 5.3. Polar lobes of three molluscan species. (*a*) section of trefoil stage of *Dentalium*. (*b*) section of trefoil stage of *Bithynia*. VB vegetal body. (*c*) scanning electron micrograph of 4-cell stage of *Crepidula*. The former lobe is visible as a puckered region on the D blastomere. (Photographs kindly provided by Dr M.R. Dohmen.)

like *Ilyanassa*, gastropods. *Crepidula* and *Bithynia* differ in that their polar lobes are much smaller (Fig. 5.3(*b*),(*c*)), while *Lymnaea* and *Patella* have no polar lobe at all. In the latter forms the early cleavages are symmetrical, forming a four-cell stage in which all blastomeres are the same size. The D blastomere can only be named retrospectively once the dorsoventral polarity has become established at the time of mesentoblast formation at the sixth cleavage.

The description of the early stages is also applicable to annelid worms, which undergo a similar spiral cleavage and gastrulation and form a trochophore larva. The description for polychaetes such as *Nereis* and *Chaetopterus* is really very similar to *Ilyanassa* except that the polar lobes are less pronounced. The trochophore is the usual end point for embryological experiments on annelids and it is composed of a pretrochal region bearing an apical tuft, the ciliated girdle called the prototroch, and a posttrochal region, which later generates the entire segmented part of the body from a growth zone. A trochophore of *Chaetopterus* is shown in Fig. 5.4(*a*). In oligochaetes and leeches there is a cytoplasmic rearrangement, called an *ooplasmic segregation*, following fertilization at which clear *pole plasms* (or *teloplasms*) are segregated to the animal and vegetal poles. Both pole plasms enter the D macromere, and in later cleavages they mix before being partitioned into both the first somatoblast (2d) and the mesentoblast (4d). In the leech *Helobdella triserialis*, the pole plasms contain many

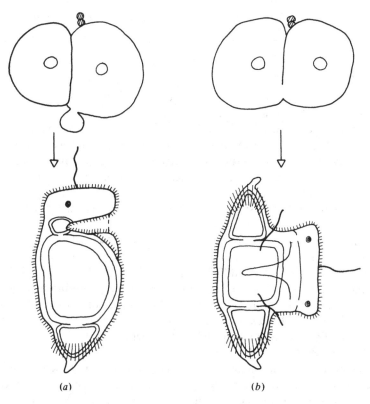

Fig. 5.4. Production of a mirror duplicated larva of the polychaete *Chaetopterus* by compression at the first cleavage. This distributes the lobe material equally between the first two blastomeres. (*a*) normal trochophore, (*b*) 'Janus larva'. (After Titlebaum, 1928.)

microtubules and their segregation is inhibited by microtubule inhibitors such as nocodazole (Astrow, Holton & Weisblat, 1989). On the other hand, in the oligochaete *Tubifex* the plasms contain dense actin filament networks and segregation is inhibited by cytochalasin (Shimizu, 1982). In these groups, and particularly the leeches, the spiral cleavage is somewhat modified. The synchrony of cleavage is lost early and the C and D macromeres divide before the others. The micromeres are very small and the first somatoblast and mesentoblast correspondingly big. The first somatoblast divides to form four *ectoteloblasts* on each side and the *mesentoblast* forms one mesoteloblast on each side. These are stem cells which end up in the posttrochal region and go through repeated divisions to lay down bands of cells forming the segmented part of the body. In leeches the cell lineage is clearly homologous but the nomenclature of cells is different: the ectoteloblasts being known as N, O, P, Q and the mesoteloblast as M. Detailed studies of the fates of these stem cells have recently been made using lineage labels (Weisblat, Kim & Stent, 1984) and we shall return to this topic below when considering segmentation.

Role of the egg cytoplasm

Many experiments on small invertebrate embryos have involved the use of actinomycin D. This is an inhibitor of RNA synthesis and the idea is that events which occur in the presence of actinomycin do not require new transcription of genes but are carried out by components formed before the treatment began. It has been shown that *Ilyanassa* will develop normally up to gastrulation in the presence of concentrations of actinomycin which block virtually all transcription (Newrock & Raff, 1975) so it is probable that most protein synthesis at the early stages is directed by maternal mRNA.

Mainly because of the properties of the polar lobe (see below) molluscan eggs have been extensively studied by electron microscopy and this has revealed different features in different species (Dohmen & Verdonk, 1979). The lobe of *Ilyanassa* includes some unusual double membrane vesicles containing dense material. The lobe of *Dentalium* contains 'multisheet vesicles'. The lobe of *Crepidula* has many folds on the plasma membrane and that of *Bithynia* contains an RNA-rich 'vegetal body' which appears during oogenesis (Fig. 5.3(b),(c)).

An important feature of the early development which is clearly under maternal control is the orientation of the spiral cleavage. Normally the third cleavage is dexiotropic, the fourth laevotropic, the fifth dexiotropic and so on. Nearly all adult snail shells are right-handed spirals when viewed from the tip. However, occasionally an individual or local race of left-handed snails may be found and their embryos undergo a laevotropic third cleavage, dexiotropic fourth and so on. Breeding experiments using right- and left-handed variants of the freshwater snail *Lymnaea* have shown that the handedness of the shell is determined by a single gene and that right-handed is dominant over left-handed. However, the handedness of an individual is not determined by its own genotype but by that of the mother (Boycott *et al.*, 1930). As we have discussed in Chapter 2, a maternal effect implies that the feature under consideration is laid down before fertilization, presumably during the development of the oocyte. Since the left-handed symmetry is apparent as early as the third cleavage this seems quite reasonable. Interestingly, the dominance of right over left is not only genetic but is also shown in cytoplasmic transfer experiments. Freeman & Lundelius (1982) showed that laevotropic eggs of *Lymnaea* could be made dexiotropic after an injection of cytoplasm from normal eggs, but the transfer of cytoplasm in the reverse direction was without effect. Moreover, in the sinestral forms spontaneous dextral phenocopies are not unusual. Presumably there is some defect in a component of the cytoskeleton which causes the spindles to tilt in the opposite direction from usual, although the molecular basis remains undiscovered.

REGIONAL ORGANIZATION

Given that most of the early protein synthesis takes place on maternal templates, that there is some inhomogeneity of ultrastructure, and that at least one aspect of pattern, the orientation of spiral cleavage, is determined during oogenesis, it seems reasonable to ask to what extent the fertilized egg, or even the oocyte, is already partitioned into different regions. This would presumably be accomplished by the localization of cytoplasmic determinants, each of which could regulate the activity of the nuclei which come within their influence in such a way as to produce the correct arrangement of cell types in the normal early embryo. The answer to this question has been sought for mollusc embryos and for the other types considered in this Chapter by four main kinds of experiment:

1. Redistribution of egg cytoplasm by low-speed centrifugation, mechanical compression or cytochalasin treatment.
2. Removal of parts of the egg cytoplasm.
3. Removal of blastomeres or portions of blastomeres and study of the development of the remainder of the embryo.
4. Study of the development of isolated blastomeres.

Egg cytoplasm

Low-speed centrifugation can stratify the egg cytoplasm to give bands of lipid droplets, soluble cytoplasm and yolk granules, arranged perpendicular to the centrifugal direction. This treatment has been shown to have no effect on the subsequent development of fertilized eggs of several molluscs and annelids including *Chaetopterus*, *Crepidula*, *Cumingia* and *Dentalium* (review: Raven, 1966). This implies either that there is no regionalization of cytoplasm before the first cleavage or that if there is then the factors responsible are not affected by centrifugation. It is often thought that the 'cortex' or cytoplasm just below the plasma membrane is not redistributed and could therefore be the site of regionalization.

Guerrier *et al.* (1978) were able to cause some lobe material to enter each blastomere of *Dentalium* either by mechanical bisection of the trefoil stage or by treatment with cytochalasin B. This effectively gives a CD + CD configuration instead of AB + CD, and the resulting veligers have many structures duplicated. Similar experiments on annelids such as *Chaetopterus* have generated quite good mirror symmetrical double embryos (see eg, Guerrier, 1970 and Fig. 5.4) and according to Henry & Martindale (1987) these are *double dorsal duplications*. They bear an obvious resemblance to the double dorsal embryos produced by organizer grafts in amphibia and suggest that the signal from the D macromere may, like that from the organizer, be some kind of gradient. It was in fact suggested by Wilson as early as 1929 that what was localized in the D quadrant was a determinant for an organizer.

When an unfertilized egg of *Dentalium* is cut in two then it is possible to fertilize both halves and, as in sea urchins, obtain a certain amount of development despite the fact that the vegetal half is haploid (Wilson, 1904a; Render & Guerrier, 1984). In such experiments the animal halves lack not only the posttrochal region, as might be expected from the fate map, but also lack the apical tuft. By contrast, the vegetal halves are capable of developing into trochophores of normal proportions but reduced size. This shows three things: first that the animal cytoplasm is not essential for formation of the normal pattern, secondly that the vegetal cytoplasm is necessary for the formation of both vegetal and animal structures and thirdly that the sizes of parts, including the polar lobe itself, can be scaled down in proportion to the size of the entire embryo.

Clement (1968) was able to separate the properties of the cortex and the mobile cytoplasm in the vegetal hemisphere of fertilized eggs of *Ilyanassa* by a regime of centrifugation which enabled the preparation of vegetal halves containing the nucleus but excluding the dense yolk granules. When allowed to develop these vegetal halves could form normal or nearly normal veligers of reduced size in a similar way to vegetal halves from unfertilized eggs. The animal halves formed defective larvae similar to those produced by animal halves of unfertilized *Dentalium* eggs.

Taken together these results suggest that, at least in the lobe forming species, the morphogenetic properties of the D blastomere lineage may be specified by some localization in the vegetal cortex of the unfertilized egg which remains in place during fertilization and is shunted into the D blastomere by the early cleavages. By contrast, the amount and composition of the mobile components of the cytoplasm have little effect on the course of development.

Lobe removal

At the trefoil stage the polar lobe of *Ilyanassa* or *Dentalium* is connected to the CD blastomere by a thin stalk and so it is a reasonably simple manipulation to cut it off. This operation was first performed by Crampton (1896) and the studies have been extended more recently by Clement (1952), Atkinson (1971) and Van Dongen (1976). Lobeless embryos are quite viable and they continue to develop at the same rate as controls for many days. However their organization is greatly disturbed. At the two-cell stage the CD blastomere is no larger than AB, and at the four-cell stage D is no larger than A, B and C. Various peculiarities of the D lineage which are

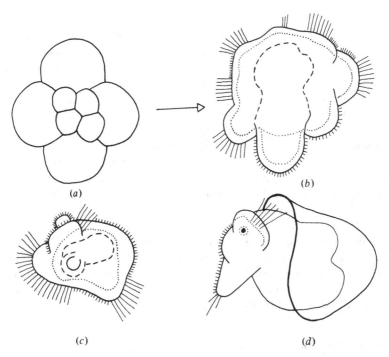

Fig. 5.5. Defect experiments on *Ilyanassa*. (*a*) an 8-cell embryo which had its polar lobe removed at the trefoil stage. It is quite symmetrical and there is no way of identifying a D macromere. (*b*) larva arising from a lobeless embryo. It has velar cilia all over but lacks most of the normal structures and possesses little regional organization. (*c*) a larva which has developed from an isolated AB blastomere. It is similar to the lobeless larva. (*d*) a larva formed from an isolated CD blastomere. This has some defects but is much more normal than the other types. (After Clement, 1952, 1956.)

apparent in normal embryos are absent in lobeless ones. In particular the cell 4d is no different from 4a,b,c. It does not form earlier, it is no larger, and it does not initiate the sequence of cleavages leading to the formation of the mesodermal bands. The lobeless embryos thus form no mesodermal bands although they do develop some muscle and mesenchyme, presumably from the second and third micromere quartets. They also have no eyes, no foot and usually no shell. They have velar cilia all over the place and the main segments of the alimentary canal other than the intestine are present although not normally organized (Fig. 5.5(*a*),(*b*)).

It is worth noting that the defects apparent in lobeless larvae are not simply of structures normally arising from the D lineage but of other parts as well, as judged by the fate map assignments given in Fig. 5.2. This reinforces the idea mentioned above that the D macromere which contains the polar lobe material is an organizer and is necessary for the normal development of the embryo because of some inductive signal which it emits. To investigate the timing of this signalling, Clement (1962) removed the D macromere at different stages by puncturing it with a fine glass needle. The results are quite complex, but the main conclusion is that removal of the D macromere at the four-cell stage produces a larva similar to the lobeless larva, lacking structures normally formed by the D lineage and also certain other structures such as the eyes and statocysts. But removal of D after formation of the third micromere quartet deletes only those structures which normally arise from the mesentoblast, and removal after formation of the fourth quartet produces no defects. The implication is that the D macromere is required as an organizer up to the 24-cell stage, as a precursor to the mesentoblast until its formation at the 29-cell stage, and is not required thereafter. Similar results have also been obtained with *Dentalium* (Cather & Verdonk, 1979).

Among annelids, lobe removal studies have concentrated on *Sabellaria*, a polychaete with a large polar lobe (eg, Render, 1983). Removal of the first polar lobe (trefoil stage) inhibits formation both of the posttrochal region and of the apical tuft. Removal of the second polar lobe (four-cell stage) inhibits the posttrochal region but not the apical tuft.

Blastomere ablation and isolation

Embryos such as molluscs and annelids are favourable to the experimentalist in several respects but particularly because single early blastomeres can be raised in isolation. According to our definitions of Chapter 2, a fragment which can self-differentiate certain structures *in vitro* is judged to be specified although not necessarily determined with respect to decisions in the developmental hierarchy which are necessary conditions for the formation of those structures. Isolation experiments were carried out by Crampton (1896) on *Ilyanassa* and Wilson (1904a,b,c) on *Dentalium*, *Patella* and *Lanice*, and Costello (1945) on *Nereis*. These authors represented their results as showing a strict mosaicism, in other words that the states of specification corresponded exactly to the fate map. However, the results are open to certain criticisms. Firstly, these authors did not really know the fate map, indeed the situation is still unsatisfactory today, and without a good fate map it is difficult to decide whether or not regulation has occurred. Secondly, they did not have antibiotics at their disposal and so their isolates often became infected and did not survive very long. This meant that they were mainly looking at very early events such as the orientation of cleavages and the positions of cilia. These are both features which will probably depend entirely on the cytoarchitecture of the egg and which will not be altered by any respecification of nuclei within the fragment. Furthermore, there is much subjectivity in the assessment of patterns whch are intermediate between perfect mosaicism and perfect regulation. If a structure is overrepresented this is often counted as mosaicism, but this begs the question of the origin of structures which may be smaller than usual but are none the less present. Whether the creatures considered in this Chapter really show a higher degree of mosaicism in isolated blastomere experiments than amphibians or sea urchins is not clear, and we shall return to this point below.

In fact, the studies by Clement (1956) on *Ilyanassa* show that isolated blastomeres do not behave in a mosaic fashion at all. CD and D blastomeres can form larvae which are somewhat abnormal in form but contain most of the structures which normal larvae contain. On the other hand AB, A, B and C blastomeres develop into larvae resembling lobeless larvae which lack shell, foot, statocysts, heart and eyes (Fig. 5.5(*c*),(*d*)). It should particularly be noted that in normal development the eyes are probably formed from the 1a and 1c micromeres, although in isolated blastomere experiments D can develop eyes while A and C cannot. Likewise Cather (1967) showed that isolated A,B,C or D blastomeres could make shell while in normal development only 2d does so. The implication is that at the four-cell stage the A, B and C blastomeres are equivalent and that they require to undergo some interaction with the D lineage to acquire their definitive states of specification.

In annelids it is also possible to obtain more or less normally proportioned larvae of reduced size by isolating the CD blastomere (*Lanice*–Wilson, 1904*c*) or the D blastomere alone (*Tubifex*–Penners, 1926). In contrast the AB cell or A + B + C give rise to severely defective larvae.

Determination of the D lineage in equally cleaving forms

The results described up to now indicate that the organizing properties of the D lineage arise from some cytoplasmic localization which is established before fertilization and which is in some way associated with the polar lobe. However, some molluscs have no polar lobe; the first two cleavages are equal and it is not possible to say which of the four blastomeres is D except retrospectively after the mesentoblast has been formed. One such species which has been investigated in some detail is the common limpet, *Patella vulgata*.

According to the morphological description of van den Biggelaar (1977), two of the macromeres remain joined by a cross furrow after the second cleavage and it is one of these which always becomes D. The first divisions of micromeres and macromeres are synchronous so that a 32-cell stage is formed before any of the divisions producing the fourth quartet. The embryos remain at this stage for some time while the macromeres protrude up within the embryo to make contact with the apical micromeres of the molluscan cross. There then appears to be a struggle between the two cross furrow forming macromeres to see which can make the most intercellular contacts and the one which does so becomes 3D. Deletion of one of the cross furrow macromeres at the beginning of the 32-cell stage does not suppress formation of the mesentoblast although if one of the cells were preprogrammed to become D then the mesentoblast would be expected to be lost in 50% of cases. So in such experimental cases it seems that another macromere can assume a central position, make the apical contacts and form the mesentoblast. On the other hand, the mesentoblast is not formed if all four macromeres are left in place but prevented from contacting the apical micromeres. This may be done in several ways: either by partial dissociation of the embryos in citrated sea water, or by deletion of the first quartet cells (van den Biggelaar & Guerrier, 1979), or, in *Lymnaea*, by brief exposure to cytochalasin B at the 24-cell stage (Martindale, Doe & Morrill, 1985).

The nature of the interaction is still unknown although it seems unlikely that gap junctions are involved. In *Patella* there is no coupling detected by injection of Lucifer Yellow until the 32-cell stage, and even then there is none between the micromeres and 3D (Dorresteijn *et al.*, 1983). Curiously, the coupling pattern for *Lymnaea* is quite different, with universal coupling during the early cleavages. In both species the first uncoupling is associated with differentiation, initially of the trochoblasts and later of the cephalic plates and shell gland (Serras & van den Biggelaar, 1987).

Do the results on *Patella* and *Lymnaea* indicate that these species have a wholly different mechanism of early development from the polar lobe forming types such as *Ilyanassa* and *Dentalium*? One is reluctant to accept this because we all like to feel that similar molecular events should underly homologous macroscopic processes. It is possible, for example, that the localization associated with the polar lobe is not a determinant which restricts the potency of the lineage but simply a bias in the cytoarchitecture which guarantees that the D macromere always makes the appropriate contacts and always forms the mesentoblast. On the other hand, it is also possible that there is such a determinant and it is activated at the one cell stage in some species and at the 32-cell stage in the other. In fact *Lymnaea* embryos contain an RNA rich ectosome in all four macromeres rather similar to the vegetal body of *Bithynia* (van den Biggelaar & Guerrier, 1983). It is the resolution of this type of question which is so important in forming a molecular approach to early development and only once it is answered shall we know where to look for the significant regulatory molecules.

Segmentation in the leech

Annelids are the segmented animals *par excellence* and the leech has been adopted by the laboratory of D. Weisblat and G. Stent to assist in understanding the mechanism of segment formation. The species mainly used for this work is *Helobdella triserialis* which has an egg diameter about 500 microns and develops to the hatching stage in about 6 days. In leeches the spiral cleavage is somewhat modified. The micromeres are very small and after the formation of the first micromere quartet, cleavage of the D lineage leads the rest. Regions of clear *teloplasm* are segregated to the animal and vegetal poles of the egg after fertilization and these enter the D blastomere, becoming mixed in the process. The D blastomere cuts off a small micromere then divides more equally to form an animal DNOPQ cell (homologous to 2d) and a vegetal DM cell. The DNOPQ cell cuts off three small micromeres, then divides equally in the medial plane to give right and left NOPQ cells. These sequentially bud off four *ectoteloblasts*: N, O, P, and Q. The DM cell cuts off two small micromeres to become the 4d homologue, then divides equally in the median plane to give right and left M cells or *mesoteloblasts*. The ten teloblasts (five on each side) bud off long bandlets of cells to form *germinal bands* (Fig. 5.6). As these are elongating they envelop the embryo such that the N bandlets end up midventral with O, P and Q arranged alongside and M within.

It is possible by low-speed centrifugation to redistribute the teloplasm such that, in a proportion of cases, it enters the C blastomere in similar amount to D. In such cases the timing and character of C cleavages are rather similar to those of D and lead to the production of extra teloblasts (Astrow, Holton & Weisblat, 1987). This suggests that the teloplasm is or contains a determinant that confers stem cell quality on the cells which inherit it, but we know nothing of its constitution except that it is rich in microtubules.

The fate map of the leech has been studied by injection of lineage labels into each of the teloblasts and into some of the other cells as well (Weisblat, Kim & Stent, 1984). The labels used are HRP, fluoresceinated dextrans and some tailor-made fluorescent oligopeptides. The chief results concern the composition of the segmental repeating unit; leeches having 32 segments of which the 21 abdominal segments show a high degree of serial homology. It turns out that individual segments are not clones or even polyclones. The progeny of one bandlet cell spans two or three segments, overlapping with the progeny of preceding and following cells in the bandlet, but populates exactly the same structures and cell types in different segments (Fig. 5.7). As far as tissue type is concerned, there is a remarkable lack of clonal restriction. All of the ectoteloblasts contribute both to the ventral nerve cord and to the epidermis, and the O teloblast even contributes cells to the nephridial tubules. The mesoteloblast contributes mainly to muscles, nephridia and connective tissue, but also a few interneurons to the nerve cord. Again, these patterns are the same between segments. Initially, the patterns were reported as being invariant between different individuals, and this remains true in general. However, it is now known that the O and P bandlets may switch over such that O cells follow P fates and vice versa

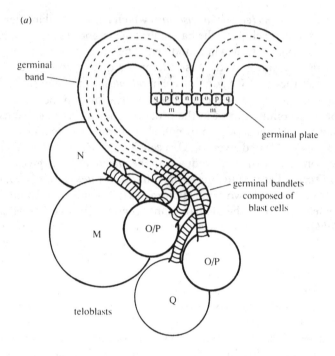

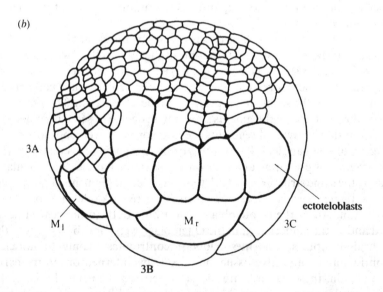

Fig. 5.6. (a) Nomenclature of teloblasts and germinal plate cells in the leech *Helobdella*. (b) formation of germinal bands in *Clepsine*, to show the relation of teloblasts to the rest of the embryo.

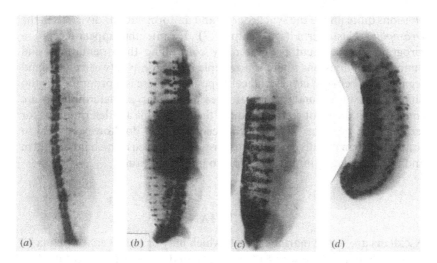

Fig. 5.7. Segmentally repeating contributions of the different ectoteloblasts in *Helobdella*. (*a*) N teloblast injected, (*b*) O/P teloblast injected, O pattern obtained, (*c*) O/P teloblast injected, P pattern obtained, (*d*) Q teloblast injected. All specimens anterior up, (*a*)–(*c*) ventral view; (*d*) lateral view. The most anterior body segments are unlabelled since they were laid down before the time of the injection. (Photographs kindly provided by Dr D. Weisblat.)

(Weisblat & Blair, 1984). Also the contributions of micromeres to a transient surface layer called the provisional epithelium are not invariant.

It is ironical that the creatures for which we have good, modern, fate map information (*Helobdella* and *Caenorhabditis*) are not very suitable for micromanipulations such as grafts or blastomere isolations. Experimental intervention in the leech has so far been restricted to ablation of particular cells *in situ*. This is done either by injecting them with DNAse or by injecting with a fluorescent lineage label and then irradiating with a high intensity of the exciting wavelength. Ablation of individual teloblasts does result in predictable, mosaic type, deficiencies in the nervous system with the exception that the O cell can compensate for ablation of P by taking its place (Weisblat & Blair, 1984). However, defects in the epidermis are filled in by cells from the other teloblasts (Blair & Weisblat, 1984), and cell death can be provoked if part of a bandlet is caused to move caudally beyond the normal limit of formation of the 32 segments (Shankland, 1984). Evidently, as in so many other cases, the leech started its laboratory career as a mosaic embryo and has become regarded as more regulative as more work has been done on it.

With regard to the mechanism of segmentation the one thing which seems very clear is that it is totally different from that found in *Drosophila*, since pattern arises through repetition of a series of stereotyped cell divisions quite unlike the synchronous and uniform nuclear divisions in the *Drosophila* blastoderm (see Chapter 7). Despite the appearance of a 'program' for segment formation by controlling the orientations and timing of cell divisions, there is no simple relationship between clones and segments or between clones and cell types, so if there is a program it is too difficult for us to understand at present. As far as determinants are concerned, it does not seem as though the teloplasms are determinants for terminal cell types or for organizing centres. They do, however, seem to confer the stem cell division characteristics on the cells which inherit them and this must be an interesting topic to pursue in future.

ASCIDIANS

Ascidians are sessile marine animals which may be mistaken for plants by the uninitiated. They are classified along with the vertebrates in the phylum Chordata because they have free-living larvae called tadpoles whose morphology is obviously similar to that of vertebrates, having a dorsal neural tube, a notochord and a muscular tail. All the species used for embryological research are simple ascidians which have an obligatory sexual cycle. Compound ascidians, in which many individuals are joined in a colony, may reproduce either sexually or by budding.

Embryonic development is exceptionally rapid. The eggs of the species considered here: *Styela, Halocynthia, Ciona, Boltenia, Ascidia* and *Phallusia* are 100–300 microns in diameter and develop to hatching tadpoles in about 24 hours, the exact rate depending on species and temperature. The larva itself will only swim for a few hours and then settles and commences metamorphosis into the adult.

NORMAL DEVELOPMENT

The adults of all the species considered are hermaphrodite although most are cross fertilized. The eggs are shed before maturation and the polar bodies extruded at the animal pole after fertilization. Three cytoplasmic zones are visible: a clear 'ectoplasm' in the animal hemisphere, a subcortical 'myoplasm', and a yolky 'endoplasm' in the interior. The ectoplasm is largely formed from the contents of the germinal vesicle. The *myoplasm* consists of a submembranous microfilament network attached to an underlying filamentous lattice, and, in *Styela* and *Boltenia*, pigment granules (Jeffery & Meier, 1983). After fertilization there is some considerable rearrangement of the egg cytoplasm (*ooplasmic segregation*) recently described in detail for *Phallusia* (Sardet *et al.*, 1989). In the first phase the myoplasm is drawn rapidly down to the vegetal pole region by a cortical contraction. It then moves up to what is now the prospective posterior side of the egg in association with the sperm pronucleus. This draws it into a

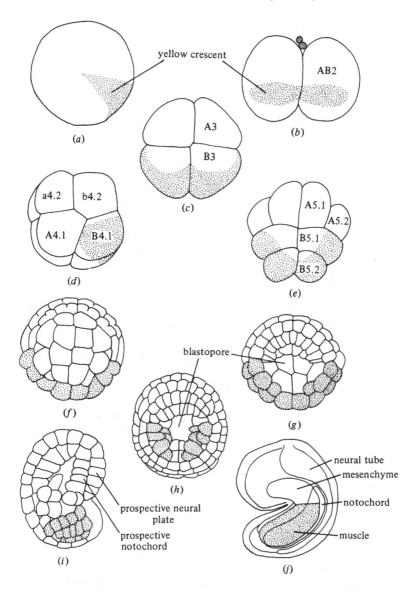

Fig. 5.8. Normal development of *Styela*. The pigment of the yellow crescent is shown stippled. (*a*) fertilized egg, lateral view, (*b*) 2-cell stage, posterior view, (*c*) 4-cell stage, vegetal view, (*d*) 8-cell stage, lateral view, (*e*)–(*h*) vegetal views, gastrulation commences in (*g*), (*i*) parasagittal section through late gastrula, (*j*) early tadpole, side view. (After Conklin, 1905*a*.)

crescent shape which because of the associated pigment granules is known as the *yellow crescent* in *Styela* and the orange crescent in *Boltenia*. The first phase can be inhibited by microfilament-disrupting reagents such as

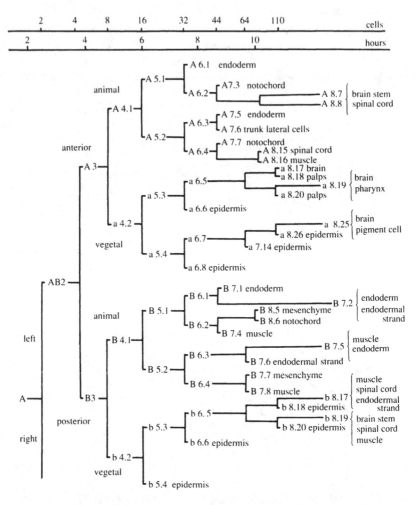

Fig. 5.9. Cell lineage of the ascidian *Halocynthia*, (After Nishida, 1987.)

cytochalasin, and the second phase by microtubule reagents such as colcemid (Sawada & Schatten, 1989).

Diagrams of the developmental stages of *Styela* are shown in Fig. 5.8 and the early cell lineage for *Halocynthia* in Fig. 5.9. The first cleavage is medial and bisects the yellow crescent. Because of the bilateral symmetry of the organism, blastomeres on both sides are given the same name and where it is necessary to distinguish them the left side names are underlined. So, for example, both of the first two blastomeres are called AB2; AB2 on the right and AB2 on the left. The second division is frontal, separating two anterior cells called A3 from two somewhat larger posterior cells called B3, which include the yellow crescent. The third cleavage is equatorial and separates

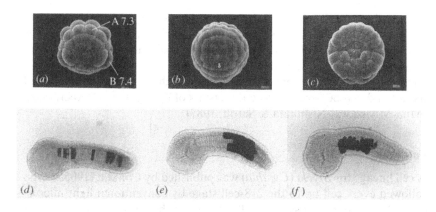

Fig. 5.10. *Halocynthia roretzi*, (*a*)–(*c*) scanning electron micrographs of embryos, anterior up. (*a*) 64-cell stage, vegetal view, showing cells A7.3 and B7.4 (*b*) 110-cell stage animal view, (*c*) 110-cell stage, vegetal view. (*d*)–(*f*) results of HRP injections. (*d*) A7.3 injected, notochord labelled, (*e*) B7.4 injected, muscle labelled, (*f*) b7.15 injected, epidermis labelled. (Photographs kindly provided by Dr H. Nishida.)

four animal cells (a4.2 and b4.2) from four larger vegetal cells (A4.1 and B4.1). It is worth attempting to remember the names of these blastomeres as many experiments are done on the eight-cell stage. The nomenclature of cells, introduced by Conklin, involves keeping the letter appropriate to the eight-cell stage and numbering all the products of each cell generation from 1 upwards. This is somewhat confusing since cells change their numbers each generation, for example A5.2 will become A6.3 and A6.4, but as with the Wilsonian system for the *Spiralia* it is well established in the field. The synchrony of cleavages ceases by the 16-cell stage although the timing and orientation of each particular division is the same in all individual embryos. The animal and vegetal cell layers continue to divide and a blastocoel forms between them. Dye coupling becomes apparent from the 32-cell stage indicating an increase in the density of gap junctions.

Gastrulation commences at the 64-cell stage as an invagination of the vegetal layer. At this stage most of the invaginating region is prospective endoderm but includes also the prospective notochord anteriorly and the prospective muscle and mesenchyme posteriorly. The animal layer, consisting of prospective neural plate and epidermis, spreads down and constricts the blastopore into a T shape with the bar anterior (Fig. 5.10(*c*)). As this T closes it is gradually moved posteriorly and the notochord and neural plate move round to what was the vegetal side and now becomes the dorsal side.

Eventually the mesenchyme and muscle end up lateral to the notochord and the remnant of the blastopore has moved right round to the posterior. The neural plate forms a tube and sinks below the epidermis and the final result is a young tadpole with the basic chordate body plan. A variety of monoclonal antibodies have been made, which recognize differentiated tissues and can be used to score the results of embryological experiments (Mita-Miyazawa, Nishikata & Satoh, 1987).

Fate map

A cell lineage for *Styela* (*Cynthia*) was published by Conklin (1905*a*), who followed every cell up to the 218-cell stage by conventional light micros-copy. That this was possible at all was due to the several visibly different cytoplasmic zones set up after fertilization which are passively parcelled out to the blastomeres. Although it was a magnificent achievement for the time, Conklin's lineage was not entirely accurate and has now been superseded by HRP injection studies carried out on *Halocynthia roretzi* by Nishida and Satoh (Nishida & Satoh, 1983, 1985; Nishida, 1987; see Fig. 5.10).

The resulting lineage is shown in Fig. 5.9. At the eight-cell stage the epidermis arises entirely from the animal tier, the notochord and endoderm entirely from the vegetal tier, the brain mainly from a4.2 and the muscle predominantly from B4.1. However, all cells at this stage give rise to more than one tissue. At the 64-cell stage, 44 cells have become allocated to produce only one cell type. By the 110-cell stage, which is the latest stage at which HRP injections could be carried out, 94 cells are so allocated. Like *Caenorhabditis*, ascidians have an almost invariant cell lineage as between individual cases. However, there is some indeterminacy in the notochord with regard to the relative positions of the derivatives of the precursors A6.2 and A6.4. There are also two cases where equivalent left and right side cells give rise to different products in an indeterminate way: either of the a8.25 pair produces the ocellus and the other the otolith; and either of the b8.17 pair produces spinal cord and the other some endodermal strand cells. Although there are several asymmetrical divisions in the lineage, the ascidian differs considerably from the annelid and the nematode patterns because there are few asymmetrical divisions which occur in a reiterated way giving a sequence of similar daughter cells from a parent stem cell.

Much of the experimental embryology of ascidians concerns the determi-nation of muscle cells and so it is particularly important to understand their origins in normal development. Nishida and Satoh showed that the larva of *Halocynthia* has 21 muscle cells on each side and that three of the four blastomeres on each side of the eight-cell stage contribute to them. B4.1 contributes 14, A4.1 contributes 2 and b4.2 contributes 5. The muscle precursors at later stages are shown in Fig. 5.10. In *Ciona* and *Ascidia* most aspects of the cell lineage are very similar but b4.2 forms only 2 muscle cells.

Interestingly, Zalokar & Sardet (1984) traced the normal fate of the mitochondria rich myoplasm in *Phallusia* using the fluorescent dye $DiO(C_2)_3$ and found that it segregated into exactly those cells shown by Nishida and Satoh to be the muscle precursors in *Ciona* and *Ascidia*.

Conklin and many of his successors evidently believed that all the visible cytoplasmic regions in the zygote, of which he distinguished six, were associated with determinants which caused the cells inheriting them to develop into the appropriate structures. However, we now understand that the fate map, even where it is at single cell resolution and where it is very similar between individuals, tells us nothing about cell commitment. Readers of this book will now be well aware that evidence about commitment must be obtained from other types of experiment and so we shall go on to consider how good is the evidence for the existence of a muscle determinant, and what its molecular nature might be.

REGIONAL ORGANIZATION

The profound ooplasmic segregation following fertilization suggests that if there are any cytoplasmic determinants preformed in the oocyte then they should be translocated to new positions in the fertilized egg. It has repeatedly been shown that eggs can have their contents stratified by centrifugation without affecting normal development. However, this only moves components such as yolk granules, pigment granules and mitochondria. The subcortical microfilament network of the myoplasm is resistant (Jeffrey & Meier, 1984).

Reverberi & Ortolani (1962) found that reasonable twins could be produced by fertilizing halves of the unfertilized egg arising either from meridional or from equatorial cuts. The half without the egg nucleus is of course haploid but this is compatible with a certain amount of development. After fertilization the ability to produce a complete pattern becomes more restricted and is shown only by vegetal fragments (see also Bates & Jeffrey, 1987b; Bates, 1988). These results indicate that in ascidians, as in other embryos, it is possible to produce normally proportioned larvae of reduced size. It is also possible to make double size larvae by fusion of embryos at the two cell stage, providing that the fusion is carried out in such a way that the egg axes of the two embryos are aligned (von Ubisch, 1938). It thus appears that the sizes of the parts can adapt to the overall size of the organism over a four-fold volume range. In at least some species it seems probable that any determinants become localized after fertilization, perhaps in the course of the visible cytoplasmic rearrangement.

Deletion experiments on early cleavage stages were carried out by Chabry (1887) and Conklin (1905b). They concluded that if one of the first two blastomeres was killed, a lateral half tadpole was formed from the remaining one, and if one or two cells of the four-cell stage were killed then the larva lacked cells normally formed by those blastomeres. Although these results are often quoted as showing the mosaic character of ascidians, it should be noted that the killed blastomeres remain attached, and under such circumstances amphibian blastomeres can also form partial larvae. Interestingly, Nakauchi & Takashita (1983) showed that defective larvae from half-blastomeres were able to metamorphose into normal adults.

A classic series of blastomere isolation experiments were carried out on *Ascidiella* and *Ascidia* by Reverberi & Minganti (1946). They isolated the four pairs of equivalent cells from the eight-cell stage and found the following pattern of differentiation following culture *in vitro*.

a4.2   epidermis only
b4.2   epidermis only
B4.1   muscle, mesenchyme, endoderm, a little intestine.
A4.1   endoderm, notochord.

(See also Fig. 5.11.)

Brain and sense organs, which normally arise predominantly from the a4.2 pair, will only develop when these blastomeres are cultured in combination with the anterior vegetal pair A4.1. These authors regard the results as evidence for a neural induction which occurs in later embryonic development, as in amphibian embryos. At the time, it was thought that the other structures corresponded exactly to the fate map, but it is now clear that all the tissues show a more restricted representation than would be expected in a strictly mosaic situation. The formation of muscle has been subjected to detailed study and will now be considered.

Determination of the muscle lineage

The centre piece of ascidian embryology is the determination of muscle cells and the role of the myoplasm, in *Styela* co-extensive with the yellow crescent, in this process. Until recently the story was simple. The myoplasm contained a muscle determinant. It was segregated to the B4.1 cells during the first three cleavages and the progeny of these cells went on to form muscle in an autonomous manner. Unfortunately, things have become less simple as more knowledge has accumulated.

The markers used to study muscle formation are all terminal differentiation products and include muscle specific actin (Tomlinson, Bates & Jeffrey, 1987), a myosin-like protein (Nishikata *et al.*, 1987*a*), and acetylcholinesterase (Meedel & Whittaker, 1983). The muscle-specific actin appears to be encoded by a maternal mRNA and synthesis of the protein commences by the 64-cell stage. For the other markers synthesis of mRNA starts by the 64-cell stage or shortly afterwards, based on the development of resistance of protein synthesis to actinomycin D treatment. From Fig. 5.9 it will be apparent that the 64-cell stage is the first at which *any* cells of

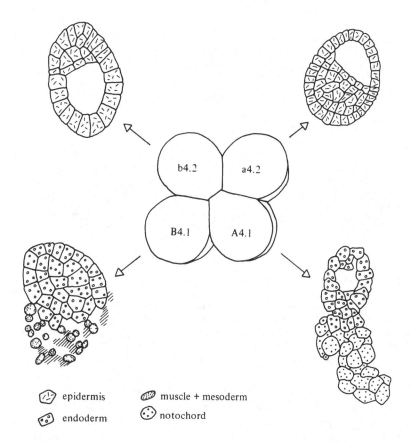

Fig. 5.11. Blastomere isolation experiment in an ascidian. Pairs of equivalent blastomeres from the 8-cell stage were cultured in isolation and the tissues formed are indicated by the stipples. (After Reverberi & Minganti, 1946.)

the muscle lineage have become allocated. All the clonal descendants of B7.4 and B7.8 will form muscle but there are other lineages in which allocation of purely myogenic cells is yet to come.

If embryos are reared in cytochalasin B it is possible to inhibit cleavage and cell movements while permitting continued nuclear division. Such embryos are referred to as *cleavage-blocked*. Whittaker (1973) using *Ciona*, and Satoh (1979), using *Halocynthia*, showed that in embryos cleavage blocked at different stages, acetylcholinesterase appeared only in cells of the B4.1 muscle lineage (Fig. 5.12). At the time it was thought that all muscle was normally formed from B4.1, but now it is known that some caudal muscle cells also arise from A4.1 and b4.2 and so we might well ask why these cells do not also express the enzyme under cleavage block.

Recent studies of isolated blastomeres have confirmed the result of

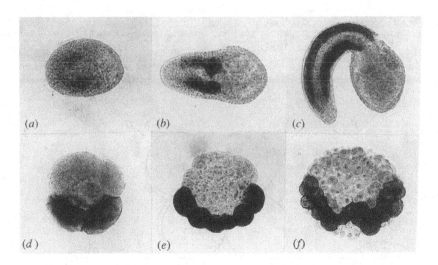

Fig. 5.12. Development of acetylcholinesterase in the ascidian *Halocynthia*. (*a*)–(*c*) normal development: the enzyme is demonstrated by a histochemical reaction which shows it to be localized in the tail muscle. (*a*) Neurula, (*b*) tailbud, (*c*) tadpole. (*d*)–(*f*) Embryos cleavage arrested by cytochalasin B. (*d*) arrested at 8-cell stage, (*e*) at 32-cell stage, (*f*) at late gastrula. The enzyme appears on schedule in most but not all of the cells destined to contribute to the tail muscle. (Photographs kindly provided by Dr N. Satoh.)

Reverberi and Minganti that muscle is formed by the B4.1 pair (Whittaker, Ortolani & Farinella-Ferruzza, 1977). A little muscle is also formed by isolated A4.1 pairs in *Ascidia* but not in *Ciona* (Meedel, Crowther & Whittaker, 1987), and at low frequency by b4.2 pairs of *Halocynthia* (Nishikata *et al.*, 1987a). However, muscle arises much more readily in three-quarter embryos from which the B4.1 pair have been removed than from the isolated A4.1 or b4.2. Indeed, Reverberi and Minganti themselves observed some cases of muscle development in such three quarter embryos. It therefore seems likely that muscle formation by descendants of A4.1 and b4.2 requires inductive interactions with surrounding cells at some time following the eight-cell stage. The failure of these blastomeres to form muscle in the cleavage blocked embryos may be because cell movements are inhibited and the correct spatial contacts have not been formed.

Because of the conviction that the yellow crescent cytoplasm contained a muscle determinant, a number of attempts have been made to introduce it into other, non-muscle forming cells. Whittaker (1980) produced an abnormal distribution of cytoplasm at the eight-cell stage of *Styela* by compressing embryos at the four-cell stage. This prevents the usual equatorial cleavage and means that four cells inherit the yellow crescent cytoplasm instead of the usual two. These pressed embryos develop abnormally after release of pressure. However, if they are cleavage blocked by cytochalasin B then 25% produced acetylcholinesterase in three or four of the blastomeres containing yellow crescent material, while cleavage blocked controls form the enzyme only in the B4.1 pair. At the time this seemed a clear case of nuclear reprogramming by a cytoplasmic factor, similar to the pole plasm graft in *Drosophila*, but its significance has been somewhat reduced by the discovery that derivatives of the neighbouring cell pairs of the eight-cell embryo also normally form muscle. Also there have been some other attempts with less impressive results. Deno & Satoh (1984) obtained only about 2% of positive cases following injection of myoplasm into A4.1 cells of *Halocynthia*, then cleavage blocking and subsequently staining for acetylcholinesterase. Bates (1988) pricked and squeezed fertilized eggs of *Ciona* or *Styela* to produce embryos of about half normal volume but containing all the myoplasm. These develop into small but normally proportioned larvae, and cleavage block experiments showed that the number of muscle cells was not increased even though the myoplasm had been distributed to more cells than usual.

The converse procedure of ablating the myoplasm has been attempted by Nishikata *et al.*, (1987*b*). A series of monoclonal antibodies were prepared against isolated myoplasm. When injected into fertilized eggs these did indeed reduce the proportion capable of forming muscle. The most effective antibody recognized a non-stratifiable component of the myoplasm.

Further information about muscle determination has been gained from the use of cytochalasin combined with the DNA synthesis inhibitor aphidicolin (Satoh & Ikegami, 1981). When the inhibitors are added to early stages both cleavage and nuclear replication are inhibited. If they are added before the 76-cell stage then no cells express acetylcholinesterase, while if they are added at or after this stage, which is still before the time of transcription, the enzyme later appears in at least some of the cells of the muscle lineage. Satoh & Ikegami showed that the cells in which the enzyme appeared were those which had undergone seven or more rounds of DNA replication by the time the inhibitors were added. It is possible to know this because the cell division is markedly asynchronous by this stage. By the early gastrula, some prospective muscle cells are in the seventh, some in the eighth and some in the ninth cell generation.

So what does all this data add up to concerning the role of the myoplasm? There is an excellent correlation between the segregation of the myoplasm and the muscle cell lineage in normal development. Probably the subcortical microfilament network is the important part of the myoplasm rather than the mitochondria rich region. The cleavage block and blastomere isolation experiments show that the muscle formed from the B4.1 cells requires neither cleavage nor correct contacts with other cells, while the muscle formed from A4.1 and b4.2 may require either or both of these. The aphidicolin experiments suggest that a minimum number of DNA replications is obligatory in all the lineages. The myoplasm transfer experiments have not been particularly successful. We are left feeling that the myoplasm is probably a necessary condition for muscle formation although perhaps not a sufficient one. There are no reports of muscle being induced in the total absence of the myoplasm, but then no worker on ascidians would have believed this possible or attempted to devise such an experiment. As to the molecular nature of the determinant, we really have no idea. There is reasonable evidence that the muscle-specific actin is synthesized off maternal mRNA, since it can be made in anucleate egg fragments (Tomlinson, Bates & Jeffrey, 1987). But it has not been shown that this mRNA is localized in the myoplasm. The other muscle markers seem to be synthesized off the zygotic genes, so if there is a single regulatory molecule in the myoplasm it would be required both to activate a set of nuclear genes and to activate one or more maternal mRNAs.

Other possible determinants

When early embryonic stages are cleavage blocked, they come to consist of a few large cells each containing multiple nuclei. In *Ciona* terminal differentiation markers characteristic of more than one cell type can be co-expressed in the same multinucleate cell (Crowther & Whittaker, 1986). Whittaker has argued that this means that several determinants are present in the fertilized egg and that they act independently as positive regulatory elements. On the other hand, similar studies in *Halocynthia* have not shown co-expression of epidermal and muscle markers (Nishikata, Mita-Miyazawa & Satoh, 1988), and this is also true of *C. elegans* (Cowan & McIntosh, 1985).

A limited amount of work has been performed on three putative determinants in addition to the muscle determinant. These concern the control of endodermal alkaline phosphatase, of the epidermis and of the dorsal axis. Alkaline phosphatase is normally expressed only in the endoderm and in cleavage blocked embryos shows an appropriately restricted pattern (Whittaker, 1977). Since it is unaffected by treatment of embryos of any stage with actinomycin D, it was thought that the mRNA was transcribed, and perhaps localized, in the oocyte. However it is now known that anucleate egg fragments will not make alkaline phosphatase, but if they are fertilized by a sperm to form andromerogones, then they will do so (Bates & Jeffrey, 1987a). This makes it probable that the mRNA is, in fact, synthesized during embryonic development.

The formation of two epidermis-specific antigens has been studied by Nishikata and others (1987) in cleavage blocked embryos and in isolated blastomeres. Expression is restricted to the a4.2 and b4.2 blastomeres in both types of experiment, which is formally similar to the muscle story. The existence of an 'axial determinant' has also been proposed by Bates & Jeffrey (1987b). They showed that if a small volume (5–15%) of cytoplasm was extruded from the vegetal pole of eggs between the first and second phases of the ooplasmic segregation, then the resulting larvae were radialized, failing to gastrulate or form a dorsal axis. They have also shown that the formation of axial structures can be inhibited by ultraviolet irradiation before the second ooplasmic segregation (Jeffrey, 1990). Although the workers concerned believe deeply in determinants in all these cases, the evidence is obviously less good than for muscle, and even the muscle determinant has its uncertainties and ambiguities.

CAENORHABDITIS ELEGANS

C. elegans is a small soil nematode which was selected by S. Brenner in 1965 as a suitable experimental organism with which to solve the problems of development and the nervous system. It is probably fair to say that this rich promise has yet to be fulfilled although a number of interesting embryological results have emerged in recent years.

The technical advantages which the worm offers are chiefly those favourable for genetic analysis. The worms will grow on agar plates on lawns of *E. coli*, they take only three days to reach sexual maturity, they have a small genome size and they can be frozen in liquid nitrogen for long-term storage. Most individuals are hermaphrodite, with a few males of XO chromosome constitution arising as a result of non-disjunction at meiosis. The hermaphrodites are self-fertilizing, and this means that a mutant stock can be established from one individual and that an F_1 mating is not required to obtain homozygosity. One individual can produce a large number (300) of progeny. It is possible to produce genetic mosaics of some parts of the genome by spontaneous loss during development of free chromosome fragments present as duplications, and it has on occasion been possible to use this method to establish in which cells a particular gene must be active in order to do its job. The main disadvantage of *C. elegans* is that it is not very suitable for micromanipulation: the embryos are very small (30×60 microns) and are surrounded by a tough shell, which must be broken open to allow the entry of inhibitors and other reagents.

As for the ascidians, development is exceptionally rapid taking about 14 hours to hatching at 22 °C. During the first six hours there is rapid cell division and after this there is almost no cell division but considerable differentiation and shape change. The eggs are laid at about the 30-cell stage. The larva hatches with 558 cells and passes through four further stages separated by moults, by which time the cell number has reached its adult total of 959 somatic cells (hermaphrodite) plus an indeterminate number of germ cells. In keeping with the treatment of other organisms in this book the focus will remain on early development although in the case of *C. elegans* this does seem rather more arbitrary than for vertebrates or insects, and there is no suggestion that there are not interesting things to study during the larval stages.

NORMAL DEVELOPMENT

Mature oocytes pass through the spermatheca and become fertilized, usually by the hermaphrodite's own sperm or perhaps by sperm from a male which has been received and stored in the spermatheca. The sperm entry point is posterior although it is not known whether this polarity is caused by the sperm or is pre-existing. The egg contains a cortical microfilament network and a number of randomly dispersed P granules which can be visualized using a monoclonal antibody and which will later segregate to the germ line (Strome & Wood, 1983). The polar bodies are extruded at about 20 minutes and the shell is secreted around this time. There is then a period of considerable cytoplasmic re-arrangement including a pseudocleavage (Fig. 5.13(b)) during which time the female pronucleus migrates towards the posterior. The first cleavage is asymmetrical and generates an anterior AB cell and posterior P_1 cell (Figs 5.13(d), 5.15(b)). By this time the P granules have all collected in the posterior and are segregated to the P_1 cell (Fig. 5.15(f)). This segregation, together with the pronuclear migration and pseudocleavage, are inhibited by microfilament inhibitors.

The P_1 cell is a stem cell which undergoes three more asymmetrical divisions to produce EMS, C and D blastomeres (Fig. 5.14). In each cell cycle the P granules coalesce in the region which will become the daughter P cell, despite the fact that AB and EMS arise anteriorly while C and D arise posteriorly. The final P_4 cell is the direct germ cell precursor and undergoes only one more division in embryonic life to form two similar daughters. In the nematode *Ascaris* studied at the turn of the century by Boveri, there is a systematic loss of genetic material ('chromatin diminution') in the somatic cell lineages, but this does not occur in *C. elegans*.

AB divides to give ABa and ABp, and at the rhomboid four cell stage the dorsoventral polarity becomes apparent, with ABp dorsal and EMS ventral (Fig. 5.13(e)). EMS undergoes an unequal division to E and MS which then divide more or less equally as do the progeny of AB, C and D. 'Gastrulation' commences at the 28-cell stage with the migration into the interior of Ea and Ep (Fig. 5.13(g)) and continues with the ingression of P_4 and most of the precursors for the internal organs until the eventual closure of the ventral cleft at about the 350-cell stage. Most zygotic transcription starts at around 100 cells and the specific products so far identified include a gut specific esterase which appears in the 4–8 intestinal cells present at the 100–150 cell stage (Edgar & McGhee, 1986). Autofluorescent rhabditin granules have also been used as markers of intestinal differentiation, appearing at the 200–300 cell stage. There is no distinctive stage of neurulation, most of the nervous system lies ventrally and in this region the neuroblasts become covered by circumferentially spreading hypoblast while dorsal and lateral neuroblasts invaginate individually.

C. elegans is famous for its invariant cell lineage. This has been followed in its entirety by direct observation using Nomarski interference contrast

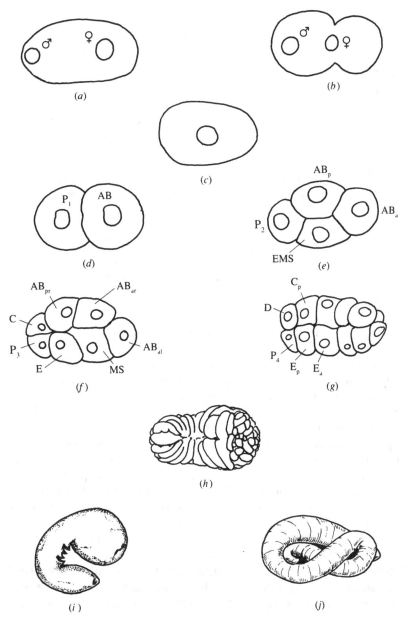

Fig. 5.13. Normal development of *C.elegans*. (*a*) after fertilization, showing pronuclei. (*b*) pseudocleavage, pronuclear migration. (*c*) nuclear fusion. (*d*) 2-cell stage. (*e*) 4-cell (rhomboid) stage. (*f*) 8-cell stage. (*g*) 28-cell stage, beginning of gastrulation. (*h*) ventral view prior to elongation. (*i*) mid-elongation, formation of embryonic sheath over surface. (*j*) late elongation, formation of surface ridges. (*a*)–(*g*) are drawings of Nomarski photographs and hence represent just one optical section, anterior is to the right.

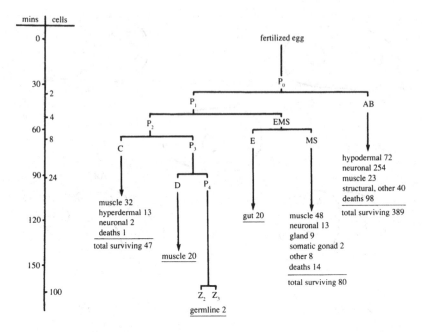

Fig. 5.14. Early cell lineage of *C. elegans*, anterior to right as in Fig. 5.13.

microscopy (Sulston *et al.*, 1983). As in the leech and the ascidian where modern methods have been used to extend the cell lineage work of the old masters, it has turned out that development is less clonal than formerly supposed. Of the six founder cells shown in Fig. 5.14, only three produce a single cell type: E produces only gut cells, D produces only muscle cells and P_4 produces only germ cells. The others produce mixtures which do not respect the traditional germ layer subdivisions, although the progeny of AB are predominantly ectodermal and of MS and C are predominantly mesodermal. The cell lineage has revealed another class of behaviour in addition to stem cell divisions and clonal divisions. These are repeated sublineages, where cells separated in time or space may undergo an identical series of divisions to give a particular arrangement of differentiated types. The observation of such sublineages has been important in the discovery of homeotic and heterochronic mutations. About one cell in eight cells undergoes programmed death during embryonic development, although they are small cells making up only 1% of total biomass.

Developmental genetics

One of the features of *C. elegans* which has gripped the imagination of many is the existence of mutations which specifically alter the observed lineages of particular cells (Sternberg & Horvitz, 1984). They are often homeotic,

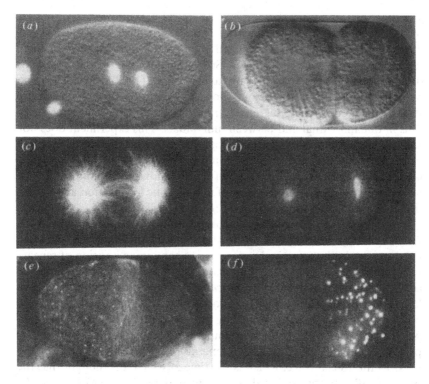

Fig. 5.15 Behaviour of cytoskeletal components during the first cell division of
C. elegans. (*a*) nuclear division shown by DAPI fluorescence, (*b*) first cleavage
produces a larger AB and smaller P_1 cell, (*c*) microtubules shown by fluorescent
antibody, (*d*) centrosomes shown by fluorescent antibody, (*e*) microfilaments
stained with fluorescent phalloidin, (*f*) P granules segregated to the posterior.
Anterior is to the left in all pictures. (Photographs kindly provided by Dr S.
Strome.)

changing the behaviour of a particular cell into that appropriate to a cell in
a different position. These include recessive mutations which when homozygous, prevent the programmed cell death seen in normal development.
Some are heterochronic, advancing or retarding events in developmental
time. Some switch between patterns found in the two sexes. A particularly
interesting group are maternal effect mutants that alter the early cleavage
pattern from asymmetric to symmetric (Kemphues *et al.*, 1988). Many fall
into none of these classes and generate patterns of division that have no
normal counterpart.

Their existence, together with the invariant cell lineage itself, has led
some workers to postulate that regional specification in *C. elegans* is
controlled to some extent by a program of directed chromosome strand
modification comparable to that found in the fission yeast *S. pombe* (Klar,
1990). However, for such a mechanism to be of use in generating an

anatomy, it is necessary that a modified strand is segregated reliably to a particular one of the two daughter nuclei. Furthermore, if genes on different chromosomes are modified in the same cell cycle, then their segregation needs to be coordinated. So far the only direct evidence bearing on this matter has been obtained by Ito & McGhee (1987) who labelled one or other parent with BUdR and showed that in the early embryos the strands from one gamete segregated randomly both between the progeny cells in one individual and between individuals. This experiment does not rule out the existence of directed chromosome modification at later stages since obviously the experimenters could only look at cells which received labelled DNA strands. At the time of writing, however, opinion seems to favour the traditional mechanisms of cytoplasmic localization and induction to explain most of the data on regional specification.

Two particularly interesting homeotic mutations are called *glp-1* and *lin-12*. These map near each other on the same chromosome and the gene products are both transmembrane proteins containing a region of repeats with sequence homology to the mammalian epidermal growth factor (EGF) and the *Drosophila* gene *Notch*. *glp-1* is a maternal effect recessive lethal. Offspring of homozygous mothers lack the anterior part of the pharynx which is normally formed by the blastomere ABa (Priess, Schnaebel & Schnaebel, 1987). When a temperature sensitive allele is used, the period required at the permissive temperature is found to extend from the four- to the 28-cell stage, which is the time at which it is thought that the pharynx cells are being induced by progeny of the EMS cell (see below). *glp-1* is also necessary during larval development for the continued mitosis of the germ cells under the influence of the distal tip cell. *lin-12* comes as both dominant (overexpressing) and recessive (underexpressing or null) alleles. These have reciprocal homeotic effects on developmental decisions at several points in both embryonic and postembryonic life (Greenwald, Sternberg and Horvitz, 1983). An example is given by the following case of two cells which form one of the few known embryonic equivalence groups.

lin-12	*ABplapaapa*	*ABprapaapa*
dominant	ectoblast	ectoblast
wild type	ectoblast	neuroblast
recessive	neuroblast	neuroblast

These mutations cause additional transformations in postnatal life, for example, in the gonad and in the part of the ventral hypodermis which forms the vulva. In each case the time of action deduced from *ts* mutations is the same as the time of induction deduced from microbeam ablation of the signalling cells. Mosaic analysis has shown that for the postembryonic functions of both *glp-1* and *lin-12* the gene activity is required in the responding rather than the signalling cells, so despite the EGF homology, the protein products are thought to be receptors (Seydoux & Greenwald, 1989).

REGIONAL ORGANIZATION

Cytoplasmic localization

It is possible to cause portions of the cytoplasm of the uncleaved egg to become extruded from the egg shell after burning a small hole in the shell with a laser microbeam (Laufer & von Ehrenstein, 1981; Schierenberg, 1985). Up to 40% of the cytoplasm can be extruded from the anterior end and up to 20% from the posterior end without affecting the pattern of early cleavages or the development of muscle and gut cells, although the worms do not develop normally to hatching. Removal of more than 25% from the posterior end abolishes the unequal cleavage. Conversely if the P_1 nucleus is extruded at the two-cell stage, and the P_1 cytoplast fused to the AB cell by a judicious laser burn, then asymmetric division can be restored. This suggests that the capability for asymmetric division is carried by the posterior cytoplasm, rather as in the case of the annelid pole plasms mentioned above. This property is not due to the P-granules, as they are still uniformly distributed at early postfertilization times when it is possible to abolish the stem cell property by extrusion of posterior cytoplasm. The polarity of the P cells changes between P_1, which cuts off EMS anteriorly, and P_2 which cuts off C and D posteriorly. Interestingly, an extruded early P_2 blastomere lacking its own anterior cytoplasm cleaves unequally while a similar fragment from later in the cell cycle cleaves equally (Schierenberg, 1987). So it really looks as though the stem cell quality is shunted from posterior to anterior during this period. A brief pulse of cytochalasin D about three-quarters of the way through the first cell cycle can produce quite severe effects on the asymmetric division, and yield polarity reversals or double anterior or double posterior cell configurations. So the stem cell behaviour is in some way dependent on microfilaments (Hill & Strome, 1990). Further progress in this area may come from following up the molecular nature of the products of maternal effect genes which seem to be responsible for the asymmetrical early cleavages. There are several of these *par* genes and the mutants show symmetrical early cleavages, do not segregate their polar granules correctly, and are severely deficient in intestinal differentiation, although some other cell types can differentiate normally (Kemphues *et al.*, 1988).

Cleavage block experiments similar to those done on ascidians can be carried out by prolonged exposure to cytochalasin B or D. The nuclei continue to divide to some extent but the cells do not cleave and remain identifiable by their size, shape and position, and may produce specific differentiation markers (Cowan & McIntosh, 1985; Edgar & McGhee,

1986). To some extent markers are produced according to the normal fate map. Thus rhabditin granules and/or esterase arises in P_1, EMS or E; hypodermal antigens are produced in P_0, P_1, AB, and P_2, but not EMS; and paramyosin is produced by P_1 and P_2. However, a single cleavage arrested cell never produces more than one marker, suggesting some sort of mutual exclusion of programs of gene activation. Also some cells do not produce markers where they might be expected to, so P_0 does not produce gut or muscle markers, and AB and EMS do not produce muscle.

A limited amount of data has been obtained in the course of these experiments concerning the developmental capabilities of isolated blastomeres, since the manoeuvres used to permeabilize the egg shell may also cause the destruction of some cells. Isolated P_1 or E cells have been shown to produce gut markers and AB cells can acquire this ability if they are fused to P_1 cytoplasts (Schierenberg, 1985). This experiment represents the nearest thing to a cytoplasmic transfer experiment of the type which is really necessary to prove the existence of a determinant. We are left feeling that there may exist in *C. elegans* determinants for asymmetric cleavage and for gut formation. But in neither case is the evidence really conclusive and we do not know whether the localizations occur during oogenesis or at some later stage.

Laser ablation

A number of studies using laser ablation of single cells or small groups of cells has not, on the whole, revealed regulative behaviour (Sulston & White, 1980). The invariant cell lineage and the lack of regulation have usually been interpreted as showing that regional specification in this organism does not involve induction. As discussed in Chapter 2, such results are certainly consistent with this view, but do not prove it. Ablation experiments are reasonably convincing in excluding named individual cells as signalling centres, since if these are removed and the neighbouring cells remain unaffected then they presumably did not need any inductive signals from the ablated cell after its death. However, it is not so good for showing that cells are not *responding* to influences in their environment because a creature with a very low cell number and a very rapid tempo of embryonic development is unlikely to be able to fill in gaps by regulative growth. It is possible that there are many potential cases in which one cell would substitute for its neighbour if their positions were exchanged, but such experiments are not easy to perform in *C. elegans*.

The laser ablation work itself has revealed the *equivalence groups* (see also Chapter 2). These are pairs or larger groups of cells with the same competence, which acquire their final fates by some sort of cell–cell interaction. In some cases the specification is controlled by a signalling cell outside the equivalence group. If the signalling cell is ablated, then all of the equivalent cells assume the fate of the most distal cell in the group. If a proximal cell of the group is ablated then the others become promoted in a serial manner depending on their distance from the signalling cell. The best worked out example of this is the equivalence group responsible for the formation of the hermaphrodite vulva which arises in postembryonic development under the influence of the anchor cell (Sulston & White, 1980) but there are also a few equivalence groups in the embryo. This phenomenology seems exactly similar to that of an embryonic field under the control of an organizer, but scaled down so that the whole is only a few cells in size.

Induction of the pharynx

An example in early development where the importance of induction has been demonstrated by more or less classical embryological methods is the formation of the anterior pharynx from the AB cell. As mentioned above, isolated AB cells do not form any muscle, although in normal development the ABa cell produces muscle cells of the anterior pharynx. Priess & Thomson (1987) showed that ABa and ABp could be interchanged and give rise to normal larvae, showing that their different developmental pathways must be specified by interactions with surrounding cells. Muscle would develop from AB so long as the P_1 cell or its EMS daughter were left in place, suggesting that EMS progeny were the source of the signal. The interaction was presumed to be over by the 28-cell stage because at this time the five muscle precursors which have formed can be burnt out by laser and do not regenerate. The time interval from four to 28 cells is also the period of requirement of the *glp-1* product for formation of anterior pharynx from AB (see above).

So far *C. elegans* has not really answered any of those questions which were posed but not answered by previous embryological work on other organisms. The results summarized here show that it may have determinants for stem cell cleavage behaviour and for gut cell differentiation. It certainly has induction. There are certainly mutations of a homeotic and heterochronic character. The study of the *lin-12* and *glp-1* genes suggest that the inductions probably involve extracellular effectors binding to cell surface receptors, now a familiar theme elsewhere. But how the other things work we still do not know. The hope is that the superior genetics compared to molluscs or ascidians will result in some answers over the coming years.

GENERAL REMARKS ON THE MOSAIC EMBRYOS

One does not need much acquaintance with embryology to notice that the thoughts of many developmental biologists are dominated by the idea that regional specification is controlled by cytoplasmic determinants. This is particularly true of those who work on the types of creature covered in this Chapter, and also particularly true of Americans even if they work on other organisms. The reasons seem to be partly historical, in that it was the main hypothesis of the great American embryologists E.G. Conklin and E.B. Wilson who had such an enormous and lasting influence on the Woods Hole marine station and on invertebrate embryology generally. It is also possible that, as a rather simple idea, it appeals to molecular biologists coming in to the field. In the days of actinomycin D experiments in the 1960s, it was commonly supposed that determinants would prove to be maternal mRNAs coding for terminal differentiation products. More recently it has been thought that determinants are positive regulatory elements which activate batteries of genes required for particular cell types. This is, for example, the main thrust of the influential book by E. Davidson, *Gene Activity in Early Development*. In this section we shall try to disentangle a number of distinct ideas which have become rather confused and conflated with the concept of the cytoplasmic determinant, with the aim of trying to decide how good the evidence really is for their existence and importance.

The features of embryonic behaviour which lead people to claim the presence of determinants are five-fold:

1. An invariant cell lineage.
2. A visible regionalization of the fertilized egg cytoplasm.
3. Mosaic behaviour of defect embryos.
4. Mosaic behaviour of isolated blastomeres.
5. Successful reprogramming of blastomeres by cytoplasmic transfer.

We shall consider each of these in turn.

An invariant cell lineage

It has been known since the late nineteenth century that certain animal types have an invariant lineage, and thus a precise fate map, at least for the early developmental stages. Recently, with the introduction of Nomarski microscopy and of lineage labels, the precision of our knowledge has increased considerably. We know that *C. elegans* and the ascidians have an almost completely (not *absolutely* completely) invariant lineage. The leech started out with an invariant lineage but it now seems to be rather less invariant than the others. We also know that there are animals, like *Xenopus*, that have a topographic fate map on the large scale but cell mingling on the small scale, and others, like the zebrafish, which have total mixing, and thus no fate map, at early stages. The implication of these results for the possible presence of determinants is clear. Where there is no fate map there can be no determinants, since any determinants would end up in different places in different individuals. However, where there is a high degree of invariance between individuals then this is compatible with the presence of determinants. Of course it does not prove that they are present; it is a necessary but not a sufficient condition for their presence. It is misleading, although often done, to call an invariant cell lineage a *determinate* lineage. This implies something that cannot be known just by observing and describing the lineage.

A visible regionalization of the fertilized egg cytoplasm

This is something which was made much of by the early workers, but rather less so today since we have become accustomed to the idea that important molecules are present in small quantities and are not visible to the human eye down the microscope. All the organisms considered in this Chapter undergo an impressive ooplasmic segregation after fertilization and this may indeed be essential for later regional specification. However, in most cases the visible components can be stratified by low speed centrifugation without affecting the course of development, and it is accordingly believed that determinants are anchored to immovable structures such as the cortical cytoskeleton. This may well be true, but serves to emphasize the fact that visible regional differences are irrelevant to the existence or otherwise of determinants.

Mosaic behaviour of defect embryos and isolated blastomeres

Mosaic behaviour means that the specification map corresponds exactly to the fate map. However the parts of the embryo are cut up or re-arranged each part should behave exactly as it would in normal development. On the whole, embryos of this Chapter from which single cells have been removed or killed do behave in a mosaic manner. But again this is only consistent with the presence of determinants, rather than providing positive evidence. Consider what would be the outcome if most structures are in fact formed as a result of inductive interactions. If an entire signalling centre were destroyed then it is true that a less than mosaic result would be expected, because the surviving responding tissue would fail to form parts predicted from the fate map. This is seen in a few cases, such as ablation of the D blastomere in molluscs or of the anchor cell in *C. elegans*. However, if only part of a signalling centre were removed then there would be little or no consequences for the responding cells and the defect would be confined to the parts expected to develop from the site of the lesion. Furthermore, if all or part of a responding field were removed, we would actually expect to see a defects corresponding to the lesion, particularly for very rapidly developing species like ascidians or *C. elegans* where the time available for compensatory growth or respecification of the surrounding cells is very limited. So mosaic behaviour in response to localized ablation is not very informative about underlying mechanism, whereas non-mosaic behaviour does imply the existence of necessary intercellular interactions.

Things are a bit different when it comes to mosaic behaviour by isolated early blastomeres. If genuine this would provide some evidence against interactions and in favour of intrinsic mechanisms (either cytoplasmic localization or programmed chromosome modification). Consider again what would happen if most structures are formed by inductive signalling. In order to obtain mosaic behaviour we would have to suppose that complete prospective signalling and responding regions were taken each time a blastomere was isolated, and it seems most unlikely that this would always be the case. Hence mosaic behaviour by isolated blastomeres is unlikely to be compatible with induction although it does not absolutely rule it out.

But on this issue we must also inquire of the actual evidence: how good is it? Do isolated blastomeres really show mosaic behaviour? The answer is no. For the molluscs, nobody would now claim that they show mosaic behaviour. The earlier claims that they did were based on very early events such as the directions of cleavages and the appearance of cilia. These probably do depend on the arrangement of pre-existing structures in the egg, but are also probably not related to the program of gene activation and repression which occurs later on. For the ascidians there is certainly muscle formation from isolated B4.1 cells, but the results fall far short of complete mosaicism since neither the neural tube nor the parts of the tail musculature not derived from B4.1 will develop in isolated blastomeres. For *C. elegans* the gut esterase does seem to come on in isolated E cells but technical difficulties and the lack of suitable markers have so far prevented a very satisfactory study of isolated blastomeres. What we see is not mosaicism but a certain degree of regional localization: certain regions like the molluscan D lineage, or the ascidian vegetal cytoplasm are necessary for development of the body plan, certain others like the E cell of *C. elegans* or the 4.1 cells of ascidians can form particular terminal cell types in the absence of later interactions with other parts of the embryo. In annelids and in *C. elegans* certain regions of zygote cytoplasm seem to be able to confer stem cell qualities on the blastomeres that inherit them. These are all very interesting phenomena and do suggest the existence of determinants for certain features of early body plan organization. However, the situation is really not very different from that in amphibians or sea urchins where there is also some early regional localization but there is good evidence that most of the body structures are formed by induction.

Cytoplasmic transfer

The existence of significant regional differences in the zygote cytoplasm is suggested by the results of blastomere isolation experiments, but the real proof of the existence of a *determinant* can only come from the transfer of cytoplasm from one cell to another and the demonstration that the host cell is reprogrammed to form structures appropriate to the donor cytoplasm. The best experiments of this type undoubtedly come from *Drosophila* and involve the pole plasm, and more recently the *bicoid, nanos* and *Toll* products (see Chapter 7). The results with the group of organisms considered here have been less good, perhaps because of the technical difficulties of the manipulations, perhaps because the determinants tend to be associated with immovable parts of the cytoplasm, and perhaps because there are fewer determinants around than the workers in the field normally think. The best examples are perhaps the transfer of stem cell character to the C blastomere of the leech by manipulations that displace the pole plasm, and to the AB blastomere of *C. elegans* by fusion with a P_1 cytoplast.

The general conclusion is that this group of 'mosaic' organisms are not really much more mosaic than the amphibians or sea urchins. For all of these creatures the more investigation has been done the less invariant the cell lineages have seemed, the less rigidly clonal is the relationship between early blastomeres and later tissues or structures, and the less real mosaic behaviour is found. Conversely, the more investigation is done the more interactions are discovered. So we must conclude that it is wrong to suppose that all or most of the larval anatomy is prefigured by an arrangement of determinants in the zygote. Indeed, this is probably impossible since it would require such a fabulously detailed subdivision of the egg cytoplasm.

On the other hand, there is a reasonable body of evidence that significant regional localizations in the zygote cytoplasm are necessary to establish early features of the body plan: certain body regions, some signalling centres and some controls over the program of cell divisions. The nature of these localizations is still one of the most important problems of experimental embryology. In no case outside *Drosophila* is the molecular basis of the localization yet known, and there is, at the time of writing, still no evidence that genetic regulatory factors are localized in the zygote. Although this may very well turn out to be the case, it is also possible that some localizations are based on the cytoskeleton, or on multiple metabolic steady states and that their effects on gene activity are indirect. Of course the fact that the determinants have not yet been found is not good evidence against their existence in these creatures. To find them requires that several problems be overcome, chiefly the small quantities of material available, and the complexity of the existing bioassays. These problems *have* been overcome in *Drosophila* which now provides the best evidence for determinants, two of which are known to be localized mRNAs (*bicoid* and *nanos*) while two others are states of receptor activation (*Toll* and *torso*) (see Chapter 7).

To date there is no evidence for the alternative intrinsic type of mechanism for regional specification which is a program of sequential chromosome modification. However, this has not yet been looked for very assiduously, and since most things which have been thought of in biology are usually discovered somewhere or other in the end we shall not be surprised to find this added to the list of known mechanisms at some time in the future.

GENERAL REFERENCES

Jeffery, W.R. (1983). Messenger RNA localization and cytoskeletal domains in ascidian embryos. In *Time, Space and Pattern in Embryonic Development*, pp. 241–259. New York: Alan Liss.

Verdonk, N.H., Van den Bigelaar J.A.M. & Tompa, A.S. (1983). *The Mollusca*, vol 3: Development. New York: Academic Press.

Wood, W.B. (1988). *The nematode* Caenorhabditis elegans. Cold Spring Harbor Laboratory.

6

Models for Man: the mouse and the chick

THE MOUSE

Although rabbits were used to some extent in the early days and rats are still used for certain purposes, the mouse has become the mainline organism for mammalian experimental embryology. This has led to the terms 'mouse' and 'mammal' being used almost interchangeably, but it should be remembered that the morphology of early development can differ considerably in other mammals, not least in the human. Unlike the other creatures considered in this book, mammals are viviparous and the resulting inaccessibility of the postimplantation stages of development causes severe problems for the embryologist. For the early, preimplantation, stages the embryos are located first in the oviduct and then the uterus of the mother. During this time mouse embryos can be collected and kept *in vitro* in reasonably simple media and most experiments have involved manipulation of these early stages. Where later development is essential to the result the preimplantation embryos are transplanted into the uteri of 'foster mothers' who have been made 'pseudopregnant', and thus receptive to the embryos, by previous mating with sterile males.

Since the *in vivo* manipulation of postimplantation stages is very difficult, and their culture *in vitro* is still only feasible for periods less than 3 days, most mouse embryology is really 'pre-embryology' dealing not with the formation of the embryo body plan but of various extraembryonic membranes which are segregated during the pre- and peri-implantation phase and which are necessary to the support and nutrition of the embryo proper. However, there has been a considerable increase in information about patterns of gene expression in the post-implantation embryo, using probes derived from other fields of investigation, such as oncogenes or genes important in the early development of *Drosophila*. It is also probable that the difficulties of micromanipulation will ultimately be circumvented using transgenic technology (see below), so more 'real' mouse embryology is to be expected in the future.

The embryos of mammals, birds and reptiles are unusual in the animal kingdom in that they undergo a considerable amount of growth during

171

development. Most other types of embryo develop in isolation from the mother and if they grow at all do so only to a relatively small extent at the expense of the yolk in the egg. By contrast, mammalian embryos after implantation obtain nourishment from the maternal blood stream, and birds and reptiles lay large eggs containing a vast amount of yolk which is gradually utilized to nourish the foetus. The extensive growth and the associated early formation of extraembryonic structures have two important implications. The first concerns proportion regulation. We have seen that in all embryos considered so far the size of structures depends on the size of the whole embryo. For example, in animals where twinning is possible following separation of the first two blastomeres, the twins approach normal proportions but at half the normal size. But this is not the case for the mouse, where separation of the early blastomeres causes the proportion of inner cell mass to trophectoderm to be reduced. The second is the question of the relationship between growth and the successive steps of the developmental hierarchy. If growth and decision-making can be separated then it should be possible to cultivate cells in particular states of determination for indefinite periods. Teratocarcinoma cells have long been regarded by some as cells arrested at an early developmental stage, and recently more use has been made of 'embryonic stem cells' (ES cells) which are grown directly from embryos. This offers a new dimension of opportunity for it provides access to much larger quantities of material than can be obtained by the collection of mammalian embryos themselves.

NORMAL DEVELOPMENT OF THE MOUSE

Mice mate in the night and so the age of the embryos is often expressed as days and a half, ie, a $7\frac{1}{2}$ day embryo is recovered on the eighth day after the mating night. However, there is some scatter in mating times and in developmental stages for a known gestation time, so age is only an approximate guide to stage. Ovulation occurs a few hours after mating and fertilization takes place at the upper end of the oviduct, the second polar body being expelled 2–3 hours later. The egg is surrounded by a vitelline membrane and a thick zona pellucida composed of mucopolysaccharide secreted by the ovarian follicle cells and the total diameter including the zona is around 100 microns.

The course of normal development up to the body plan stage is shown in Fig. 6.1. The first few cleavages are very slow and in contrast to the embryos considered in the previous Chapters expression of the zygotic genome commences as early as the two-cell stage. The first cleavage occurs about 24 hours after fertilization and the second and third cleavages, which are not entirely synchronous, follow at intervals of about 12 hours. In the early eight-cell stage the shapes of individual cells are still clearly visible but they cease to be visible when the whole embryo acquires a more nearly spherical

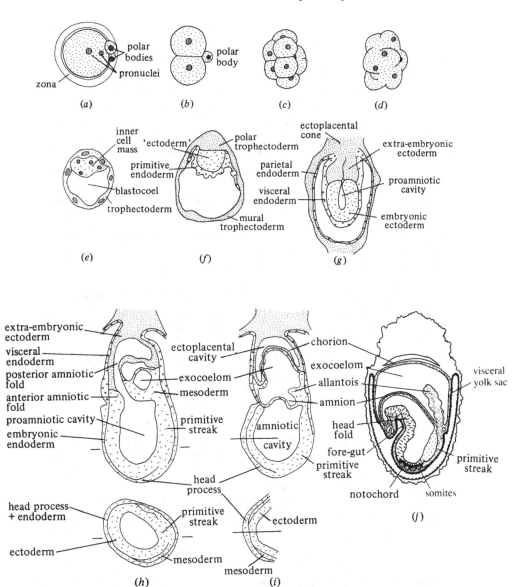

Fig. 6.1. Normal development of the mouse. (*a*) fertilized egg, (*b*) 2-cell stage, (*c*) 8-cell stage, (*d*) 8 cell stage after compaction, (*e*) blastocyst, about 3.5 days, (*f*) implantation stage, about 4.5 days, (*g*) egg cylinder, about 6.5 days, (*h*) primitive streak stage, (*i*) later primitive streak, closure of amniotic cavity, about 7.5 days, (*j*) head fold stage, about 8.5 days. There is no growth before implantation but the conceptus increases in size by a factor of about 6 from (*f*) to (*j*). (Adapted from Snell & Stevens, 1966, except (*d*) by author and (*j*) from Beddington, 1987.)

shape in a process called *compaction*. This consists of a flattening of blastomeres to maximize intercellular contacts, mediated by the calcium dependent adhesion molecule E-cadherin (also called L-CAM or uvomoru-lin). The cells themselves become polarized in a radial direction. This is most obviously apparent from the appearance of Con-A binding microvilli on the outer surfaces (Fig. 6.5) but also involves a variety of changes in the cell interior. Gap junctions are also formed at this stage and allow electrical coupling and diffusion of low molecular weight dyes throughout the embryo. At the earlier stages there is only communication between sister cells via their mid-bodies although these may sometimes persist for more than one cell cycle.

The embryo is called a *morula* from compaction until about the 32-cell stage. During this period, point desmosomes and tight junctions appear, creating a permeability seal between the inside and outside of the embryo, and a fluid filled blastocoel begins to form in the interior. This is about 3 days after fertilization and around the time that the embryo moves from oviduct to uterus. The cavity expands the embryo into a *blastocyst* (Fig. 6.1(*e*)) which consists of an outer cell layer called the *trophectoderm* and a clump of cells attached to its interior, the *inner cell mass* (ICM). At the 60-cell stage, about one-quarter of the cells are found in the ICM and three-quarters in the trophectoderm.

From $3\frac{1}{2}$ to $4\frac{1}{2}$ days both the ICM and the trophectoderm diversify (Fig. 6.1(*f*)). The ICM delaminates a layer of *primitive endoderm* on its blastocoe-lic surface. Despite its name this layer is not now thought to contribute to the embryonic (definitive) endoderm but rather to extraembryonic mem-branes. The trophectoderm becomes divided into a *polar* component, which overlies the ICM, and a *mural* component which makes up the remainder. While the polar trophectoderm continues to proliferate, the mural trophec-toderm becomes transformed into *giant cells* in which the DNA continues to be replicated but without mitosis.

At about this stage the embryo hatches from the zona and becomes implanted in a uterine crypt, which is only competent to receive the embryo during a short period about 4 days from copulation. The trophectoderm then becomes known as the *trophoblast* and soon stimulates proliferation of the connective tissue of the uterine mucosa to form a *decidual swelling*. From this stage onward the embryo receives a nutrient supply from the mother and can begin to grow in size and weight.

The next stage of development is known as the *egg cylinder* (Fig. 6.1(*g*)). Cells derived from the primitive endoderm move out to cover the whole inner surface of the mural trophectoderm and start to secrete an extracellu-lar basement membrane known as Reichert's membrane containing lami-nin, entactin and type IV collagen. These cells are called the *parietal endoderm*. The remainder of the primitive endoderm remains epithelial and forms a layer of *visceral endoderm* around the egg cylinder. The cells of this layer somewhat resemble the later foetal liver being characterized by the synthesis of α-foetoprotein, transferrin and other secreted proteins. The inside of the egg cylinder consists of the primitive ectoderm from which the entire embryo will develop, and extraembryonic ectoderm derived from the polar trophectoderm. This extraembryonic region has now become the *ectoplacental cone* and as it proliferates it produces further layers of giant cells which move around and reinforce the trophoblast. In contrast to the differentiated extraembryonic tissues, the primitive ectoderm from which the entire embryo will be derived remains visibly undifferentiated.

At about 7 days the anteroposterior axis of the future embryo becomes apparent with the formation of the *primitive streak* at one edge of the embryonic ectoderm. It should be remembered that the primitive ectoderm forms a deep cup and so appears U-shaped in a section. The streak marks the posterior end of the future embryonic axis which will extend across the ectoderm in the manner shown in Fig. 6.1(*h*). The streak is a region of cell movements which are not well understood but which result in the formation of the definitive endoderm and mesoderm. As will be mentioned below, the gene *int-2* is expressed in the early primitive streak (Wilkinson *et al.*, 1988) and the early mesoderm is characterized by the presence of vimentin and of a cell surface carbohydrate related to the blood group I-antigen (Fenderson, Eddy & Hakomori, 1988).

The amniotic fold forms as an outpushing of the ectoderm and mesoderm at the junction of primitive streak and extraembryonic ectoderm. The side of this fold nearer the embryo becomes the *amnion* and the side nearer the ectoplacental cone becomes the *chorion* (Fig. 6.1(*i*)). The fold pushes across the proamniotic space and divides it into three: an amniotic cavity above the embryo, an exocoelom separating amnion and chorion, and an ectoplacental cavity lined with extraembryonic ectoderm. These cavities are lined on their luminal surfaces by complex carbohydrates bearing terminal groups of the *N*-acetyl lactosamine type. Into the exocoelom grows the allantois, which is a projection of tissue from the posterior end of the primitive streak. By 7½ days a *head process* appears in the definitive endoderm between the anterior extremity of the primitive streak and the anterior edge of the embryo. The mid part of this becomes the notochord and the remainder part of the gut lining. The mesoderm becomes somites and lateral plate and the ectoderm becomes epidermis and neural plate, all arranged in the typical vertebrate body plan. By 8½ days the embryo has elongated somewhat in length and a massive head fold has formed at the anterior end, mainly composed of the anterior neural tube. For the purposes of understanding the experimental work on the mouse embryo we do not need to describe its development any further than this general body plan stage which shows obvious homology with other vertebrate embryos.

It should be stressed that the details of the above description hold only for the mouse. The course of development of mammals varies somewhat between species and in particular the arrangement of the extraembryonic membranes can differ considerably. In the human embryo it should be noted that the amniotic cavity forms within the ICM shortly after the blastocyst has commenced implantation, and that the embryonic axis develops within a flat plate of primitive ectoderm rather than an egg cylinder.

Fashionable genes

At the time of writing there is a considerable industry of *in situ* hybridization and antibody staining of normal embryos to reveal the spatial patterns of expression of genes which are thought to be of developmental importance. Pre-eminent among these genes are those bearing the homeobox, a sequence first identified as a common feature of homeotic genes in *Antennapedia* and *Bithorax* complexes of *Drosophila*. The homeobox codes for a DNA binding domain and the protein products of genes containing the homeobox are transcription factors which regulate the levels of activity of other genes. There are four clusters of homeobox genes in the mouse genome which are thought to be derived from a single cluster in the insect–vertebrate common ancestor. Many genes are expressed in early development, the earliest detectable transcripts appearing at $7\frac{1}{2}$ days. The spatial patterns of expression become established and readily detectable by *in situ* hybridization over the following 5 days, depending on the gene in question (Holland & Hogan, 1988). None shows a stripy pattern like the pair rule or segment polarity genes of *Drosophila* but most show some restriction to a particular part of the anteroposterior axis of the embryo. In particular, the anterior boundaries of expression are usually quite sharp. Most homeobox genes are expressed in the central nervous system but many are expressed in other tissues as well. Intriguingly the anatomical order of expression, in terms of anterior boundaries, is the same as the spatial order in which the genes are arranged on the chromosome within each cluster (Graham, Papalopulu & Krumlauf, 1989; Duboule & Dollé, 1989). This is also true in *Drosophila* although it is thought probably to be of evolutionary rather than developmental significance. In other words, the original evolution of genes expressed at successively more posterior body levels may have proceeded by gene duplication from one end of the cluster and this arrangement has been maintained throughout subsequent evolutionary time despite splitting or duplication of the entire cluster. At the time of writing we do not know the function of any of the homeobox genes in the mouse, but the most popular view is that the different combinations of gene activity are the epigenetic codings for different anteroposterior body levels. One piece of evidence for this is that the earliest genes to come on, at $7\frac{1}{2}$–8 days, are expressed at the same anteroposterior levels in ectoderm and mesoderm, and so seem to reflect position in the body rather than tissue type. Conversely, some of those expressed late in development are confined to particular histologically defined tissues. It is currently hoped that their functions will become clear when the genes can be ablated by homologous recombination (see below), but it may be that the amount of redundancy in the mouse genome will defeat this approach and a much longer biochemical road will have to be taken.

Other fashionable classes of genes are the 'finger' genes such as *Krox-20*, with a sequence homology to the prototype Transcription Factor IIIA of *Xenopus*; 'pax' genes with homology to the *Drosophila paired* box; and various oncogenes and growth factors identified by biological assays. Numerous studies on the expression patterns of these genes are under way and it seems likely that many of them have functions in embryonic development although more often in the stage of organogenesis rather than primary body plan formation. Among the oncogenes may be mentioned *int-2* which has a sequence homology to fibroblast growth factor (FGF) and is expressed in the primitive streak of the $7\frac{1}{2}$ day embryo (Wilkinson *et al.*, 1988). bFGF is thought to be a mesoderm inducing factor in *Xenopus* (see Chapter 4) and it has been shown that the *int-2* protein has some mesoderm inducing activity when tested on *Xenopus* ectoderm (Paterno *et al.*, 1989). Whether it also plays this role in normal mouse development we do not know. Also expressed in the primitive streak is the *brachyury* (T) gene (Wilkinson, Bhatt & Herrmann, 1990). This codes for a nuclear protein and its absence in the mutant causes death because the allantois is not properly formed.

int-2 is expressed again at a somewhat later stage in rhombomeres 5 and 6 of the hindbrain, while *Krox-20* is expressed in rhombomeres 3 and 5. The rhombomeres are bulges existing in the hindbrain of all vertebrates which have assumed some significance recently as it has been realized that they are the basis for the segmental arrangement of cranial motor nerve roots and sensory ganglia (Lumsden & Keynes, 1989) and are also allocation territories, or 'compartments' (Fraser, Keynes & Lumsden, 1990). The rhombomeres are, in addition, spatial domains for expression of the homeobox genes (Wilkinson *et al.*, 1989) so may represent a fundamental segmental component of the vertebrate body plan.

A final class of fashionable genes are not genes at all but regulatory elements which can be revealed by the 'enhancer trap' method, a procedure first introduced for *Drosophila* and later used for the mouse. As applied to the mouse, a ß-galactosidase reporter gene equipped with a weak and non-tissue specific promoter is introduced into the germ line by injection of DNA into the male pronucleus of the fertilized egg. This is a standard method for making transgenic mice (see below). The expression of the reporter gene depends on integration near a host enhancer or promoter and so an interesting spatial pattern of expression suggests that the construct has integrated near a regulatory element of some developmental importance (Allen *et al.*, 1988). In view of the relative paucity of developmentally significant genes known in mammals, this method yields a surprisingly high proportion of potentially interesting patterns. A variant of this method is the 'gene trap' in which a ß-galactosidase gene is introduced into embryonic stem cells and, if it is expressed, then the cells are introduced into embryos to create a strain containing the insert. Its expression pattern is then the same as the gene into which integration occurred (Gossler *et al.*, 1989).

Fate map

Various methods of prospective labelling have been applied to the mouse embryo in attempts to gather information about the normal fates of different regions. Among the passive methods are the injection of horse-radish peroxidase (HRP) and fluorescent dextrans into individual blasto-meres; and the performance of orthotopic grafts between embryos labelled and unlabelled with ^{3}H-thymidine. Labels of this type have the disadvantage of being diluted by growth, which is very rapid in the mouse embryo after implantation, although they are useful for short-term studies or for the preimplantation stages.

A heritable label which has been in use for some time involves isozymes of the enzyme glucose phosphate isomerase (GPI). GPI exists as two allelic forms, a and b, which have different electrophoretic mobilities on starch gels. Since different mouse strains are homozygous for different alleles it is possible to reconstruct embryos in which a particular region or cell type is of one variant and the remainder of the other. The embryos can then be implanted in foster mothers and allowed to grow. After a while they are recovered and dissected and each part of the embryo and its surrounding membranes examined for its proportions of the a and b isozymes. An advantage of this method is that GPI is an enzyme which is present in all tissues, but there are also some disadvantages, in particular that the spatial resolution obtained is only as good as the dissection, and it is not easy to detect proportions of one isozyme below a few percent. Heritable labels which can be identified in sections are now being made by transgenic technology (Thomson & Solter 1988; Beddington *et al.*, 1989 and Fig. 6.2) and it is likely that these will represent an improvement. In principle they can combine a capacity for *in situ* detection with presence in all tissues and no dilution. Some studies have also been carried out in which embryos are infected with defective retrovirus carrying an enzyme label such as ß-galactosidase (eg, Sanes, Rubenstein & Nicholas, 1986). Here the idea is that a clone can be marked at a known stage, although not a known position, and that all its progeny can be visualized later on with the appropriate histochemical method.

Injection of HRP or fluorescent labels into blastomeres between the eight- and 32-cell stages is instructive as regards the normal origins of

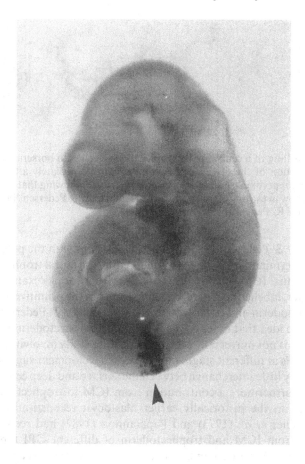

Fig. 6.2. A cell label made with transgenic technology. The dark staining patch derives from a graft of a single somite made at the head fold stage. The donor was of a transgenic strain which uniformly expresses a bacterial ß-galactosidase gene, so donor-derived cells can be visualized with a histochemical reaction for ß-galactosidase. (Photograph kindly supplied by Dr R. Beddington.)

trophectoderm and ICM (Pedersen, Wu & Balakier, 1986). After compaction the blastomeres are polarized, with a microvillous exterior domain and a smooth interior domain. About half the ⅛ blastomeres divide meridionally to yield two polar cells, and about half divide tangentially to yield one inner apolar cell and one exterior polar cell. At the 16-cell stage, about one-quarter of the outer cells divide tangentially to yield an inner apolar cell. After this stage outer cells produce only pairs of outer cells. So by the 32-cell stage there are between 8–12 inner apolar cells and it is these which become the inner cell mass. In expanded blastocysts, single labelled cells of the polar trophectoderm give rise to clones of postmitotic cells in the mural trophec-

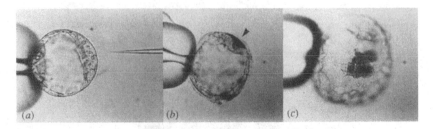

Fig. 6.3. Labelling of a central polar trophectoderm cell with horseradish peroxidase. (*a*) Injection of the HRP, (*b*) Peroxidase stain immediately after injection (arrowhead), (*c*) peroxidase stain after two days culture, showing that the labelled progeny are now in the mural trophectoderm. (From Cruz & Pedersen, 1985, kindly supplied by Dr R. Pedersen.)

toderm (Cruz & Pedersen 1985; Fig. 6.3), consistent with the polar region serving as a germinative or stem cell zone feeding the mural trophectoderm. When individual ICM cells are injected and the blastocysts examined 24 or 48 hours later, labelled cells are frequently found in the primitive ectoderm, primitive endoderm and polar trophectoderm (Winkel & Pedersen, 1988). However, the idea that inner cells contribute to trophectoderm in normal development is not universally accepted. Another study involving labelling of surface cells at different stages in blastocyst development suggested that there was very little interchange between the surface and deep cells (Dyce *et al.*, 1987). Furthermore a contribution from ICM to trophectoderm was not detected in the historically earlier blastocyst reconstitution experiments. Gardner *et al.* (1973) and Papaiannou (1982) had reconstructed blastocysts from ICM and trophectoderm of different GPI type, reimplanted them in foster mothers, recovered the conceptuses at a later stage, dissected them and analysed each part for GPI isoenzymes. The conclusion from this was that the ectoplacental cone, the secondary giant cells and the extraembryonic ectoderm of later stages were derived solely from the trophectoderm. The entire foetus, the amnion, allantois and extraembryonic endoderm were derived from the inner cell mass. The majority view, then, remains that ICM and trophectoderm do not exchange cells after the expanded blastocyst stage in normal development.

For the postimplantation embryo, Tam & Beddington (1987) performed orthotopic grafts of various regions of the primitive streak from $7\frac{1}{2}$ and $8\frac{1}{2}$ day embryos, using ^{3}H-thymidine as a label. The hosts were cultured *in vitro* for a further day and the labelled cells localized by autoradiography. The anterior part of the streak tends to populate the notochord and somites along the entire length of the body. Even more labelling of somites and some of neural plate is obtained from anterior grafts somewhat lateral to the midline. The middle of the streak populates mainly lateral plate mesoderm in the posterior half of the body. The posterior part of the streak populates mainly mesoderm of the amnion, visceral yolk sac and allantois. Somewhat similar results have been obtained by Lawson & Pedersen (1987) by injection of individual endoderm cells with HRP at primitive streak stages. These studies are consistent with the idea that the primitive streak in the mouse works in approximately the same way as the chick in that cells from the surface layer move laterally towards and then through the streak to become definitive endoderm and mesoderm. Whether there is an equivalent to the node regression and anteroposterior sequence of laying down of axial structures is not yet clear. All the studies show a substantial degree of mixing between labelled and unlabelled cells showing that the fate map is a rather fuzzy statistical construction. This represents a sharp contrast to the creatures considered in the previous Chapter where fate maps were precise down to the single cell level, but represents only a moderate increase of indeterminacy over the situation in *Xenopus* where there exists a reproducible topographic projection but with some cell mixing in the short range.

Chimaerism and mosaicism

Among mammals an animal composed of cells which are genetically dissimilar is called a 'mosaic' if it has arisen from a single zygote, and a 'chimaera' if it has arisen from some experimental or natural mixture of cells from different zygotes (McLaren, 1976). Genetic mosaics should not, of course, be confused with 'mosaic development' an entirely different concept discussed in Chapter 2. Mouse mosaics and chimaeras have been used a great deal in experimental work on developmental genetics and organ development. Most of this is outside the scope of this book but one relevant feature is the degree of cell mixing which is revealed both in chimaeric and mosaic animals.

In all female mammals one copy of the X chromosome becomes heterochromatic and inactive and thus ensures that all female mammals are naturally mosaic for X-linked heterozygous loci. X-inactivation occurs earlier in the extraembryonic ectoderm and endoderm than in the epiblast itself (Monk & Harper, 1979). In these extraembryonic tissues the *paternal* X-chromosome is preferentially inactivated. Inactivation occurs in the embryo itself at about the primitive streak stage and is apparently random with respect to maternal or paternal origin. X-inactivation is one of the features, together with ultrastructure and early synthesis of terminal differentiation products, which indicates a precocious development of the extraembryonic parts relative to the embryo itself. As for the distinction between maternal and paternal chromosomes, a differential methylation of key loci during gametogenesis would provide an attractive explanation (Monk, 1986).

The inactivation of one copy of the X chromosome produces a mosaic with respect to any heterozygous locus, since some cells will express the gene in question and others will not. Many studies have examined the number and distribution of clones arising from this cause and found a surprisingly fine grained arrangement, for example, there are about 9000 coherent clones per retinal epithelium in the eye (Deol & Whitten, 1972; West, 1976). Similar results have been obtained by identifying the progeny cells of each member of an aggregation chimaera. In preimplantation chimaeras in which one component is labelled with ^{3}H-thymidine the labelled patch remains coherent (Hillman *et al.*, 1972; Garner & McLaren, 1974) so the processes causing the mixing must happen later on. Possible explanations, which would have a cumulative effect, are cell mixing during gastrulation and clone breakdown during organ formation when many organs are formed by extensive convolutions of neighbouring epithelial and mesenchymal tissue layers.

SPECIFICATION OF EARLY BLASTOMERES

Isolated blastomeres and giant embryos

Since embryos of various amphibian, insect and sea urchin species will form twins when divided at early stages, it should not come as any surprise that the same is true of early mammalian embryos. However, there is a difference. The other forms can produce twins which are at least approximately normally proportioned while mammalian blastocysts resulting from blastomere separation are not. This probably arises from the unique geometry of the first developmental decision ('inside versus outside') which is in turn perhaps related to the fact that the early decisions are concerned with the specification of extraembryonic parts rather than regions of the embryo itself.

Following earlier work using rats and rabbits, Tarkowski (1959) destroyed blastomeres of the two- and four-cell stage mouse embryo and studied the subsequent development of the remaining cells both during the pre-implantation stages and also after reimplantion into foster mothers. When the cell numbers in the blastocysts derived from $\frac{1}{2}$ blastomeres were counted it was clear that the proportion of cells in the ICM relative to trophectoderm was lower than normal. Early postimplantation embryos were half-sized but they caught up by $10\frac{1}{2}$ days and those that developed to term were normal. The development of isolated $\frac{1}{4}$ and $\frac{1}{8}$ blastomeres *in vitro* was studied by Tarkowski & Wroblewska (1967). They showed that it was possible to obtain more than one blastocyst from the products of a single embryo but that these blastomeres, particularly from the eight-cell stage, frequently formed empty vesicles of trophectoderm. It therefore seemed as though all blastomeres had the same state of specification at early stages but the smaller the embryo the less likely it was to develop an ICM. This led to the idea that the first developmental decision occurred in the morula and that cells became specified to form ICM and trophectoderm depending on whether they lay in an inside or outside position.

Further evidence for the equivalence of early blastomeres came from experiments in which giant embryos were produced by the aggregation of two or more normal embryos at the 8- or 16-cell stage (Tarkowski, 1961; Mintz, 1965). These have a higher proportion than normal of cells in the ICM of the blastocyst. According to Rands (1985) the figures for percentages of cells in the ICM are 24% for normal embryos, 17% for $\frac{1}{2}$ embryos and 45% for 4x embryos thus approximately paralleling the volume/ surface ratio. Giant embryos formed by aggregation of morulae adjust to a normal size from $5\frac{1}{2}$ to $6\frac{1}{2}$ days, during formation of the egg cylinder, and do so by a lengthening of the cell division cycle in all regions (Lewis & Rossant, 1982).

Specification of the ICM and trophectoderm

The most direct evidence for the importance of position in the establishment of the ICM and trophectoderm comes from experiments of Hillman *et al.* (1972). They labelled donor embryos with ^{3}H-thymidine and then assembled morulae in which the labelled cells were either 'outside' or 'inside' a mass of unlabelled ones (Fig. 6.4). A labelled blastomere can be put outside simply by sticking it to another embryo at the four- or eight-cell stage. In order to put it inside, it has to be surrounded by several unlabelled embryos and thus eventually incorporated into a giant blastocyst. When the blastocysts were examined by autoradiography the label was usually in the trophectoderm if the labelled blastomere had been outside, and in the ICM if the labelled blastomere had been inside. In the latter case it is possible to force an entire early embryo to become part of the ICM of a giant blastocyst. This suggests that up to a certain stage all cells can become either ICM or trophectoderm depending on their position.

How is a difference of position converted into a difference of cell state? Following the energetic work of the group of M. Johnson, it is now generally believed that the two cell states are acquired as a result of cellular polarization which normally occurs at the eight-cell stage and is associated with compaction. In normal development it is known that microvilli arise on the outer surfaces of the blastomeres during compaction, and that a proportion of fourth cleavages are tangential, yielding large outer polar cells and smaller inner apolar cells. The microvillous part of the membrane can be visualized by the binding of fluorescein conjugated concanavalin A. It has been shown that $\frac{1}{8}$ blastomeres can become polarized in the same way *in vitro* especially if they are attached to another cell. They then usually divide at right angles to the polar axis to give a polar and an apolar cell as in the intact embryo (Ziomek & Johnson, 1980; Fig. 6.5). So the first step in

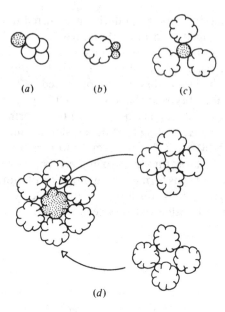

Fig. 6.4. 'Inside–outside' experiment of Hillman *et al.*, 1972. In (*a*) and (*b*) early blastomeres from a labelled embryo are attached to the outer surface of another. These tend to become part of the trophectoderm. In (*c*) a labelled blastomere is surrounded by three unlabelled embryos, and in (*d*) an entire labelled embryo is surrounded by 14 unlabelled embryos. In both these situations the labelled cells tend to contribute to the inner cell mass.

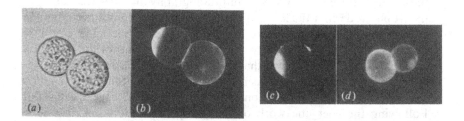

Fig. 6.5. Polarization of mouse embryo blastomeres. (*a*) and (*b*) show two $\frac{1}{8}$ blastomeres which are derived from an isolated $\frac{1}{4}$ blastomere. The cell has divided and become polarized, as visualized with fluorescein conjugated Con A ((*a*) phase image, (*b*) fluorescent image). The polarization is more apparent for the left-hand cell. (*c*) and (*d*) show that when an isolated $\frac{1}{8}$ blastomere divides it can generate dissimilar daughters. (Photograph kindly provided by Dr M. Johnson.)

the establishment of ICM and trophectoderm is a self-polarization of the blastomeres dependent on cell contact and the second step an asymmetrical cell division. This is somewhat reminiscent of the situation in the invertebrate embryos discussed in the previous Chapter in which a cytoplasmic determinant becomes localized in the fertilized egg and then partitioned to a subset of blastomeres by the early cleavages.

At the 16-cell stage the polar and apolar cells clearly bear surface features that stabilize their normal positions since when they are disaggregated, selected, labelled and mixed, the apolar ones sort themselves to the interior of the aggregate (Surani & Handyside, 1983). On the other hand, their commitment is not irreversible. Aggregates of all polar or all apolar cells of the $\frac{1}{16}$ stage will still form complete blastocysts (Johnson & Ziomek, 1983). So at this stage they are specified but not determined.

The mechanism of compaction and the associated cell polarization are still not satisfactorily understood although a great deal of work has gone into the problem. Compaction can be inhibited by a number of types of interference including deprivation of Ca, addition of antibodies to E-cadherin (the Ca-dependent adhesion molecule), addition of cytochalasin to disrupt microfilaments, or inhibition of gap junctions by injection of two-cell stage embryos with antibody to gap junction protein. However, none of these treatments inhibit cell polarization although they may decouple the plane of division from the axis of polarization. Some discrimination between the processes is provided by the application of protein synthesis inhibitors. These allow compaction if applied after the late two-cell stage, coupling if applied after the mid four-cell stage and polarization if applied after the late 4 cell stage (Levy *et al.*, 1986).

It is generally thought that by the 64-cell expanded blastocyst stage, the ICM and trophectoderm have both become determined and are not interconvertible. Handyside (1978) and several subsequent workers have investigated the properties of inside cells isolated by immunosurgery, a technique in which an anti-mouse antibody and complement are used to destroy the outer layer of cells. From morulae the inner cells will form complete blastocysts, from early blastocysts they will sometimes do so, and from later blastocysts they never do so. This is of course at variance with the claim of Pedersen's group that there is a continual recruitment of cells from ICM to polar trophectoderm in normal development, even in quite late blastocysts.

Nuclear transplantation and imprinting

The first reports of nuclear transplantations in the mouse embryo suggested that ICM nuclei could support the development of enucleated eggs, thus rather resembling the situation in amphibians. However, subsequent work in which karyoplasts from different stage embryos were fused to enucleated eggs with Sendai virus has shown that only nuclei from one- and two-cell stages will support development reliably (McGrath & Solter, 1984a). The early loss of totipotency may be due to the early onset of zygotic transcription in the mouse.

It has been known for some time that parthenogenetic embryos develop poorly. This used to be ascribed to homozygosity which led to the expression of recessive lethal genes. But more recent studies involving nuclear transplantation have shown that this is not the case. Heterozygous androgenetic diploids, made by substitution of another male for the female pronucleus, develop at most to the six-somite stage and have a large trophoblast relative to embryo (Barton, Surani & Norris, 1984). Partheno-genetic diploids, made by activation of the egg and suppression of the second polar body, develop slightly further but have poorly developed extraembryonic tissues (Kaufman, Barton & Surani, 1977). Nuclear trans-plantations have shown that the essential condition for normal develop-ment is for the egg to contain one paternally derived and one maternally derived pronucleus (McGrath & Solter 1984b; Surani, Barton & Norris, 1984). In chimaeric embryos containing some normal and some andro- or parthenogenetic cells, the androgenetic cells tend to be found in the trophoblast and yolk sac and the gynogenetic cells tend to be found in the embryo and yolk sac mesoderm. This suggests that there is some difference between the state of homologous genes on chromosomes from the two parents such that the paternal copy is more readily activated in the trophoblast and the maternal copy in the embryo.

Similar conclusions have come from genetic experiments in which mice are produced which have both copies of an individual chromosome from the same parent. This is done by setting up crosses in which complementary non-disjunctions are selected from chromosome sets containing Robertso-nian translocations (see Cattanach, 1986 for explanation). Not all chromo-somes behave in this non-equivalent way, but several do. This difference between homologous chromosomes, now known as *imprinting*, is thought to be due to DNA methylation, and support for this idea comes from the discovery that several transgenes have been found to be differentially methylated in sperm and eggs, and sometimes also differentially expressed depending on whether they entered the zygote through the maternal or paternal line (eg, Swain, Stewart & Leder, 1987). The transgenic studies also show that the genes in question have their methylation state altered accordingly during gametogenesis if they enter an animal via the opposite sex parent.

Although a fascinating biological phenomenon, it is not clear at present that imprinting has anything to do with regional specification of the body plan. It does represent an example of heritability of an epigenetic character by means of DNA methylation, and it is quite possible that other DNA methylations or demethylations occuring during development encode cellular states and represent the molecular basis of determination. This would be an alternative to the autocatalytic feedback loops discussed in Chapter 3, but at present positive evidence is lacking.

LATER DEVELOPMENTAL DECISIONS

Formation of primitive endoderm

As we have seen, ICMs isolated from the very earliest cavitating blastocysts can reform an outer layer of trophectoderm. At later stages, around $3\frac{1}{2}$ days, the state of commitment of the cells has changed and isolated ICMs in culture will now produce a layer not of trophectoderm but of primitive endoderm over their entire outer surface (Handyside, 1978; Hogan & Tilly, 1978). This process has not been studied in nearly so much detail as the formation of the trophectoderm, but the similar geometry ('inside versus outside') might suggest a similar mechanism. In reconstructed blastocysts a reduction of the number of ICM cells can lead to conceptuses in which primary endoderm derivatives are present but in which there is no embryo (Gardner, 1978). Some reports have suggested that primitive endoderm or its derivatives can be formed from the ectoderm at later stages, but the experiments concerned rely heavily on the assumption that immunosurgery can remove every endoderm cell from an isolated inner cell mass, and this does not seem to be the case (Gardner, 1985). As with the formation of the trophectoderm, there seems to be little or no regulation of proportions in the short term although adjustment can occur later so long as some cells of each population are present.

Once formed, the primitive endoderm cells can be distinguished visually from the primitive ectoderm by their 'rougher' appearance. When fragments or single cells from the primitive endoderm are injected into blastocysts of different GPI type, they are found to contribute to the visceral and parietal yolk sacs but not to the actual embryo of later stages (Gardner & Rossant, 1979). In the visceral yolk sac it is the endoderm rather than the mesoderm that is labelled. By contrast, the smooth cells were found to contribute to the embryo, the amnion and the mesoderm of the visceral yolk sac. These differences imply that the primitive endoderm becomes determined as a precursor to certain extraembryonic structures around the time of its appearance as a visible cell layer. This is compatible with the finding from the fate mapping studies that in normal development all three germ layers of the embryo arise from the primitive ectoderm and that the primitive endoderm forms only extraembryonic structures.

Mural and polar trophectoderm

An environmental difference also probably underlies the subdivision of the trophectoderm into mural and polar regions, the polar tissue forming above the ICM. Presumably the ICM emits some substance which inhibits the formation of giant cells and maintains the stem cell character of the polar region. There are two pieces of evidence for this. Firstly, trophectoderm which develops without an ICM, such as that from $\frac{1}{4}$ or $\frac{1}{8}$ blastomeres, is mural all over. Secondly, when an extra ICM is introduced, as in the rat-mouse chimaera experiments of Gardner & Johnson (1973), a polar region forms over the second ICM as well as over the first.

Visceral and parietal endoderm

The visceral endoderm is the epithelial tissue which surrounds the egg cylinder and which becomes displaced from the embryonic region by the definitive endoderm. From about $6\frac{1}{2}$ days it is characterized by synthesis of α-foetoprotein and other secretary proteins. Parietal endoderm consists of dispersed cells lining the inner surface of the trophectoderm from about the time of implantation and makes various basement membrane components. Hogan & Tilly (1981) found that the visceral endoderm would maintain its properties when cultured in isolation but in the presence of extraembryonic ectoderm (itself becoming trophoblast giant cells) would become parietal type. They proposed that the visceral endoderm was a germinative or stem cell region whose descendent cells would switch to a parietal type when they came into contact with extraembryonic ectoderm. Gardner (1984) introduced a new heritable cell label consisting of cytoplasmic malic enzyme, for which null alleles exist in certain mouse strains, and which is well expressed in the extraembryonic membranes. It was shown, by injection of wild-type cells into null blastocysts, that clones in the visceral endoderm remained coherent while those in the parietal endoderm became highly dispersed. Cockroft & Gardner (1987) later showed that visceral endoderm cells from early egg cylinders populated mainly the parietal endoderm when introduced into blastocysts. In fact the colonization pattern is indistinguishable from that of primitive endoderm cells from younger embryos. By the time that the specific biochemical markers appear on the visceral endoderm (about $6\frac{1}{2}$ days) very few cells gave rise to clones. Although these studies do not entirely concur in timing with those of Hogan & Tilly, if the two pieces of work are put together it seems probable that the early visceral endoderm represents the same cell state as the primitive endoderm, that it does for a period act as a source of cells for the parietal endoderm, and that the differentiation pathway adopted by the cells depends on whether they are in contact with embryonic or extraembryonic ectoderm. An analogous situation exists for F9 teratocarcinoma cells (see below).

Egg cylinder ectoderm

We have seen that twins can be produced experimentally by separation of the first two blastomeres. But twinning has also been observed to occur by subdivision of the ICM in embryos cultured *in vitro* and can be induced by injecting the mothers with the cytotoxic drug vincristine on day 7 of gestation (Kaufman & O'Shea, 1978). This implies that twins can arise from disruption of the embryo as late as the head process stage. The possibility of twinning in the gastrula indicates that the specification of parts in the transverse axes of the body is still labile at this stage.

By comparison, it is thought that human monozygotic twins usually arise from division of the ICM (70–75%), less often from blastomere separation (25–30%) and most infrequently from division of the primitive streak (1%). These figures are arrived at on the basis of whether the twins share a common placenta and amnion (Hamilton & Mossman, 1976).

N. Skreb and colleagues have studied the self-differentiation of explants from rat embryos transplanted to ectopic sites in the adult animal (review: Svajger *et al.*, 1986). Mouse embryos were not used because under such conditions they tend to form teratocarcinomas. It was found that, prior to gastrulation, the isolated embryonic ectoderm could form tissues characteristic of all three germ layers, although there is little difference in behaviour at different anteroposterior levels. By the head fold stage, gut was no longer formed. Isolated endoderm from the head fold stage failed to differentiate unless accompanied by mesoderm, in which case it could form several types of gut epithelia. Isolated mesoderm from head fold stages formed only brown adipose tissue, but if accompanied by endoderm produced also cartilage, bone and muscle. These experiments represent the nearest thing in mammalian embryology to the construction of a specification map by culture of fragments in a neutral medium. Of course the neutrality of the medium in this case is highly questionable, but the results do suggest that even major subdivisions of the body plan do not become specified until gastrulation, and that the maturation of tissues in the various organs may involve persistent inductive interactions between the germ layers.

EMBRYONIC STEM CELLS AND TERATOCARCINOMA

ES cells

Several earlier workers managed to establish cultured cell lines from early mammalian embryos, but the first lines showing developmental pluripotency were produced by Martin (1981) and Evans & Kaufman (1981) from ICMs and from delayed blastocysts respectively. These so-called 'embryonic stem cells' (ES cells) proliferate when grown on feeder layers of irradiated cells and will differentiate into *embryoid bodies* in the absence of feeder cells. Embryoid bodies consist of an outer layer of cells resembling the primitive endoderm, and an inner complement of ES cells. Recently, the requirement for feeders has been obviated by the discovery of a differentia-

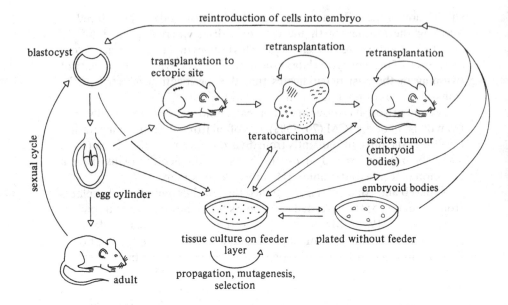

Fig. 6.6. Interrelationship between embryos, ES cells and teratocarcinomas.

tion inhibitory activity (DIA, also called 'leukaemia inhibitory factor'), which can substitute for them (Smith *et al.*, 1988). ES cells can be established from any strain of mouse and are usually of normal karyotype. When implanted into syngeneic mice ES cells form tumours containing several differentiated tissue types; and, more significantly from the developmental point of view, when ES cells are injected into blastocysts they will colonize the resulting embryos giving a high frequency of chimaerism (Bradley *et al.*, 1984). In favourable cases the chimaerism extends to the germ cells, making it possible to breed intact mice from cells which have been grown in culture (Fig. 6.6).

We shall consider below some of the implications of these techniques for the future, and deal here just with the developmental properties of ES cells themselves. Their very existence shows that it is possible to disengage growth and determination: ES cells have been passaged as many as 250 times without loss of the ability to repopulate embryos. The pattern of chimaerism, involving as it does a contribution to the primitive endoderm as well as the primitive ectoderm derived tissues, suggests that the cells are in the same state of developmental commitment as normal ICM cells (Beddington & Robertson, 1989), even though a number of lines have been isolated from delayed blastocysts in which further developmental steps may have occurred. So ES cells represent a source of ICM available by the gram rather than by the microgram, and even in the era of recombinant DNA technology there are many types of investigation that benefit from this sort of abundance of starting material. Potentially they could also be used to investigate the exact requirements for endoderm formation and the subsequent developmental steps, to the degree that they occur in culture. Such work has been anticipated to some extent by the use of teratocarcinoma cells, but ES cells do seem much more embryo-like and thus less vulnerable to objections than teratocarcinoma cells.

Teratocarcinoma

Most of the thoughts about the nature of ES cells and the uses to which they might be put were anticipated by work on teratocarcinomas. These are malignant and transplantable tumours which consist of several types of tissue and also contain undifferentiated cells. The undifferentiated cells, which will grow in tissue culture, are considered to be the stem cells for the tumour and are often called embryonal carcinoma (EC) cells. Three types of teratocarcinoma can be distinguished: spontaneous testicular, spontaneous ovarian and embryo derived (Stevens, 1980). Spontaneous testicular teratocarcinomas arise in the testes of foetal male mice of strain 129. They are thought to arise from primordial germ cells and the well known F9 cell line is of this type. Spontaneous ovarian teratocarcinomas arise in females of LT mice at about 3 months of age. They are derived from oocytes which have completed the first meiotic division and undergo approximately normal embryonic development as far as the egg cylinder stage. Embryo-derived teratocarcinomas can be produced by grafting early mouse embryos to extrauterine sites in syngeneic hosts, usually the kidney capsule or testis. The embryos become disorganized and produce various adult tissue types together with proliferating EC cells.

Different teratocarcinoma cell lines differ greatly in their properties: some need feeder cells to grow and others do not. Some will grow as ascites tumours in the peritoneal cavity and others will not. Some will differentiate *in vivo* or *in vitro* while others will not. Most lines have an abnormal karyotype although a few are normal or nearly normal. Most of the differences have probably arisen by selection of variants during establishment of the tumour and during culture, although it may to some extent also reflect a heterogeneity of the parent cell type. Chimaeric mice can be produced by injection of some types of teratocarcinoma cells into blastocysts (Brinster, 1974), but this does not work nearly as well as with ES cells. It does, however, have some theoretical interest since it shows that at least one sort of tumour can be made to revert to normal behaviour if it is placed in the appropriate biological environment.

At one time, teratocarcinomas were enthusiastically sold to the scientific world as more or less identical to embryos, but enthusiasm has waned as more and more differences have emerged in expression of individual genes between real embryos and EC cells. However, many interesting biological experiments have been performed with them and a few of these are worthy of mention within our context of early body plan formation. F9 cells, after removal from their feeders, can be induced to form a primary endoderm-like tissue by treatment with retinoic acid (Strickland & Mahdavi, 1978). In the presence of cAMP much of this becomes parietal endoderm. On the other hand, if cells are allowed to aggregate and are then treated with retinoic acid, they form a visceral type endoderm around the exterior (Hogan et al., 1981). This is somewhat reminiscent of the way in which different embryonic environments can direct the pathway of differentiation of the primitive endoderm although whether retinoids or cyclic AMP are really involved in vivo we do not know. Work with a human teratocarcinoma line, NT2/D1, has shown that the sensitivity of homeobox genes to induction by retinoic acid runs in the same sequence as the normal expression of these genes from posterior to anterior (Simeone et al., 1990). This, together with similar results on mouse cells and on Xenopus embryos, has led to the suggestion that normal anteroposterior pattern may be controlled by a gradient of retinoic acid with a high point at the posterior end.

Teratocarcinomas have also been used to examine the possible growth requirements of early embryo cells. Certain lines, such as PC13, secrete growth factors of the fibroblast growth factor (FGF) class, and while they do not respond themselves, the factor(s) are needed for continued proliferation of the differentiated cells arising from retinoic acid treatment. In turn, the differentiated cells secrete insulin-like growth factor II, which is needed for survival, but not growth, of the parent EC cells (Heath & Rees, 1985). This sort of reciprocal dependence is reminiscent of situations in vivo. For example, the polar trophectoderm seems to maintained in a proliferating state by the proximity of the ICM, and in turn the trophectodermal vesicle provides a suitable environment for survival of the ICM. Once again, we shall have to wait to find out whether FGFs or IGFs are really at work in vivo but the probable role of FGF as a mesoderm inducing factor in Xenopus makes this an interesting issue.

TRANSGENIC MICE

The creation of mice with an altered genetic constitution has become a vast industry. Although at present it has only peripheral relevance to the study of early development, the relevance will doubtless increase as the techniques improve. Much work with transgenic mice is directed toward understanding the normal functions and regulation of known genes or substances and there are three methods currently used to introduce genes. Firstly, the DNA can be injected directly into one pronucleus of the fertilized egg. This leads to a reasonable yield of transgenics with a moderate chance of integration in the germ line. Normally, integration is at a random position and may comprise many tandem copies (Brinster et al., 1985). Secondly, the gene can be inserted into a retrovirus which is then introduced into postimplantation embryos by various routes (Jaenisch *et al.*, 1981). Thirdly the gene can be introduced into ES cells by transfection or by retroviral vector infection and 'the cells then incorporated into blastocysts as described above (eg, Robertson *et al.*, 1986).

At present, all three methods are mainly directed toward 'overexpression experiments', in which one is looking at the effects of too much of the product in question or its expression in ectopic locations. It is known that, despite the abnormal chromosomal location of transgenes, they can show correct temporal and regional expression so long as sufficient flanking DNA sequences are included. For example an α-foetoprotein transgene was expressed in visceral endoderm, yolk sac, foetal liver and gut, but not in the adult animal and this closely resembles the normal expression pattern (Krumlauf *et al.*, 1985). A very large amount of work has been performed in which the flanking sequences of the genes are modified in order to identify the particular regulatory sequences responsible for control of their expression and to identify the proteins which bind to these sequences. There have also been some attempts to understand the normal functions of genes whose functions are presently unknown. For example, uniform overexpression of the Hox1.1 gene caused abnormalities anterior to the normal expression limit, consistent with a role in anteroposterior specification (Kessel, Balling & Gruss, 1990).

Although most transgenic work up to now has involved uniform overexpression, other techniques are being developed which will be more useful to the embryologist in the future. By making the appropriate molecular constructs, it is possible to express the coding region of one gene under the control of the promoter of another and this holds out the promise of performing discriminating embryological experiments on the hitherto rather inaccessible early postimplantation stages. One possibility is to ablate a particular region of the embryo by expressing a toxin under the control of a suitable promoter. A model for this approach has been provided by partial ablation of the pancreas in transgenic mice containing the Diphtheria A toxin under the control of the elastase enhancer (Palmiter *et al.*, 1987). Another possibility is to express a potential morphogen ectopically and alter the pattern of surrounding tissues accordingly. These applications will, of course, depend on the availability of a suitable library of stage and position-specific promoters. Most currently available promoters are for genes expressed during terminal differentiation, but it is likely that more useful ones will soon become available, for example from the homeobox genes.

The converse of the overexpression experiment is the creation of mice in which the expression of a particular gene, for example *int-2* or a homeobox gene, is abolished, with the hope that the null phenotype will reveal the gene's function. This is done by transfecting ES cells with a construct containing a defective copy of the gene and arranging a selection in tissue culture which selects for homologous recombination with the endogenous gene and against random integration elsewhere in the genome (eg, Mansour, Thomas & Capecchi, 1988). The desired cell clone is then injected into blastocysts and if germ line chimaerism can be attained then so can mice heterozygous for a null allele of the gene in question. When such mice are mated, 25% should be homozygous null, and might have phenotypes which would be informative about the normal function of the gene in question. An early example of this approach is the ablation of the proto-oncogene *wnt-1*, which results in defects in the mid- and hindbrain (McMahon & Bradley, 1990). This interesting line of work represents an almost exact reversal of the story in *Drosophila* where interesting developmental functions were identified by their phenotypes and the genes were subsequently cloned (see next Chapter). Its successful application really depends on there being a relatively low level of genetic redundancy in the developmental program, so if one component is taken away its function will not be covered by other components.

HOW TYPICAL IS THE MOUSE?

There is no doubt that research in mammalian embryology is fuelled by the potential practical benefits which may accrue from the ability to manipulate embryos of domestic animals and, subject to rules which society may properly impose, human embryos. Indeed, embryo transfer in cattle and *in vitro* fertilization in humans bear witness to the success of this programme to date. This means that technical feats such as the production of transgenic mice, or the achievement of homologous recombination of modified genes in ES cells, may receive a disproportionate amount of publicity compared to the results of embryological research proper. However, we are more concerned here with the contribution made by mammalian embryology to the science generally, and in particular we should ask the question: is there anything different about the mouse from other experimental models?

The obvious technical difficulties of micromanipulation of mouse and other mammalian embryos have led workers to be rather more inventive than they might have been if they had worked on *Xenopus* or invertebrate embryos. The early adoption of teratocarcinoma cells to bypass the shortage of embryonic material for biochemical work is an obvious example of this, as was the early programme of searching for molecular markers of different regions during normal development. But although the techniques used may have been somewhat different, as far as the substantive results are concerned the mouse seems to fit rather well into our system of thought about how regional specification works.

As in many other types of embryo the early blastomeres are equivalent and totipotent, there being no evidence for any regionalization prior to the polarization of blastomeres at the eight-cell stage. The first subdivision into

ICM and trophectoderm seems to be intimately associated with this polarization and thus bears an obvious similarity to other examples of cellular polarization followed by asymmetrical division, such as is found in the egg after fertilization in many other species. Although in the mouse this process is not usually *called* a segregation of cytoplasmic determinants there is no particular reason to think that polarization and asymmetrical division happens very differently in the mouse from *Xenopus* or annelids. Subsequent regional distinctions (mural versus polar trophectoderm; primitive ectoderm versus primitive endoderm) are set up in small groups of cells in response to environmental signals. The distinct states are initially labile (specification) but soon become stable and clonally heritable (determination). These events do not seem to differ significantly from inductive interactions found elsewhere. It was shown clearly for the first time by the mouse experiments, and has now been confirmed in *Xenopus*, that *single cells* can become determined. Determination is not therefore a property only of grafts containing many cells. A certain amount of stress has from time to time been laid on the fact that the decisions of early mouse development are all binary ones, since binary decisions are sometimes thought to have an intimate relationship to cell division. However, the spectacular creation of giant embryos by aggregation of morulae and their size regulation shortly after implantation argues against any rigid connection between the steps of determination and the cell cycle, as does the very existence of ES cells.

THE CHICK

The early chick embryo resembles the mouse in several respects. The early blastoderm becomes divided into two layers somewhat like the ICM, and the gastrulation movements occur through a primitive streak which forms at the posterior end. The embryo undergoes extensive growth during development and devotes a substantial proportion of its early tissue to extraembryonic structures, some of which: the amnion, chorion and allantois, are clearly homologous to those of mammals.

For the experimentalist it has the great advantage over the mouse that the embryo is accessible at all stages following cleavage. Early blastoderms can be cultured *in vitro* for long enough to form a recognizable primary body plan, or can be manipulated *in ovo* and kept alive until later stages. Furthermore it is possible to explant small pieces of tissue onto the chorioallantoic membrane of advanced embryos, where they become vascularized and will grow and differentiate in isolation. For these reasons our knowledge of the normal morphogenetic movements, the fate maps and the early determinative events are superior to our knowledge of the postimplantation stages of mammalian development. The studies of induc-

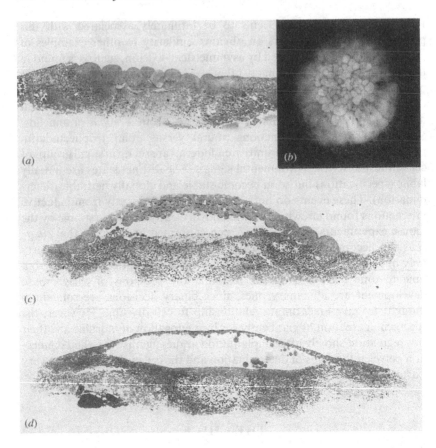

Fig. 6.7. The earliest stages of chick embryo development. (*a*),(*b*) cleavage (stage III), (*c*) formation of subgerminal cavity (stage V), (*d*) thinned blastoderm (stage X) showing the first islands of primary hypoblast. (Photographs kindly provided by Dr H. Eyal-Giladi.)

tive interactions show that the situation is generally similar to that in amphibians although as we shall see there are some important differences as well.

NORMAL DEVELOPMENT

At the time of oocyte maturation the future embryo consists of a patch of cytoplasm 2–3 mm in diameter at one edge of an enormous yolk mass. After fertilization the early cleavages take place in the oviduct over about 5 hours during which time the yolk becomes surrounded by albumen and shell membranes. Cleavage produces a circular blastoderm initially one cell thick and later several cells thick. The cleavage pattern is very variable from one embryo to the next and the blastomeres at the ventral and lateral faces of the sheet remain open to the yolk for some time (Fig. 6.7(*a*),(*b*)). After this, the egg spends about 20 hours in the uterus undergoing slow rotations driven by uterine peristalsis while the calcareous shell forms around it. When the blastoderm consists of a few hundred cells, a cavity opens beneath it (the subgerminal cavity) (Fig. 6.7(*c*)). Cells are shed from the lower surface of the blastoderm into this cavity and probably die so that by the end of the uterine period the central region of the blastoderm has thinned to an organized epithelium one or a few cells thick (Fig. 6.7(*d*)). Because of its translucent appearance this is known as the *area pellucida*. The outer, more opaque, part of the blastoderm is called the *area opaca* and the junctional region the *marginal zone*. Note that the marginal zone of the chick embryo is not a homologous structure to the marginal zone of an amphibian. The blastoderm consists by now of about 60 000 cells and it is at this stage that the egg is laid. A lower layer of cells, the *hypoblast*, then develops, partly by ingression of small groups of cells all over the area pellucida and partly by spreading of cells from the deep part of the posterior marginal zone. The advancing cell sheet incorporates the islands of already invaginated cells as it goes to form an integrated layer. The islands are sometimes called *primary hypoblast* and the cell sheet *secondary hypoblast*. The hypoblast contributes only to extraembryonic structures and may perhaps be homologous to the primitive endoderm of mammals. The upper layer of cells now becomes known as the *epiblast*.

The stages of body plan formation are depicted in Fig. 6.8 showing top views, and Fig. 6.9 showing transverse sections. A condensation of cells called the *primitive streak* arises at the posterior edge of the area pellucida and elongates until it reaches the centre (Figs. 6.8 (*a*),(*b*); 6.9(*a*)). Cells from the epiblast migrate into the streak and pass through it to become the mesoderm and the endodermal part of the lower layer. There is probably quite a high degree of random cell mixing involved both before and during this stage which is usually regarded as gastrulation. The area pellucida gradually changes from a disc to a pear shape and a further condensation called *Hensen's node* appears at the anterior end of the primitive streak (Fig. 6.8 (*b*),(*c*),(*d*)). This contains the presumptive notochord cells, some of which migrate anteriorly to form the *head process*, that part of the notochord lying within the head. The remainder of the node moves posteriorly and as it does so the principal structures of the body plan appear in its wake: the notochord in the midline, the somites on either side of it, and the neural plate in the overlying epiblast.

By about one day of incubation, the anterior end of the embryo is marked by an uplifting of the blastoderm called the head fold, and one somite and the anterior neural folds have appeared in the track of the regressing node (Fig. 6.8(*e*)). Somites form in anteroposterior sequence from the segmental plates which flank the notochord and neural tube. The visible event of

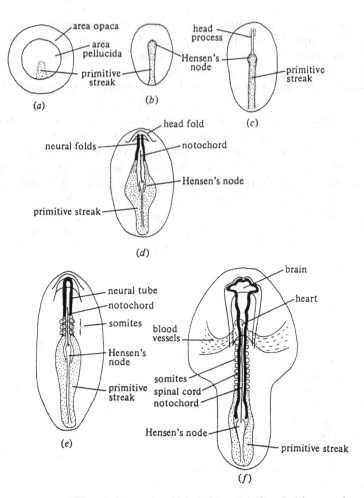

Fig. 6.8. Body plan formation in the chick embryo. (*a*),(*b*) primitive streak stages, (*c*) head process, (*d*) neurulation, (*e*) somite formation, (*f*) early tailbud.

segmentation corresponds to a transition from a mesenchymal morphology to epithelial spheres of tightly apposed cells. By about 36 hours there are ten somites and the neural tube has closed to form three brain vesicles (Fig. 6.8(*f*)). During these stages of overt body plan differentiation, some workers have described a segmented pattern in the mesoderm, not only in the somites themselves but extending right from the head fold to the node. These structures are called 'somitomeres' and are best visualized in scanning electron micrographs (review: Jacobson, 1988). They have been described in a wide variety of vertebrate embryos, not just in the chick, but their reality is still a matter of scepticism in some quarters. Although node regression and the formation of the posterior part of the body continues for

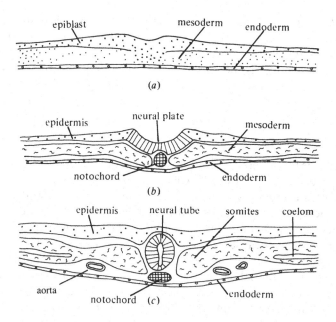

Fig. 6.9. Body plan formation in the chick embryo: transverse sections. (*a*) primitive streak, (*b*) neural plate stage, (*c*) neural tube stage.

some time, this stage, reached after about 2 days of incubation, marks approximately the junction between early and late development since the general body plan has been laid down and the formation of individual organs is about to begin. Hatching of the chick does not occur until much later, after 21 days of incubation.

Some biochemical and immunological markers have been collected for the early chick. Two-dimensional gels show some hypoblast but no epiblast specific spots, recalling the precocious differentiation of the extraembryonic tissues in the mouse (Lovell-Badge *et al.*, 1985). Much attention has been given to the cell adhesion molecules: the calcium dependent adhesion molecule E-cadherin (also called L-CAM or uvomorulin) which is required for compaction of the eight-cell mouse embryo is also found in the chick. It is ubiquitous at early stages then is lost from the forming neural plate during node regression, and then from most other tissues except the epidermis. It re-appears in the segmental plate during somitogenesis, with the highest concentrations in the region of mesenchymal–epithelial transition. The neural cell adhesion molecule N-CAM is also found all over early on and is then lost from tissues other than the neural tube. Both molecules reappear in many tissues at later stages (Rutishauser & Jessell, 1988).

Chick development is described by several stage series, the most important being that of Hamburger & Hamilton which is featured in most embryology textbooks, and that of Eyal Giladi & Kochav (1976) which divides the period before the primitive streak into 14 stages designated by Roman numbers. Landmark stages are V, at which the subgerminal cavity is formed; X at which the area pellucida is complete; and XIII at which the hypoblast is complete. The appearance of the primitive streak is H & H stage 2 and the primary body plan stage shown in Fig. 6.8(*f*) is H & H stage 10.

Fate map

Although the blastoderm of the newly laid egg contains about 60000 cells, it is estimated that only about 500 will contribute to the embryo proper, the rest forming extraembryonic structures as in mammals (Spratt & Haas, 1960). The prospective regions in the chick blastoderm have been studied for almost as long as those of the amphibian but the fate maps are not as accurate, perhaps because there is more random mixing of cells and so the fate maps have an inherently lower resolution. The three techniques which have been used for labelling are vital staining, application of carbon particles and orthotopic grafts labelled with tritiated thymidine. Each worker has disagreed to some extent with the others and so in Fig. 6.10 are shown 'compromise' fate maps devised by Waddington (1952) for the early blastoderm and a vital staining experiment of Nicolet (1970) on the head process stage. More recent cell labelling studies have emphasized the importance of the posterior marginal zone of the pre-streak stage as a source of cells (Stern, 1990). Like the area pellucida itself, this region contains two cell layers; the upper layer making a substantial contribution to the ectoderm of the streak, and the lower layer to the secondary hypoblast.

The movements of cells into and out of the primitive streak have been followed using orthotopic grafts labelled with tritiated thymidine (Nicolet, 1971). Initially the streak arises from the posterior $\frac{1}{4}$ of the area pellucida and then extends to the centre by active stretching. The labelling index of the streak is no higher than elsewhere, so although it is growing it cannot be regarded as a special growth zone, but rather as a locus of cell movements. The principal movement is a migration of epiblast cells from both sides which enter and move through the streak into the meso- or endoderm. However, they probably do not invaginate through the midline as often suggested in textbook figures, because there is in the midline a population of cells derived from the posterior marginal zone which do not contribute to the mesoderm (Stern, 1990). As in the mouse, the embryonic endoderm as well as the mesoderm is derived from the epiblast and the hypoblast cells become pushed to the outer rim of the lower layer.

During the phase of node regression, the node itself is the prospective region for the notochord along the entire length of the body and the region around the node for the remainder of the axial parts. The neural plate arises from epiblast anterior to the node and the somites and lateral plate from

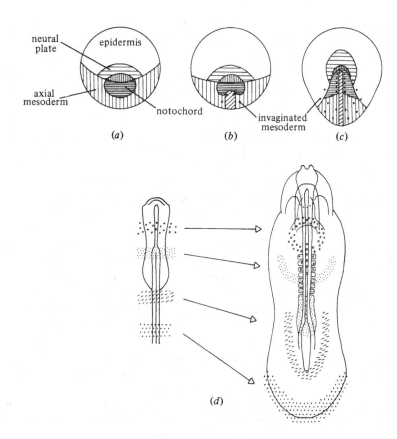

Fig. 6.10. (*a*)–(*c*) Fate maps of the chick epiblast during establishment and elongation of the primitive streak (after Waddington, 1952). (*d*) a marking experiment by Nicolet (1970) on the head process stage showing that the posterior part of the primitive streak does not normally contribute to the body axis.

epiblast and mesoderm posterior to the node. So the primitive streak does not map in a one-to-one manner onto the later anteroposterior body plan, in fact the posterior half of the streak is destined to become extraembryonic mesoderm. This should be remembered when assessing experiments on isolated posterior half streaks.

REGIONAL ORGANIZATION

AP polarity and fragmentation of the blastoderm

As long ago as 1828, von Baer propounded a rule which enabled the anteroposterior axis to be predicted in the majority of eggs. This states that if the egg is horizontal with the pointed end to the right then the tail of the embryo should be towards the observer. It was shown by Kochav & Eyal-Giladi (1971) that the polarity depended on the direction of rotation in the uterus. The embryo and yolk do not rotate along with the outer surface of the egg in the uterus but are none the less tipped in the direction of rotation and, in fact, a simple tipping of the blastoderm at the critical period of 14–16 hours of uterine life was enough to establish the polarity with the posterior end up. The critical period coincides with the formation of the area pellucida by shedding of cells into the subgerminal cavity, and these would presumably tend to congregate at the future anterior end. Once polarized, the intact blastoderm cannot be repolarized by tipping in the other direction, however fragments can be repolarized if they do not contain the original posterior edge, suggesting that the polarization consists of some fairly local covert change at the posterior edge.

An extensive study involving fragmentation of the blastoderm *in ovo* was conducted by Lutz (1949) using the duck. In this species the egg is laid at a slightly earlier stage than that of the chick. He found that up to four complete embryos could be produced by cutting (Fig. 6.11). Spratt & Haas (1960, 1961) obtained similar results with the chick showing that an embryo could arise from as little as an eighth of a blastoderm or as much as three blastoderms fused together. The size of the embryos themselves did not vary by a factor of 24, but more like a factor of 3 in volume. This is another indication that, as we saw for the mouse, the proportions within the embryo are accurately controlled but the proportion of tissue devoted to extraembryonic as opposed to embryonic tissues is not.

Lutz and many subsequent workers showed that when the blastoderm was cut parallel to the future axis both of the twins would be of normal orientation, but if the cut is perpendicular to the axis only the posterior twin would be normally oriented and the anterior one would be randomly oriented (Fig. 6.11). We have seen that the entire blastoderm can become polarized at an early enough stage merely by tipping, and the cutting experiments show that this polarity is located only in the posterior region, since anterior fragments do not share it. Presumably we are here dealing with another *symmetry breaking process* in which the system is poised to become polarized in any orientation so that any slight reproducible bias in the environment will make the process deterministic (see Chapter 3). This is reminiscent of the dorsoventral polarization of the fertilized amphibian egg, although it should be remembered that the amphibian becomes polarized as a single cell while the chick becomes polarized as a blastoderm of several hundred cells. So although both processes may share the general dynamical character of symmetry breaking they are unlikely to be the same at the molecular level.

In the course of their experiments on induction (described below) the group of Eyal-Giladi have also carried out many isolation experiments on embryos at stage X–XIII, during the period of hypoblast formation

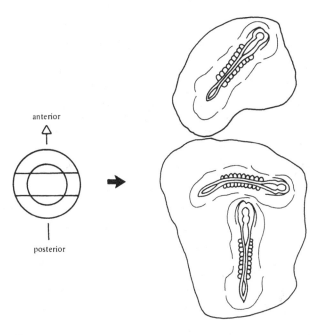

Fig. 6.11. Formation of triplets from a single duck blastoderm cut into three pieces. The posterior embryo has the original orientation but the anterior and middle embryos do not. (After Lutz, 1949.)

(review: Eyal-Giladi, 1984). Unfortunately the degree of differentiation reached is poor in those experiments involving smaller pieces if tissue, such as isolated epiblast fragments, so there is still some uncertainty about exactly which parts can form which structures. With this proviso, the results are as follows (Fig. 6.12). Explanted hypoblasts or pieces of area opaca do not differentiate into any recognizable embryonic structures at all. Explanted parts of the area pellucida denuded of hypoblast or taken prior to the arrival of the hypoblast will not form a primitive streak but do form some embryonic tissues, in particular some blood cells, some mesenchyme and perhaps some muscle. A whole area pellucida or a part thereof including the hypoblast will produce a primitive streak, as will the complementary parts of the blastoderm, consisting of the area opaca plus the marginal zone. The isolated marginal zone itself will also produce a streak although its differentiation capacity is improved by inclusion of the area opaca. In some cases it is possible to obtain two streaks from the same embryo, one from the isolated area pellucida and the other from the complementary ring of surrounding tissue.

Isolation experiments of tissues from later stages have usually involved culture on the chorioallantoic membrane of late embryos, since this

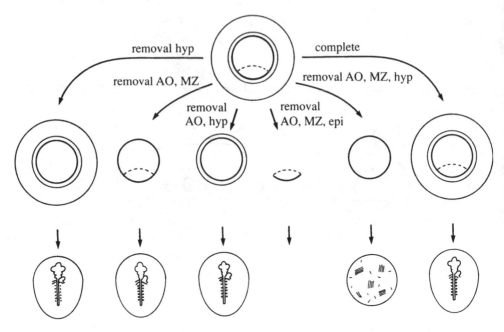

Fig. 6.12. Results of isolation experiments carried out on the early chick blasto-derm. AO: area opaca, MZ: marginal zone, hyp: hypoblast, epi: epiblast.

provides a richly vascular environment. In a classic study Rawles (1936) cultured a large number of fragments from head process stage embryos. Although often quoted in textbooks as evidence for late regulation, this interpretation depends on tissues obtained in only small proportions of cases, and when only reasonable proportions of cases are counted the results become very similar to Nicolet's fate map, thus showing essentially mosaic behaviour. This suggests that by the head process stage the decisions involved in the specification of the major tissue types of the body have already taken place, although much potential is evidently packed into the small region around the node.

Isolated node regions transplanted to the area opaca or the chorioallan-toic membrane will self-differentiate into axes containing notochord, neural tube and somites, of a character posterior to the level reached at the time of explantation (Spratt, 1955). This suggests some sort of connection between node regression and specification of anteroposterior levels of the body plan, but as in amphibians we have little other experimental evidence relevant to this process. Early nodes, from H & H stage 4, but not from later stages, can completely regulate after removal (Grabowski, 1956). This occurs only if the gap closes, if it remains open there are always defects in the final pattern. If the node were regarded as a signalling centre then this would be a significant phenomenon because as we have seen in Chapter 3 a simple LSDS model cannot regenerate its source region while a Gierer–Meinhardt type model can.

Induction

Studies on induction in the chick have suffered in the past from a failure to use a reliable cell label which can distinguish between donor and host tissues. Without such a label it can be difficult to decide which structures have really been induced from the host, and which are formed by self-differentiation of the graft. Although in the early work (review: Waddington, 1952) no suitable labels existed, we have now possessed tritiated thymidine and the quail label for many years. Both of these need to be introduced by grafting a labelled fragment into an unlabelled host. Tritiated thymidine has already been mentioned in connection with fate mapping, its main disadvantage being that the label becomes diluted out by continued growth, limiting the useful labelling period to 1–2 days. Cells of the quail were shown by Le Douarin to contain a heterochromatic blob in the interphase nucleus which could be stained with the Feulgen reaction to provide a positive label heritable on cell division (see Le Douarin, 1982). This is ideal for experiments on induction and it is surprising that it has not been more widely used.

We shall start by considering the origin of the non-axial type of mesoderm. As shown by the isolation experiments discussed above, isolated epiblast fragments will self-differentiate non-axial type mesodermal tissues (blood cells, mesenchyme and muscle) in the absence of the hypoblast. Stern & Canning (1990) have shown that the precursors to this mesoderm consist of cells identified by an antibody called HNK-1 (originally found to bind to neural crest cells of later stages) which are seen to be scattered across the epiblast from stage XII. The cells can be labelled with anti-HNK-1 conjugated to colloidal gold particles and their position identified at a later stage of normal development. It appears that at least the first formed mesoderm and definitive endoderm consist exclusively of these cells and that the negative cells are left in the ectoderm. If the positive cells are ablated by treatment with the antibody and complement at stage XIII then no mesoderm forms, suggesting that by this stage they cannot be regenerated from the negative cells. In a *Xenopus* embryo the formation of these tissues would correspond to the induction of ventral mesoderm, thought perhaps to be due to FGF. In the chick it seems that no spatially restricted induction is involved since the positive cells arise all over the epiblast and later sort out from the negative cells.

When we consider the formation of the axial mesoderm, there is more evidence that induction is involved. The primitive streak normally originates on the posterior side of the area pellucida epiblast. But it has been shown by many workers starting with Waddington in the 1930s that a second streak can be induced elsewhere if the hypoblast is rotated through 90 or 180 degrees. It is also known that the ability to form a streak can be restored to a piece of isolated epiblast by coculture with hypoblast (review: Eyal-Giladi, 1984). Although an important caveat concerning these experiments is that no cell label was used, it is generally agreed that a primitive streak can be induced by the hypoblast or at least the posterior part of it. The active region is specifically the secondary hypoblast, the part of the hypoblast formed by spreading of cells from the marginal zone. This was shown by Azar & Eyal-Giladi (1979) who found that regenerated hypoblast was not active if it consisted solely of cells invaginated from the epiblast, but was active if it contained cells from the marginal zone. There seems to be an inhibitory effect created by an existing streak which antagonizes the formation of another. We have seen that many streaks can arise from a stage XIII blastoderm that is cut into fragments, but the intact blastoderm forms only one. According to Eyal-Giladi & Khaner (1989) a fragment of posterior marginal zone can suppress the streak initiated by a fragment of lateral marginal zone if it exceeds it in size by a factor of 1.5. Such repressive effects are reminiscent of the Gierer–Meinhardt activator–inhibitor model discussed in Chapter 3, although because of the general failure to label cell populations in these experiments it cannot be excluded that we are seeing a competition for mesoderm *cells* between streaks rather than a diffusible chemical inhibition. In an interesting recent experiment Mitrani & Shimoni (1990) showed that epiblasts deprived of the posterior marginal zone and thus unable to form a streak, had this ability restored when treated with activin. As we have seen, the activins are good candidates for the inducer of the organizer in *Xenopus* (Chapter 4).

The epiblast competence to form a streak is maximal at stage XIII and decays by H & H stage 3. During this period the primitive streak is elongating and there remain some hypoblast cells underlying it, but later on when the streak's development has become autonomous the lower layer beneath the streak has become composed entirely of definitive endoderm. Both the signalling capacity of the hypoblast and the competence of the epiblast are maximal at the posterior end although this may simply be an effect of differential exposure: since the hypoblast develops from the posterior margin, an epiblast isolated in the stage range X–XIII will have already received some stimulus at its posterior end. Exposure to a further uniform stimulus such as a sheet of reaggregated hypoblast cells would then be expected to evoke one streak in the posterior, which is what is found (Mitrani & Eyal-Giladi 1981).

So in general the induction of the streak by the hypoblast seems to parallel closely the induction of the amphibian organizer by cells of the dorsovegetal region of the morula, except that the competent tissue is not a uniform sheet of ectoderm but a population of pre-committed mesoderm cells intimately mixed with their ectodermal progenitors.

The next inductive interaction is a regionalization of the mesoderm under the influence of the node. Bellairs (1963) showed that at the head process stage posterior third blastoderms would not form somites unless a thin strip of anterior streak material was included. Nicolet (1970) grafted tritiated thymidine labelled nodes to such posterior isolates and found that somites were formed from host tissue while the grafts differentiated mainly into notochord. Hornbruch *et al.* (1979) grafted quail nodes at varying distances from the host streak of complete head process stage chick blastoderms. They found that when the graft was far away it self-differentiated into notochord, somites and some neural tissue but when it was close to the host streak then extra somites could be induced from host tissue. These experiments are all quite similar and all suggest that the mesoderm can be raised in its state of commitment from the level of non-axial mesoderm up to the level of segmental plate under the influence of the node. This bears an obvious similarity to the dorsalizing activity of the amphibian organizer discussed in Chapter 4 and a gradient model for the process was advanced by Hornbruch *et al.* (1979). There may, however, be a difference in that within the node it may be prospective neural rather than prospective notochordal tissue which is most active. This was suggested by experiments of Fraser (1960) in which the nodes were modified in various ways to minimize notochord formation, such as by punching a small region from the centre, or by scraping off the lower cell layers.

The third inductive interaction in the avian, as in the amphibian embryo, is neural induction, formerly misnamed 'primary embryonic induction'. In the 1930s Waddington reported that neuroepithelium could be induced from the epiblast of the area pellucida and area opaca in response to implants of primitive streak. He even obtained two cases of induction from the embryonic shields of rabbit embryos after implanting chick primitive streak (review: Waddington, 1952). These experiments were performed without cell labels but Grabowski (1957) repeated them using vitally stained nodes from definitive streak stages and found that the neural tissue was indeed derived from the host. The interaction has been further investigated by Gallera (1971) using tritiated thymidine labelled grafts. Inductive activity initially lies in the anterior primitive streak and later in the region just anterior to the regressing node. It disappears by about H & H stage 8 (four somites). The competence of the ectoderm disappears before this, at the about H & H stage 5 (head process). About four hours of contact are required between the inductor and the area pellucida epiblast in order to obtain a response, and the signal will pass across a Millipore filter with a nominal pore size of 0.45 microns. There is little evidence for the existence of anteroposterior regional specificity as is found in neural induction in

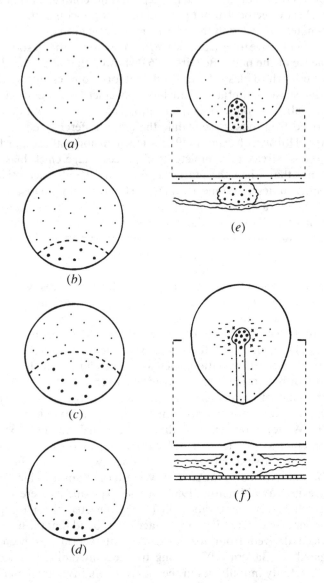

Fig. 6.13. Sequence of inductive interactions in the chick embryo.
Small dots represent non-axial mesoderm cells, large dots organizer cells, crosses pre-segmental plate cells, dashes pre-lateral plate cells, vertical hatch endoderm cells. (*a*) stage X. Non-axial mesoderm cells appear all over area pellucida. (*b*) stage XI. Limit of hypoblast shown by dashed line. Some non-axial mesoderm cells now promoted to organizer cells by a signal form the hypoblast. (*c*) stage XII. As the

amphibians, but this may simply be because it has not been looked for, the appropriate inducer for such experiments being tissues from different levels of the mesoderm at the 1–2 somite stage.

In Fig. 6.13 the information reviewed above is used to construct a feasible model for early regional specification in the chick. At about stage X the non-axial mesoderm cells are produced, later becoming visualizable with the HNK-1 antibody. As the hypoblast spreads from the posterior it induces some or all of the overlying non-axial mesoderm cells to become organizer cells. The organizer cells emit a freely diffusible inhibitor which spreads throughout the area pellucida and antagonizes this process, thus limiting the final size of the primitive streak and suppressing the formation of streaks elsewhere. By stage XIII a cluster of organizer cells has become established covertly in the posterior region of the epiblast. By H & H stage 2 this arrangement has become overt with the organizer cells aggregated into a primitive streak and attracting non-axial mesodermal cells in from the sides. Those mesodermal cells which penetrate to the lower layer become converted by the change of environment to endoderm. Those which enter the middle layer remain mesodermal. If they end up near the dorsal midline they become raised in level of commitment to segmental plate, if they are somewhat further away they are raised up to lateral plate, and if they are further away still they remain in the mesodermal ground state which later differentiates to extraembryonic mesoderm and blood islands.

This model may well not be correct, but, as with the models presented in Chapter 4 for the amphibian and sea urchin, the attempt is simply to propose a scheme which is compatible with most of the existing data and which can serve as a guide for further experiments.

Segmentation

We have suggested that the segmental plate (the presomite mesoderm) is induced from the competent field of non-axial mesoderm by organizer cells concentrated in Hensen's node. A recent clonal analysis of the segmental plate showed that the this process continues during node regression. Stern *et al.*, (1988) injected the fluorescent label RDA into single cells at different anteroposterior levels of the segmental plate. In the anterior two thirds,

proportion of organizer cells increases so the inhibition builds up and fewer become induced. (*d*) stage XIII. Organizer cells begin to aggregate. (*e*) H & H stage 2. Cluster of organizer cells now visible as primitive streak. Non-axial mesoderm cells migrate through streak to form mesoderm and definitive endoderm. (*f*) H & H stage 4. Hensen's node is composed of organizer cells. Mesoderm cells that entered the lower layer have become endodermal. Mesoderm cells adjacent to the organizer have become promoted to segmental plate.

they found that clones were later confined to the somites. But in the posterior third they found that clones often populated both the somites and other tissues derived from the lateral plate mesoderm. This means that determination of cells to become somite continues to occur in the vicinity of the node, and that it occurs no earlier than about 20 hours before segmentation.

The important facts about somitogenesis are that it proceeds independently of the presence of the axial tissues, and that the sequence of somite formation is unaltered by surgical rearrangement of the segmental plate. In particular it is possible to make transverse cuts or radiation burns across the segmental plate but the process of segmentation continues on schedule on the distal side of the gap. It is also possible to reverse the anteroposterior orientation of a region of the plate and see segmentation proceeding autonomously in the graft, ie, in reverse sense to the host body (review: Menkes & Sandor, 1977). This means that the timing of segmentation must be preset in the segmental plate cells perhaps at the time of somite determination. Stern and his colleagues have argued that the cell cycle plays an important role in this timing process. Labelling studies with tritiated thymidine show that there is a partial synchrony of cycles in the segmental plate, such that neighbouring cells are likely to be at similar points in the cell cycle. There is also a spatial projection of the age structure from anterior to posterior corresponding to about two cell cycles. This means that as one moves from the last formed somite posterior to the zone of somite determination, one passes sequentially through a zone in which the most popular phase is M, then a G2 zone, then an S zone, then another G1 zone and so on through two cycles. Temperature shocks applied to the embryos give defects in the pattern of segmentation corresponding in position to those cohorts of segmental plate cells which were clustered around M phase at the time of the shock (Primmett et al., 1989). Although there are usually one or two defects per file of somites, there may sometimes be three or even four, showing that the shocks can have effects on the pool of cells not yet committed to become somite.

These workers have advanced a model for segmentation with four components: (i) segmentation is associated with a rise of adhesiveness of the cells entering the new somite, (ii) this change can only occur two cell cycles after somite determination, (iii) it can only occur in a window of the cell cycle comprising about ½ of the total cycle and spanning M phase, (iv) the discontinuity or 'gating' inherent in segmentation occurs because the first cells to reach the end of the window emit some substance which recruits all other cells in the vicinity which are also in the time window to increase adhesiveness as well. The substance must decay fairly rapidly so that it does not influence the next cohort of cells to enter the window of competence. This model explains the autonomous character of segmentation, the possibility of increasing or reducing the size of somites in abnormally sized embryos, and the effects of temperature shocks. It has much in common with the 'clock and gradient model' discussed in Chapter 4 but differs in that the hypothetical oscillator turning once per segment is replaced by the familiar cell division cycle which turns about once per seven segments.

A COMMON PROGRAM FOR VERTEBRATE DEVELOPMENT?

The general vertebrate body plan which is achieved by the end of early development is similar in amphibia, birds and mammals, and if the method of getting there in terms of the hierarchy of decisions is also similar, then we can probably conclude that any differences are superficial ones, presumably adaptations to the different modes of embryonic life of the animals concerned.

Despite the obvious differences in the morphology of the early stages, and in the gastrulation movements, the three types of induction which can be distinguished in the chick have obvious homologues in the amphibia. Streak induction corresponds to the induction of the organizer in *Xenopus*, the induction of pre-somite mesoderm from area pellucida mesoderm corresponds to dorsalization, and neural induction by the node to corresponds to neural induction by the archenteron roof. The main difference seems to be that the ventral mesoderm induction apparent in *Xenopus* does not occur in the chick, instead the non-axial mesoderm cells arise all over the epiblast and are presumably the target for the streak inducing signal since when the streak forms its meso- and endodermal parts are composed of these cells. At the biochemical level this may not be much of a difference for the same ventral inducing signal, perhaps an FGF-like molecule, may still be required and be present at such a dose that only a proportion of cells respond. However, the logic of development is somewhat altered because a spatially localized induction has been replaced by a cell sorting process, and since this is a book in which logic rather than molecules have priority we must take the difference seriously.

The similarities between the chick and the amphibian are encouraging since if these two have so much in common then it is unlikely that the mammals can be all that different. Of course many scientists have spent whole careers in medical research organizations assuming the truth of this idea. There are also some homologies with the invertebrate embryos, most notably the probable role of homeobox containing genes similar to those of the *Drosophila Antennapedia/Bithorax* complexes in the specification of anteroposterior body levels, and we shall return to the possible significance of this in the later Chapters. But some other genes from *Drosophila*, such as *engrailed*, have very different expression patterns in vertebrates. Also, although a case has recently been made that the rhombomeres of the hindbrain are homologous to the body segments of *Drosophila*, the mechanism of the principal vertebrate segments, the somites, appears to have little in common with segmentation in *Drosophila*. Finally, we should not forget the drastic differences of descriptive morphology between vertebrates and invertebrates. After all, the sequence and nature of the interactions needed to build a leech or nematode body plan would not produce a vertebrate. Where the program of regional specification is concerned, the vertebrate embryos, and particularly the amniotes, still probably have the best claim to be regarded as the 'models for Man'.

GENERAL REFERENCES

Bellairs, R. (1971). *Developmental Processes in Higher Vertebrates*. London: Logos Press.

Rossant, J & Pedersen, R.A. (1986). *Experimental Approaches to Mammalian Embryonic Development*. Cambridge: Cambridge University Press.

Waddington, C.H. (1952). *The Epigenetics of Birds*. Cambridge: Cambridge University Press.

7

The breakthrough

The last few years have seen an absolute explosion of results on the early regional specification of the fruit fly *Drosophila melanogaster*, and the work is now so advanced that we are approaching a satisfactory explanation of the developmental program itself. This has been achieved by a brilliant combination of developmental genetics, experimental embryology and molecular biology. Because of the small size of *Drosophila* and its short life cycle of two weeks, it has been intensively used in genetic research for most of the present century and over this time a large number of mutants have been collected and accurate genetic and cytological maps of the chromosomes have been compiled. Over most of this period the developmental genetics dealt only with mutants which were viable as adults but from the late 1970s it was realized that many mutations of the genes controlling regional specification would lead to grossly perturbed body plans and be lethal by late embryonic life. The raw material for the breakthrough was a near comprehensive collection of late embryo lethal mutations affecting the body plan which were isolated in massive screens mainly by Ch. Nüsslein-Volhard, E. Wieschaus, T. Schüpbach and colleagues. Another crucial element of the breakthrough was the tradition of insect experimental embryology, mainly a German subject, which had previously been conducted on insect species other than *Drosophila*. This led the *Drosophila* workers to compile accurate fate maps and to repeat and extend manipulations which had suggested the existence of organizing centres at the two poles of the egg. These embryological experiments became much more powerful when grafts could be carried out between embryos of different genetic constitution. The third, and decisive, element of the breakthrough was the technology of molecular biology. By the 1980s this had become powerful enough to clone the genes which the genetic and embryological studies had identified as being of importance in regional specification. The resulting availability of probes enabled the expression patterns of the genes to be visualized by *in situ* hybridization, and the changes of expression pattern caused by other mutations also to be visualized. So it became possible to observe directly the previously cryptic cell states which are the foundation of the body plan itself.

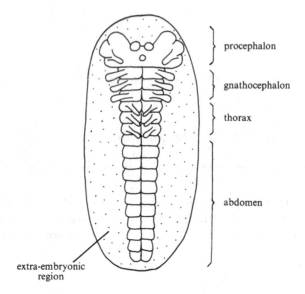

Fig. 7.1. The general body plan of an insect is exemplified by this diagram of a late germ band stage embryo of the beetle *Tenebrio*. (After Ullman, 1964.)

INTRODUCTION TO INSECTS

All adult and larval insects are built up of an anteroposterior sequence of segments which fall into the three principal body regions of head, thorax and abdomen. The basic arrangement is shown in Fig. 7.1 which depicts an embryo of the beetle *Tenebrio*, which has a more typical morphology than *Drosophila*. The prototype body plan is most clearly seen at that embryonic stage which is called the *extended germ band* and is the stage at which all insect species display their maximum morphological similarity. At this stage the head may consist of as many as six segments: three procephalic and three gnathal, the gnathal segments bearing leg bud-like appendages which later become the mouthparts. The middle part of the body is the thorax which always consists of three segments: the prothorax, mesothorax and metathorax. All of these bear a pair of legs on the ventral side and the meso and metathorax also bear a pair of wings on the dorsal side. The number of abdominal segments varies with species but is usually in the range 8 to 11. *Drosophila* does not display the procephalic head segments even in the embryo, but the three gnathal segments appear transiently, and there are the usual three thoracic segments (T1-3) and eight abdominal segments (A1-8). In *Drosophila*, as in other Diptera, only the mesothorax bears wings, the metathoracic wings being represented by small balancing structures called halteres. Although the conventional segments are dominant in the larval and adult body plan we shall see that during early

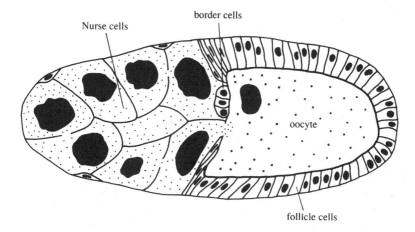

Fig. 7.2. Arrangement of the egg chamber during oogenesis of *Drosophila*. The nurse cells are connected to each other and to the oocyte by cytoplasmic bridges. (After King, 1970.)

development the most important repeating units are the *parasegments*, which have the same period but are out of phase with the later segments.

Drosophila is a *holometabolous* insect, meaning that it undergoes an abrupt and complete metamorphosis, so the egg hatches into a larva which is quite different in structure from the adult. The larva grows and passes through two moults before becoming a resting stage called a pupa in which the body is remoulded to form the adult. Much of the adult body is formed from the *imaginal discs* and the *abdominal histoblasts* which are only present as undifferentiated buds in the larva. Some other species, particularly those of the short germ type, are *hemimetabolous*, meaning that the larva resembles the adult and acquires the final adult form via a series of moults.

NORMAL DEVELOPMENT

As we shall see, events during oogenesis are quite important for regional specification in the *Drosophila* embryo. The germ cells are derived from the *pole cells* of the early embryo which migrate to the embryonic gonads. In the process of oogenesis, one germ cell divides four times to produce 16 cells interconnected by cytoplasmic bridges. Two of the cells have four bridges, and the others fewer. One of the four-bridge cells becomes the *oocyte* and the other 15 cells all become *nurse cells*. The whole cluster, called the egg chamber, is surrounded by ovarian follicle cells, which are derived from the gonads and thus of somatic rather than germ line origin (Fig. 7.2). The nurse cells become polyploid and export large amounts of RNA and protein into the oocyte. In the later stages the oocyte becomes visibly polarized in both dorsoventral and anteroposterior axes and a granular

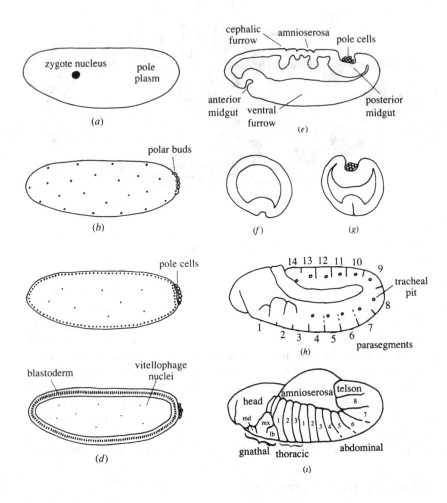

Fig. 7.3. Embryonic development of *Drosophila*. (*a*) zygote, (*b*) cleavage, (*c*) syncytial blastoderm, (*d*) cellular blastoderm, (*e*) gastrulation and germ band extension, (*f*),(*g*) transverse sections through gastrula, (*h*) fully extended germ band, showing parasegmental grooves, (*i*) retracting germ band showing segments.

pole plasm forms at the posterior end. The follicle cells assume a squamous form over the nurse cells and a columnar form over the oocyte, and secrete both the vitelline membrane and the chorion which is a tough outer coat surrounding the egg. The yolk is not made locally, but comes from the fat body via the haemolymph.

Fertilization occurs in the uterus, the sperm entering at the anterior end through a specialization of the chorion called the micropyle. Development of *Drosophila* is very fast compared with most other insects and the larvae hatch after less than 24 hours at normal laboratory temperatures. The early embryonic stages are depicted in Fig. 7.3. The initial period is called 'cleavage' but as in most insects it is actually a period of rapid synchronous *nuclear divisions* without cellular cleavage. After the first eight divisions the pole cells are formed at the posterior end, incorporating those nuclei lying within the pole plasm. After nine divisions most of the nuclei migrate to the periphery to form the *syncytial blastoderm*, those remaining internally are later incorporated into vitellophages which end up in the gut lumen. After four more nuclear divisions, during the third hour, the *cellular blastoderm* is formed. At this stage there are about 5000 surface cells, 1000 yolk nuclei and 16–32 pole cells. The pole cells divide once more before gastrulation but mitosis of the blastoderm cells slows dramatically. In what follows positions along the anteroposterior axis of the early stages will often be expressed as *% egg length* (%EL) from the posterior pole. This sometimes causes confusion, for example, the region 100–50% EL is the *anterior* half and the region 10–0% EL is the *posterior* tenth.

The columnar epithelium of the blastoderm is quite thick and most of it is destined to become part of the embryo, only a thin dorsal strip becoming the extraembryonic *amnioserosa*. Gastrulation commences at about 3 hours with the formation of a deep *ventral furrow* along much of the embryo length. This consists of a mesodermal invagination along the ventral midline, joined shortly later by invaginations of *anterior* and *posterior midgut* at the respective ends. The *cephalic furrow* appears laterally at 65% EL. Concurrent with gastrulation the germ band begins to elongate, driving the posterior end with the pole cells round to the dorsal side of the egg. By about 4 hours the first *neuroblasts* appear in the neurogenic ectoderm which is now mid-ventral, having closed over the ventral furrow.

Segmentation in the mesoderm is apparent by 5 hours, and in the ectoderm at about 6 hours, by the appearance of 15 *parasegmental grooves* and ten paired tracheal pits. This is about the stage of maximum germ band extension and is the stage at which the maximum morphological similarity is apparent between different species of insect. The *parasegments* are not exact precursors of the later segments but a segment forms from the posterior $\frac{2}{3}$ of each parasegment combined with the anterior $\frac{1}{3}$ of the next. Although their morphological appearance is transient, parasegments appear to be fundamental units for the construction of the body plan as we shall see below (Martinez Arias & Lawrence, 1985). The head displays three transient bulges corresponding to gnathal parasegments but without mesodermal segmentation.

The germ band then retracts and as it does so the epidermal grooves are translated to the position of the definitive segmental grooves which coincides with the tracheal pits. The anterior and posterior midgut fuse in the middle and the ventral nerve cord becomes segregated. Dorsal closure of the epidermis occurs at 10–11 hours, displacing the amnioserosa into the interior. At about this time the head 'involutes' into the interior and is therefore scarcely represented on the outer surface of the larva. This happens also in other Diptera but is unusual for insects in general. In later stages the Malpighian tubules are formed at the junction of posterior midgut and hindgut; muscles, blood vessels, the fat body and the gonads arise from the mesoderm; and the central nervous system is formed by complex but stereotyped cell divisions in the ganglia of the ventral cord.

The larva has no legs, its head is tucked away in the interior and it has three thoracic and eight abdominal segments. The specializations of the epidermal cuticle which are formed during late development are very important as they are the features used to assess the phenotypes of late embryo lethal mutations. On the dorsal side the region from T2–A8 is covered with fine hairs. On the ventral side are *denticle belts* on each of the thoracic and abdominal segments (see Fig. 7.8(*a*)). Each belt occupies mainly the anterior part of a segment but it also just straddles the segment boundary and extends into the posterior of the next segment. The thoracic can be distinguished from the abdominal segments both by the shapes of the denticle belts and by differences in the epidermal sensory structures, for example, the thoracic segments bear paired ventral *Keilen's organs*. Segment A8 bears the *posterior spiracles*, which are openings of the tracheal system which carry distinctive structures called *Filzkörper*. The structure of the larval head is very complex, but its most prominent component is a horny *cephalopharyngeal skeleton* secreted by the stomodeal part of the alimentary tract. This is derived from the procephalic part of the head. There are numerous other sclerites and sense organs which have been placed on the fate map and have been used to assess the phenotypes of mutants. The term 'acron' has been used to designate a presegmental region of the head, but its use is not consistent between authors. At the posterior end of the body the structures posterior to segment A8 are often called the *telson*.

The *imaginal discs* and *abdominal histoblasts* are present as small nests of cells in the first instar larva. It is not known at exactly what time the rudiments become established in the embryo but evidence from clonal analysis indicates that it may be before the time of the first postblastoderm mitosis at 4–4.5 hours.

Fate map

Prior to the migration of nuclei to the periphery there is not a definable fate map because the orientation of nuclear divisions is not the same for different individuals. So the fate maps for *Drosophila* necessarily refer to the blastoderm stage or later and in Fig. 7.4 is shown the fate map of the cellular blastoderm by Technau and Campos Ortega. The methods used to construct it were quantitative histology, injection of HRP into individual cells and defect mapping with the UV microbeam (Lohs-Schardin *et al.*,

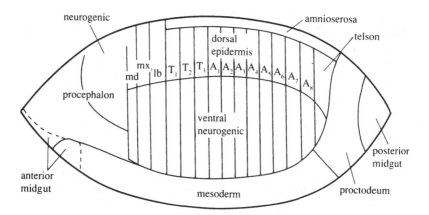

Fig. 7.4. Fate map for the cellular blastoderm of *Drosophila*, after Hartenstein *et al.*, (1985). md: mandibular, mx: maxillary, lb: labial, T: thoracic, A: abdominal segments.

1979; Hartenstein *et al.* 1985). The last is not strictly speaking a method for fate mapping but it is acceptable in this case as there seems to be very limited regulation of defects in *Drosophila*. The principal features of the fate map are as follows: prospective regions exist for all the larval segments, there being no regions of indeterminacy representing later cell mixing or later growth from a small bud. The prospective regions for the recognizable gnathal, thoracic and abdominal segments are arranged in anteroposterior sequence from 75% to 15% of the egg length. The procephalic head structures are not arranged in a simple way, but prospective regions for all of them are there in the anterior 25%. The prospective region for the anterior midgut also maps to the anterior, and for the posterior midgut, Malpighian tubules, proctodeum and germ cells to the posterior, as one would expect from the descriptive embryology. In terms of allocation of the 5000 surface cells: about 1000 will be anterior to the cephalic furrow, about 1250 will invaginate into the mesoderm and posterior gut, about 1770 will enter the ventral neurogenic region and about 920 the dorsal trunk epidermis.

Cellular commitment

Before the introduction of HRP as a cell label, studies of determination in *Drosophila* relied on the use of genetic labels which were scored in the adult organism (eg, Simcox & Sang, 1983). The general conclusion was that isolated nuclei were not determined before cellularization, while isolated cells from the blastoderm stage or later were determined to the segment but not to terminal cell type. A similar conclusion was reached from observation of labelled clones induced by somatic recombination (see below), although this method reveals allocation rather than determination. Clones induced at the cellular blastoderm were observed not to cross segment boundaries nor to cross an invisible anteroposterior boundary running through each of the thoracic imaginal discs (Garcia Bellido *et al.*, 1979). This was called the anteroposterior *compartment boundary* and is now known actually to be the parasegmental boundary of the early embryo, defined by the *wingless* and *engrailed* systems (see below). It is presumed that this boundary is inherited and maintained in each of the thoracic imaginal discs.

In situ labelling of cells by injection with HRP shows that ectodermal clones remain confined to one parasegment while mesodermal clones may contribute to several parasegments and to several tissue types, for example, somatic muscle, visceral muscle and fat body. Exchange of ectodermal cells between different dorsoventral levels shows that prospective neuroblasts still form neuroblasts even at more dorsal levels, while prospective dorsal epidermis can be caused to become neuroblasts after transplantation to the neurogenic region (Technau, 1987).

Pole plasm

The pole plasm is formed during oogenesis and was suspected of being a germ cell determinant for many years. Direct evidence that the pole plasm can program the nuclei to become prospective germ cells was obtained from the experiments of Illmensee & Mahowald (1974). Pole plasm was transplanted to the anterior end of host eggs at cleavage stage. Then the cells which formed in the region of the graft were grafted to the posterior end of a genetically distinguishable second host at cellular blastoderm stage. The second host was grown to an adult and then crossed with flies carrying the same recessive marker genotype as the first host. The production of double recessive progeny proved the presence of functional, labelled, germ cells in the chimaeric fly. This experiment has been repeated using pole plasm from unfertilized eggs and late stage oocytes, and ectopic pole cells have been obtained although it has not been shown that these can become functional germ cells. A number of maternal effect mutations prevent the formation of the pole plasm and result in sterility of the resulting offspring (viable alleles of maternal posterior group – see below). One of these, *vasa*, is known to code for a protein which is localized to the pole plasm region in the oocyte and which remains confined to germ line cells for the whole life cycle (Hay *et al.*, 1988), but we do not know whether *vasa*, or any of the others, actually has the properties of a determinant. Heterotopic grafting of HRP labelled pole cells shows that they cannot form other tissues, so in this case the determinant really does specify a cell type rather than a body region (Technau, 1987).

DROSOPHILA DEVELOPMENTAL GENETICS

The spectacular progress in the understanding of *Drosophila* development depends largely on the sophistication of the genetic methods available. The genome size is small in comparison with most animals, both in gene number and in DNA content. The generation time is short and the animals are small, so experiments involving complicated breeding protocols and large numbers of individuals can be completed in a few weeks. The genetic maps are very detailed as are the cytological maps of the giant polytene chromosomes from the larval salivary glands. All these features have been vital for the rapid transition from the discovery of a potentially interesting mutant phenotype to the possession of a cloned gene.

Embryologists have often wondered how much regional specification is already to be found in the egg. In previous Chapters we have seen some quite convincing evidence for cytoplasmic determinants laid down during oogenesis, but we are still in most cases ignorant of their molecular nature. In *Drosophila* we do not need to speculate any more since we know exactly how much information is laid down during oogenesis and we know the molecular structure, localization and mode of action of several of the key substances. This is because mutations in genes required during oogenesis manifest themselves as *maternal effects*, meaning that the structure of the embryo corresponds not to its own genotype but to the genotype of the mother. Strictly speaking when discussing maternal effect mutations one should always say 'eggs from mutant mothers' rather than 'mutant eggs' but for simplicity we will in fact refer to mutant eggs and hope that this does not cause confusion.

A genetic component which deserves introduction right at the beginning of this section is the transposable element known as the P-element. P-elements encode their own transposase enzyme and hence catalyse their own transposition from one locus in the genome to another. But the transposase is only active in the germ line of individuals resulting from the mating of a P containing male with a P free (M strain) female, called a *dysgenic cross*. Subsequent generations containing the P elements are stable to further transpositions. As we shall see, P elements are used to introduce mutations for cloning purposes, and, even more important, they are used to introduce genes into the germ line to create transgenic stocks. The gene to be inserted is cloned into a P element which lacks its own transposase and contains a label gene to enable transformants to be identified. This is injected into the posterior pole of the egg along with a helper P which produces transposase but is incapable of integration itself. If the P-element is incorporated into the genome of one or more pole cells then the resulting flies will produce some offspring containing the integrated P-element. The method has been used for various purposes which will all be mentioned below: to introduce enhancer traps, to create strains carrying reporter constructs, to rescue endogenous mutations, and to express genes in an ectopic manner.

Identification of relevant genes

The *Drosophila* genome contains about 5000 genes which can be mutated to lethality, and it has been possible to assemble a fairly complete collection of genes which, when mutated, give rise to pattern alterations in the embryo. There are several different types of screening protocol for doing this, of which a simple example is shown in Fig. 7.5. This is a screen for autosomal recessive mutations and like most screens depends on the use of a *balancer* chromosome which both suppresses recombination, carries some visible label gene, and is lethal in the homozygous condition. Males homozygous for a different visible label (a) are mutagenized and are then mated to females carrying the balancer. Each *individual* offspring fly represents one mutagenized gamete, so individual males from the F1 generation are isolated and crossed again. In the F2 generation each *tube* of flies represent one mutagenized gamete. For this second step, females are used which carry a dominant temperature sensitive lethal gene and this enables selection against offspring not carrying the balancer. When offspring with one mutagenized chromosome and one balancer are mated with each other to produce the F3 generation, it is possible to ask whether or not flies carrying the homozygosed mutant chromosome are viable. If they are, they will display the recessive label phenotype (a/a) from the mutagenized males. If they are not, then the dead embryos or larvae are examined to see whether they have any significant pattern abnormality. The original mutagenesis induces mutations on all the chromosomes, but it is only those on the balanced chromosome which can survive as a stable stock. Once obtained, mutations are sorted into groups of alleles of the same gene by complementation tests, and are genetically mapped. A mammoth screen of this type by Nüsslein-Volhard, Wieschaus and colleagues yielded a total of 140 genes, mainly zygotic lethals, which gave effects on embryonic pattern (Nüsslein-Volhard *et al.*, 1984; Jürgens *et al.*, 1984; Wieschaus *et al.*, 1984).

Another class of mutations which are important to the embryologist are maternal effect genes which have an effect on embryonic pattern because they disturb the normal formation of the oocyte. These are often identified as *female sterile* mutations in screens similar in principle to those for zygotic lethals (Schüpbach & Wieschaus, 1986). Some zygotic lethal genes also show important maternal effects. Because of the lethality, homozygotes cannot grow to flies whose fertility can be tested. The effects have been identified by the method of X-ray induced mitotic recombination using larvae which are heterozygous for the lethal mutation to be tested and also for another, dominant, female sterility gene on the same chromosome

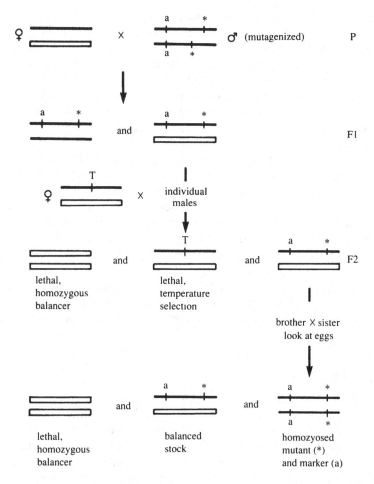

Fig. 7.5. Mutagenesis screen for autosomal recessives with embryo lethal pheno-types. The outlined chromosome is a balancer which suppresses recombination, carries visible labels, and is lethal in the homozygous condition. a is another visible label gene, T is a dominant temperature-sensitive lethal, and * represents a mutation. The eggs resulting from the F2 cross are examined for abnormalities and, if no interesting ones are found, the stock is discarded at this stage. The homozygous balancer and *ts* dominant are chosen to have their lethal effect at a later develop-mental stage so as not to interfere with the visual assessment.

(Perrimon *et al.*, 1986). In the absence of any recombination, no eggs are produced because the germ cells carry the dominant female sterility gene. X-ray induced recombination will create clones which are free from the female sterility gene and homozygous for the lethal mutation. Any maternal effect will then show up as defects in embryonic survival or morphology. Many maternal effect genes affecting early pattern are zygotic lethals and have been identified in this way, or by the equivalent procedure of pole cell transplantation into homozygous recessive female sterile embryos.

Another method of identifying important genes does not depend on mutagenesis but on the *enhancer trap* method which has already been mentioned in Chapter 6 (O'Kane & Gehring, 1987). An *E. coli* ß-galactosidase gene is introduced into the germ line by P element mediated transformation (see above). The ß-galactosidase is controlled by a weak promoter so will not be expressed unless the construct integrates near an enhancer element of the host genome. The offspring are screened directly for the expression pattern of the enzyme, and if an interesting expression pattern is seen it is presumed that the enhancer responsible normally controls some interesting endogenous gene, which can now be cloned using the insert as a starting point. The enhancer trap method has recovered many genes already known from mutagenesis as well as others which were formerly unknown.

Types of mutation

Most mutations are 'loss of function' representing the production of a smaller amount of gene product, or a gene product of reduced effectiveness. These are usually recessive and the phenotypes are called *hypomorphs*. In *Drosophila* genetics much effort goes into the creation of several alleles for each locus to obtain at least some that have lost all function, the so-called *null* alleles giving *amorphic* phenotypes. Frequently the hypomorphic alleles can be arranged in a series of increasing severity with the amorphic phenotype as the limit form. Such allelic series can often give useful information about function, particularly where the weaker alleles give something recognizable and the stronger ones do not.

Some alleles may be temperature sensitive, often because the mutation affects the thermal stability of the protein product. These are useful because they can help to establish the developmental stages at which a gene is required. This is done by shifting between the permissive and non-permissive temperatures, and if the shift to the non-permissive temperature is made before the time of requirement then most cases will be mutant, while if it is made afterwards, most cases will be normal.

Dominant mutations are sometimes *haploinsufficient*. This means that they are loss of function mutations for a gene for which a reduction in the level of the product to 50% of the wild type level is sufficient to cause a mutant phenotype. In such cases the homozygous phenotype will be more severe than the heterozygous one. More often dominant mutations are 'gain of function', resulting in the production of active gene product in positions or at times when it is not normally found. Gain of function mutations often work by inactivation of some other repressive or inhibitory function.

A huge advance was made in *Drosophila* developmental biology when it was realised that embryo-lethal phenotypes could be informative, and most of the phenotypes of genes covered in this Chapter will be described in terms of late embryo/first instar larval cuticular morphology. Recently, with the availability of probes for many developmentally significant genes, it has become quite common practice to describe phenotypes of earlier acting mutations in terms of changes in later expression patterns, for example, of the pair-rule gene *fushi tarazu*. This is even more direct than looking at larval morphology, and can preserve quantitative features which are obscured during later morphogenetic movements and cell differentiation.

Mosaic analysis

A genetic mosaic is a creature derived from a single zygote but composed of cells of more than one genetic composition (see also Chapter 6). *Drosophila* mosaics can be made in two ways: by X-ray induced somatic mutation and by the gynandromorph method. In the former the radiation can cause recombination between the individual chromatids of two homologous chromosomes. This means that when a heterozygote for a suitable label gene is irradiated, the recombination can result, after a further cell division, in the formation of a homozygous mutant clone together with a homozygous wild type clone (Fig. 7.6). The homozygous mutant clone will express the label and be visible.

A *gynandromorph* is a creature which is part male and part female, and this type of mosaic is created at an early nuclear division by loss of an unstable Ring-X chromosome from female embryos. The part composed of cells without the Ring-X are XO and hence male while those retaining it remain female. The Ring-X is often lost at the first nuclear division yielding a half-and-half embryo. Since the division plane is random, different individual embryos have the boundary between male and female in different places. Recessive label genes on the X chromosome are uncovered in the XO cells to provide the visible identification.

Both methods are used to investigate the cell-autonomy of developmental mutations. If a homozygous developmental mutant clone is marked with a cell autonomous label, usually another recessive mutation reasonably closely linked, then one can tell whether the mutant phenotype is

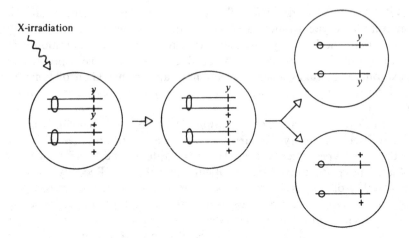

Fig. 7.6. Creation of a labelled clone by somatic crossing over. Every cell in the embryo is initially heterozygous for a label gene: *yellow*. If crossing over occurs between strands of homologous chromosomes then two homozygous clones will be generated by the next mitosis, one + / + and one y/y.

confined to mutant cells (autonomous) or is more or less widely distributed (non-autonomous). Obviously genes required for the generation and transmission of inductive signals are likely to be non-autonomous, while those responsible for the interpretation of such signals and for genetic regulation should be autonomous. These methods were somewhat limited by the quality of suitable label genes active during the embryonic stages but the recent discovery of a gene *serendipity*, expressed in all cells of the blastoderm, promises to improve the situation in the future. Another gene, *sex lethal*, is expressed in all cells of the female and will undoubtedly be used as a label in gynandromorphs.

Mosaic analysis has in the past been extensively used in adults, where there have long been reasonably good cell autonomous cuticular labels, such as *yellow* or *multiple wing hairs*. Irradiation is carried out during larval development when the rudiments for the adult cuticle, the imaginal discs, are undergoing rapid growth. This has enabled the study of the *adult phenotype* of mutations which are zygotic lethals when expressed in the whole organism but are viable as clones in a normal background. It has also, as mentioned above, been used to induce homozygous zygotic lethal mutant clones in the germ line to study their maternal effects. Mosaic analysis also led to the discovery of the famous 'compartments' in the imaginal discs, originally discovered as boundaries which clones would not cross (Garcia Bellido *et al.*, 1979). Originally a hierarchy of compartments was supposed to exist but nowadays one only hears about the anteroposterior compartment boundary which has stood the test of time and is derived from the parasegmental boundaries in the early embryo.

Cloning of genes

As will be apparent from the tables at the end of this Chapter many of the genes important for early regional specification have now been cloned and sequenced. Some genes have been cloned by 'chromosome walking' since their genetic map positions were known and clones were available from a near enough location to make this feasible (readers who are not familiar with chromosome walking should look it up in a textbook of molecular biology). Where this was not the case, a starting point for the walk was created by P element mutagenesis. When a P element integrates into or near a gene it may mutate it to inactivity and it is possible to screen for a particular mutation resulting from a dysgenic cross by methods similar to those of Fig. 7.5. A mutation caused by P element insertion can then be used as a starting point for cloning the gene. First some breeding is done to reduce the number of P elements in the genome as much as possible, while retaining the mutant. Then a genomic library from this strain is screened with the P element probe and the resulting clones are tested by *in situ* hybridization to polytene chromosomes to find which particular P element they represent. The right one can be used as the starting point for the chromosomal walk to obtain the whole of the required gene. Proof that the cloned candidate really *is* the required gene can be quite difficult, but in principle this is obtained by showing that it is not expressed in null mutants of the same locus, that several known mutants have identifiable sequence changes, or best of all, by rescuing the phenotype of the null mutant by P element mediated transformation.

Once the gene is cloned and sequenced the next step is to determine the normal expression pattern using *in situ* hybridization to different stages. This was originally done autoradiographically on sections but more recently wholemounts with non-radioactive probes have been favoured since they give particularly clear and vivid patterns and enable the whole embryo to be inspected at a glance. The next step is to make antibodies to the protein product, usually of a fusion protein expressed in bacteria, and use this to determine the protein expression pattern.

Study of the developmental mechanisms then comes to mean 'how does the pattern of gene expression at one stage cause the pattern of gene expression at the next stage?' and various standard types of experiment have been devised to find out. A large and growing literature examines the expression pattern of one gene in embryos mutant for another. This sort of experiment has provided most of the entries in the 'developmental function' column of the gene table, but it does have drawbacks. In particular if removal of gene A has no effect on gene B it is usually assumed that the expression of gene B is independent of gene A. However, it may be dependent on gene A but also on other things as well so that the removal of just one of them produces no effect. This *redundancy* is probably not too serious a problem in *Drosophila*, but may be much worse when similar experiments come to be carried out in vertebrates.

If removal of gene A does have an effect on gene B then people worry about whether the effect is *direct* or *indirect*. Ways of establishing directness are by examining DNA-protein interactions *in vitro*, or by cotransfection of the two components into *Drosophila* tissue culture cells and seeing whether the expected interaction occurs. It is worth remembering that 'direct' in a developmental program means 'one-to-one' rather than necessarily direct molecular contact. For example, if A activates B by molecular contact then it will do so *in vitro*. But if A + (some other substance present all over the early embryo) activates B then it may not work *in vitro*, but it is none the less still 'direct' in a logical sense since no more *information* is required to turn on B than the expression pattern of A.

When the effect of A upon B is examined, it is quite common to look not at the endogenous B gene product but at a construct composed of the regulatory and promoter region of B fused to a reporter gene, usually *E. coli* ß-galactosidase. The reasons for this may be simply sensitivity if the endogenous product is hard to detect. Also in many constructs the ß-galactosidase is quite stable and so 'integrates' the gene activity up to the time of examination. Most often it is because the regulatory region of B has been modified in order to find which DNA sequences are responsible for each component of the expression pattern. The strain containing the reporter construct is made by P element mediated transformation.

The homeobox

We have already met the homeobox in Chapters 4 and 6. It is a 180 nucleotide long stretch of DNA found within the coding regions of a number of genes involved in regional specification in *Drosophila*, particularly the homeotic selector genes. It codes for an amino acid sequence rich in lysine and arginine whose function is to act as a specific DNA binding domain. Homeobox containing genes code for transcription factors whose function is the positive or negative regulation of other genes, although of course transcription factors need not contain a homeobox, and other motifs such as the 'zinc fingers' are also common. The homeobox has been particularly useful to developmental biologists because it has served as a probe for finding formerly unknown genes, some in *Drosophila* but mainly in vertebrates, which have often turned out to be expressed in early development.

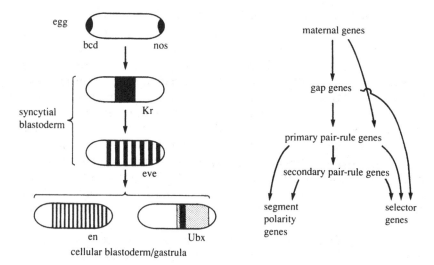

Fig. 7.7. Overview of anteroposterior specification in *Drosophila*. On the left, expression patterns of representative genes are shown for each level in the hierarchy. On the right, is an indication of the regulatory connections, although these can only grow denser as more information is gathered.

OVERVIEW OF THE DEVELOPMENTAL PROGRAM OF *DROSOPHILA*

What follows is long and complex, although hopefully it is simpler that the original literature itself. The reader who is not already familiar with *Drosophila* should hold onto the following brief summary in order to navigate through the story of the breakthrough.

One system operates in the dorsoventral axis and is responsible for the formation of mesoderm, neurogenic region and epidermis from ventral to dorsal. As a result of a series of events in the egg chamber of the mother, the product of a maternal gene *dorsal* becomes distributed in the nuclei of the blastoderm in a ventral–dorsal gradient. It activates a set of zygotic genes including *twist*, *single-minded* and *zerknüllt* which control formation of the various bands of tissue.

An independent system operates in the anteroposterior axis and is much more complex (Fig. 7.7). The first phase of specification occurs in the egg chamber with the establishment of three maternal systems. The *anterior system* is concerned with the production of a gradient of the *bicoid* protein from anterior to posterior. The *posterior system*, of which the pole plasm is an integral part, deposits the mRNA of a gene called *nanos* in the prospective abdominal region. The *terminal system* activates a cell surface receptor produced by the gene *torso* at the anterior and posterior poles.

The products of these three maternal systems divide the embryo into several zones depending on their relative concentrations or activities. Before cellularization one nucleus can affect the gene activity of nearby nuclei simply by producing a genetic regulatory protein, no receptors or signal transduction mechanisms being required. Because of the overlaps between the domains of activity, and because different concentrations of the same substance can have different effects, the maternal systems can activate a spatial pattern of zygotic gene activity which is more complex than their own. The best characterized genes activated at this stage are the *gap* genes, *hunchback*, *Krüppel* and *knirps*, and the *primary pair-rule genes*, *hairy, even skipped* and *runt*. The gap genes are expressed in one or a few domains while the pair-rule genes are expressed in stripy patterns with a periodicity of two segment widths. Their periodicity arises because many different combinations of maternal and gap genes can activate the same pair-rule gene.

Various combinations of concentrations of maternal and gap and primary pair-rule gene products then activate the *secondary pair-rule genes*, of which the best characterized is *fushi tarazu*. The overlapping periodic patterns of primary and secondary pair-rule genes lead to the activation of a repeating pattern of *engrailed* and *wingless*, which has a single segment width periodicity and causes the subdivision of the axis into parasegments. Simultaneously the combined maternal, gap and pair-rule gene product combinations activate the homeotic selector genes of the *Antennapedia* complex and the *Bithorax* complex which control the character of each parasegment and thus its subsequent pathway of differentiation.

In an attempt to aid readers with less than 640K of random access memory, the story is presented gene by gene and a comprehensive gene list is provided at the end of the Chapter for reference purposes. We shall start with the rather simpler dorsoventral mechanism and then deal with the anteroposterior mechanism later.

THE DORSOVENTRAL PATTERN

The pattern along the dorsoventral axis is relatively simple consisting at the cellular blastoderm stage of four strips committed to become, from ventral to dorsal, the mesoderm, the ventral neurogenic region, the dorsal epidermis and the amnioserosa (Fig. 7.4). Cell transplantation experiments suggest that at this stage the ventral cells are determined, but the dorsal cells are not, and can be caused to form the appropriate cell types if grafted to the neurogenic or mesodermal regions (Technau, 1987). The disposition of these four territories is controlled by a ventral–dorsal nuclear gradient of the *dorsal* gene product, and the establishment of the gradient is under the control of substances preformed during oogenesis and revealed as maternal-effect mutations. The gradient probably activates zygotic genes, at least one for each territory, which are identifiable as zygotic dorsoventral pattern mutations.

Maternal control

At least 15 maternal effect genes have been identified by the type of exhaustive screen described earlier (see gene table). The loss of function phenotypes consist for most of them in the formation of a folded tube of larval cuticle bearing the fine hairs typical of the dorsal epidermis, but lacking structures normally derived from the lateral or ventral territories of the blastoderm (Anderson, 1987). The prototype mutation of this dorsalizing class was called *dorsal* (see Fig. 7.8(*b*)). Several of the genes have alleles with different degrees of function and these can be arranged in hypomorphic series in which more and more structures are lost from the ventral side until in the amorphic embryos only the symmetrical tubes of dorsal cuticle remain. Since the amnioserosa, and not the dorsal epidermis, is the most dorsal territory in the blastoderm fate map, one might legitimately wonder why the amorphic phenotype does not consist of amnioserosa, but it does not, *dorsal* also has some dominant haploinsufficient alleles and two of the other genes (*Toll* and *easter*) have dominant, gain of function, alleles with a ventralizing effect. There are also three genes, *gurken*, *torpedo* and *cactus*, whose loss of function phenotype is ventralization.

The dorsoventral system has been studied by various types of genetic and embryological experiment. The 15 genes can to some extent be arranged in a temporal series by making double mutants of which one has a dorsalizing and the other a ventralizing effect. Whichever predominates is assumed to act later in the developmental program (eg, Anderson, Jürgens & Nüsslein-Volhard *et al.*, 1985). Some idea of the time of action can also be gained by using temperature sensitive alleles and shifting between the permissive and non-permissive temperature.

It can then be asked whether a particular gene is required in the germ line (the oocyte itself and the nurse cells) or in the soma (in particular the ovarian follicle cells). Chimaeric embryos are created by grafting, say, mutant pole cells into normal embryos. When these are grown up the females will produce offspring that are mutant if the gene was required in the germ line, or normal if it was required in the soma. Studies of this sort have shown that some are required in the germ line and some in the soma (eg, Schüpbach, 1987).

With regard to establishing the position within the early embryo at which gene products are found, and the position at which they are required, cytoplasmic transplantations have been most useful. Most of the dorsal group mutants can be 'rescued' by the injection of small amounts of cytoplasm from wild type embryos before the pole cell stage (Anderson & Nüsslein-Volhard 1984). The appearance of Filzkörper, which normally arise at a dorsolateral posterior position, is a criterion for weak rescue, and the appearance of ventral denticles is a criterion for strong rescue.

These various types of experiment have led to the following story, in

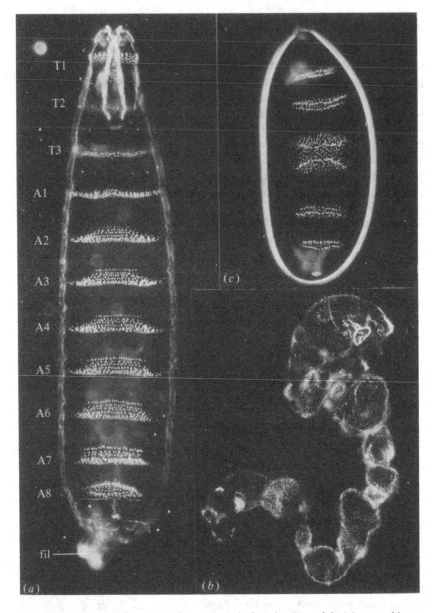

Fig. 7.8. Larval cuticle preparations: normal, *dorsal⁻*, *Bicaudal⁻* (*a*) normal larva of *Drosophila*. T: thoracic segments, A: abdominal segments, fil: Filzkörper. (*b*) embryo produced by a *dorsal⁻* female, consisting of a featureless tube of dorsal epidermis. (*c*) embryo produced by a *Bicaudal⁻* female, consisting of a mirror symmetrical duplication of the most posterior segments. (Photographs kindly provided by Dr C. Nüsslein-Volhard.)

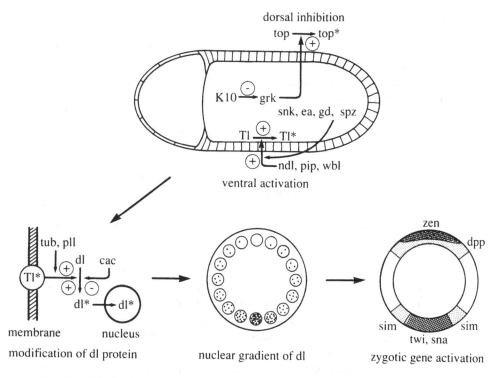

Fig. 7.9. Mechanism of dorsoventral specification. At the top are indicated the events in the egg chamber, and below the formation of the nuclear gradient of *dorsal* protein and the activation of the zygotic genes.

which 'mutant' will indicate loss of function unless otherwise stated (Fig. 7.9). The first known gene to act is called K10. The mutant has a dorsalizing effect on both the embryo and the structures around it which are secreted by the follicle cells. It codes for a DNA binding protein which is found in the oocyte nucleus around the beginning of yolk accumulation (Prost *et al.*, 1988). K10 probably inhibits the expression of *gurken* (*grk*), which is also required in the oocyte but gives the opposite phenotype to K10, producing ventralized embryos and eggshells. *gurken* is therefore presumed to be responsible for making an inhibitor of ventralization on the dorsal side. Next comes *torpedo* (*tor*), which is the *Drosophila* epidermal growth factor receptor and which is required in the follicle cells rather than the oocyte (Schüpbach, 1987). This has a similar ventralizing phenotype to *gurken* and so is presumably also involved in the inhibitory process on the dorsal side. Next come a group of genes (*pipe, windbeutel, nudel*) which are required in the follicle cells and whose mutants have dorsalizing effects. One of these is presumed to code for an extracellular factor which can be extracted from the perivitelline fluid and induces ventral structures if injected into the perivitelline space of eggs mutant for one of the preceding dorsalizing genes. This factor is activated by proteases which are products of the genes *snake* (*snk*) and *easter* (*ea*), and itself activates a cell surface receptor on the oocyte surface which is the product of the *Toll* (*Tl*) gene.

The rescue experiments showed that a special position in the system was held by *Toll*. In general the dorsoventral polarity of the hosts is not altered by rescue and it can be seen that the rescued ventral structures always arise on the ventral side of the hosts. But in *Toll*⁻ rescue the new ventral side appears instead at the position of the injection. So *Toll* activity can impart polarity to the egg (Anderson, Bokla & Nüsslein-Volhard, 1985). It is known that the *Toll* gene product itself has the characteristics of a cell membrane protein and is uniformly distributed (Hashimoto *et al.*, 1988), so presumably it is *Toll* product in an activated state that is responsible for rescue. However the rescuing activity from the wild type eggs is present both on the dorsal and the ventral side! A possible explanation for this is that in the *Toll*⁻ mutant there is an excess of the ligand in the perivitelline space. If functional, but unactivated *Toll* protein is injected into such an egg, it will all be activated by the excess of ligand, and hence the ventral structures will appear at the site of the injection. As mentioned above, *Toll* has some gain of function alleles producing an expansion of the normal ventral territory and these presumably represent receptors in a state of constitutive activation.

The activation of *Toll* on the ventral side commences shortly before, or even perhaps after, egg maturation and the subsequent events occur during the embryonic cleavage stages. *Dorsal* was the first dorsalizing mutant to be discovered and now seems to be the final step in the causal chain. It is a nuclear protein homologous to the mammalian *c-rel* oncogene (Steward *et al.*, 1988). Its mRNA is uniformly distributed in the oocyte and the protein is synthesized after fertilization. Its distribution is initially uniform as well but during the syncytial blastoderm stage it enters the nuclei preferentially on the ventral side, forming a ventral–dorsal gradient of nuclear protein (eg, Roth *et al.*, 1989). The entry to the nuclei depends on the proximity of activated *Toll*, as it does not occur in *Toll*⁻ mutants and occurs all over in *Toll* dominant ventralizing mutants. It probably involves some modification of the *dorsal* protein, which depends on the product of other late acting genes, *tube* (*tub*) and *pelle* (*pll*) and is antagonized by *cactus* (*cac*), which is the third gene of the group having a loss of function ventralizing phenotype.

Zygotic control

There are also zygotic genes whose mutants show dorsalization and ventralization, and these are now known to be regulated by the nuclear gradient of *dorsal*. Amorphic mutants of *twist* (*twi*) and *snail* (*sna*) have no mesoderm. *twist* has been cloned and the mRNA appears in the midventral region in the syncytial blastoderm (Thisse *et al.*, 1987). By gastrulation it is confined to the mesoderm of the ventral furrow and it fades by the extended germ band. *twist* is not expressed in dorsalized maternal mutants and is expressed all over in ventralized maternal mutants suggesting that it is activated, directly or indirectly, by *dorsal.*

Many zygotic genes affect neuronal development in the neurogenic regions flanking the ventral furrow. Of these some are early acting and seem to define a preneurogenic belt rather than being concerned with neuronal differentiation itself. They are possible candidates for regulation by *dorsal.* For example, *single minded* (*sim*), whose mutant phenotype is the loss of a particular subset of neurons, comes on in this region at the cellular blastoderm stage (Thomas *et al.*, 1988).

Another group of zygotic genes have loss of function alleles causing reductions in the amnioserosa and dorsal epidermis. Of these, *decapentaplegic* (*dpp*) is the most severe and *zerknüllt* (*zen*) somewhat less severe. *dpp* is a large and complex gene with many effects on later developmental stages. It has been cloned and shown to encode a protein belonging to the TGF β superfamily of cytokines. It is expressed on the dorsal side in the syncytial blastoderm, later falling in the amnioserosa and resolving into two longitudinal stripes in the dorsolateral epidermis (St Johnston & Gelbart, 1987).

zen is found in the *Antennapedia* complex and is a nuclear protein containing a homeobox. It is expressed on the dorsal side in the syncytial blastoderm and later becomes confined to the amnioserosa (Doyle *et al.*, 1986). All the maternal dorsalizing mutants show uniform expression of *zen* suggesting that it is normally repressed in the ventrolateral region by *dorsal.* The *Toll* gain of function mutant and the *cactus* ventralizing loss of function mutation both repress *zen* all over except for small patches at the two poles, and these are repressed by the maternal terminal mutant *torso* (see below). So it seems that the initial expression of *zen* is controlled by maternal genes, with activation by the terminal system and repression by the dorsal system. Later details of its expression pattern can also be affected by the other zygotic genes such as *dpp.*

Lessons from the dorsoventral system

The method of dissection of the dorsoventral system exemplifies the power of the three-pronged approach using developmental genetics, microsurgery and molecular cloning, and the same principles have governed the analysis of the more complex anteroposterior system. Although the role of the zygotic genes is not as well understood as in the anteroposterior system, it does seem as though each dorsoventral level is coded by expression of one or more of the zygotic genes and that these control, directly or indirectly, the subsequent tissue movements and cell differentiation events. The dorsoventral zygotic gene products are, then, prime candidates for epigenetic coding factors. Their expression is controlled, although perhaps not directly, by the nuclear gradient of the *dorsal* protein. Because a smooth gradient of *dorsal* leads to several discrete territories of zygotic gene expression, it is reasonable to regard *dorsal* as a morphogen although the gradient is set up by a quite different mechanism from those envisaged in Chapter 3. The polarity and steepness of the *dorsal* gradient is, in turn, controlled by the differential activation of the *Toll* protein by an extracellular ligand released from the follicle cells of the egg chamber. The activated form of *Toll* can reasonably be regarded as a cytoplasmic determinant, although again it is not quite what was expected since it does not directly regulate gene expression, but modifies the *dorsal* protein so that it can enter the nuclei.

THE ANTEROPOSTERIOR SYSTEM

Maternal information: the anterior system

Theorists are constantly having to remind some molecular biologists that one concentration gradient with one set of responses cannot specify a two- or three-dimensional pattern, and in *Drosophila* they can cite evidence to prove it. In *Drosophila* the blastoderm is an ellipsoidal monolayer and the embryo can be regarded as effectively two dimensional since the internal vitellophages do not contribute to the body plan. It turns out, as predicted, that the specification of structures along the anteroposterior axis is controlled by a system which is largely independent from that controlling the dorsoventral pattern. Once again, it was the brilliant combination of genetic analysis, experimental embryology and molecular biology led by Nüsslein-Volhard and her colleagues which has led to the solution. The basic mechanism is summarized in Fig. 7.11.

   The anterior end of the egg becomes organized in response to a gradient of the protein product of the *bicoid* (*bcd*) gene. *bicoid* is a maternal effect gene, mutations of which causes a deletion of the head and thorax with substitution of a second telson at the anterior end (Frohnhöfer & Nüsslein-Volhard, 1987). Temperature shift experiments show that, although it is the genotype of the mother and not the embryo that is important, the time of activity of the gene is during early embryonic life. This suggested that perhaps transcription of *bicoid* occurred during oogenesis and translation after fertilization.

   A number of microsurgical experiments have been done on this system which show exactly the properties expected if there is a morphogen source at the anterior end of the egg (Frohnhöfer & Nüsslein-Volhard, 1986). A phenocopy of the *bicoid* mutation can be produced by pricking the egg at the anterior end and extruding about 5% of the cytoplasm, but not by pricking elsewhere. Conversely, *bicoid⁻* eggs can be 'rescued' toward a

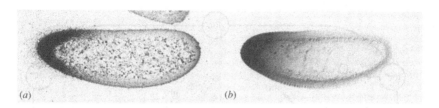

Fig. 7.10. (*a*) *In situ* hybridization of *bicoid* RNA, (*b*) antibody stain of *bicoid* protein. Note that the protein extends much further than the RNA and is graded from the anterior. ((*a*) kindly provided by Dr P. Ingham, (*b*) by Dr W. Driever.)

normal phenotype by injection of cytoplasm taken from wild type eggs. To be effective the cytoplasm must come from the anterior 5% of the egg. The position at which the injection is made determines the position of the induced anterior end, so, for example, injection of wild type cytoplasm to a central position of a *bicoid⁻* egg produces a head at the injection site, flanked by two thoraxes in mirror symmetrical arrangement.

Cloning and sequencing of the *bicoid* gene suggested that it was a genetic regulatory protein since it contained a homeobox. *In situ* hybridizations showed that the mRNA was transcribed in the oocyte and the nurse cells and became localized at the anterior end of the oocyte (Berleth *et al.*, 1988; Fig. 7.10(*a*)). In females mutant for two other genes with similar but weaker effects, *exuperantia* or *swallow*, the *bicoid* gene is transcribed as usual but the message does not become localized nearly as sharply. Study of the *bicoid* protein by antibody staining showed that it was synthesized during the period 1–4 hours of embryonic development and formed an exponential concentration gradient from anterior to posterior (Driever & Nüsslein-Volhard, 1988*a*; Fig. 7.10(*b*)). The localized mRNA is the source and, as the protein appears to have a short half life, the remainder of the embryo is the sink. The level of protein can be manipulated by changing the number of active copies of the gene in the female and this displaces in the expected directions markers such as the cephalic furrow or stripes of pair-rule gene expression (Driever & Nüsslein-Volhard, 1988*b*). Except for the fact that the gradient is set up across a syncytium instead of a sheet of cells, the behaviour is exactly as predicted by the LSDS model described in Chapter 3. The function of the *bicoid* protein gradient is to regulate the zygotic expression of the gap genes *hunchback* and *Krüppel* (Fig. 7.11), and *bicoid* also turns on *orthodenticle* (*otd*) and perhaps other genes which are required to make a head from the anterior *torso* domain.

Maternal information: the posterior system

At the posterior end of the *Drosophila* egg lies a special region of cytoplasm called the pole plasm and the evidence that this contains a germ cell determinant has been considered above. Nowadays the pole plasm has

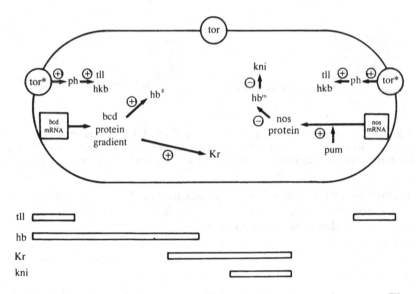

Fig. 7.11. Roles of the maternal genes which control anteroposterior pattern. The *bicoid* system leads to activation of *hb* and *Kr* by the *bcd* protein gradient. The *nanos* system leads to activation of *kni* by double repression. The terminal system leads to activation of *tll* and *hkb*.

acquired a further significance as the source of an abdominal determinant, the product of the maternal effect gene *nanos* (*nos*).

Pricking of the posterior pole and extrusion of 5–10% egg volume causes defects in the larva; not in the telson as might be expected, but in the abdomen whose prospective region is not in the extreme posterior but around 50–20% EL (Nüsslein-Volhard, Frohnhöfer & Lehmann, 1987). *nanos⁻* eggs also lose the abdomen but retain the telson. This mutant phenotype can be rescued towards normality by injection at early cleavage stages of cytoplasm from wild type eggs. Rescue only works if the cytoplasm comes from the extreme posterior of the early cleavage embryo, or from the pole cells themselves. But it is necessary to inject the cytoplasm not into the pole of the host embryo but into the prospective abdominal region (Lehmann & Nüsslein-Volhard, 1986). There are a number of other maternal effect mutants causing loss of the abdomen. Of these, *pumilio* (*pum*), like *nanos*, shows an abdominal defect with normal pole plasm, while the others, such as *oskar*, show an abdominal defect with absence of pole plasm. Their roles can be assessed to some degree from their complementation ability. Embryos from *pumilio* mothers can rescue themselves if pole plasm is transplanted to the abdominal region, so this gene is thought to be involved in transport of the active factor (Lehmann & Nüsslein-Volhard, 1987(*a*)). Cytoplasm from nurse cells of any of the mutants except *nanos* will rescue any of the other mutants. For this reason *nanos* is thought to code for the factor required for abdomen formation and the others for components of the pole plasm required to localize the *nanos* mRNA. Study of the localization of *nanos* mRNA shows that it is made in the nurse cells and localized to the pole plasm and pole cells of the early embryo.

It is now established that the posterior system works in a somewhat different way from the anterior system (Fig. 7.11). The *bicoid* protein is a real morphogen, evoking different responses at different concentrations, so its source region behaves as an organizer of pattern. The *nanos* protein is a factor which has a single function: to allow the transcription of the zygotic gap gene *knirps* (*kni*), and, in normal development, to ensure that this occurs in the prospective abdominal region. The mechanism of action depends on another gene, *hunchback* (*hb*). This is a zygotic gap gene, but is also active in oogenesis such that the egg is normally filled with a uniform concentration of maternally derived *hunchback* mRNA. Translation of this mRNA commences in early cleavage but is inhibited in the posterior by the *nanos* protein. *hunchback* protein inhibits the transcription of *knirps* so normally *knirps* becomes turned on in the posterior but not in the anterior by this mechanism. Since activation of *knirps* depends on a double inhibition it follows that the *nanos* protein would not be necessary in the absence of the maternal *hunchback* message, and indeed this is the case: embryos from double mutant *hunchback⁻/nanos⁻* mothers are normal (eg, Hülskamp *et al.*, 1989).

Bicaudal

Embryos with a mirror symmetrical double posterior phenotype have been known to originate from maternal mutations at several loci. Dominant mutations in *Bicaudal-D* are the most effective, giving symmetrical double posterior body plans joined by a plane of mirror symmetry at some level within the abdomen (Mohler & Wieschaus, 1986; Fig. 7.8(*c*)). They are not quite symmetrical as only the original posterior end forms pole cells. The gene has been cloned and codes for a cytoskeletal protein. The mRNA is normally uniformly distributed in the oocyte, but in the mutant becomes more concentrated in the anterior. It appears that the double posterior phenotype arises because both maternal *hunchback* and *bicoid* mRNAs are inactivated, hence zygotic *hunchback* fails to be activated in the anterior, and *knirps* is activated in the anterior as well as the posterior (Wharton & Struhl, 1989).

Maternal information: the terminal system

Mutations in a third group of maternal effect genes affecting pattern revealed a function which had not previously been suspected. The embryos resulting from loss of function mutations show defects at both termini with a normal pattern in between. In the head the labrum and parts of the skeleton are missing and at the posterior end segment A8 and the telson are missing. The gene *torso* (*tor*) also has a gain of function allele which causes substantial suppression of segmentation in the thorax and abdomen (Klinger *et al.*, 1988). Cloning and sequencing of the *torso* gene revealed that it was a cell surface receptor of the tyrosine kinase class, comparable to certain vertebrate growth factor receptors (Sprengler *et al.*, 1989). As with *Toll*, expression is uniform in the oocyte and so it is thought that the receptor becomes activated at the termini, the ligand probably being the product of another terminal gene: *torsolike* (*tsl*), which is required not in the germ line but in the somatic tissues of the female. In fact, the follicle cells at the termini are somewhat different morphologically from those surrounding the central part of the oocyte, and are called 'border cells'.

Another step in the pathway is thought to be represented by the product of *polehole* which is a serine–threonine kinase, also expressed uniformly in the oocyte and early embryo (Ambrosio, Mahowald & Perrimon, 1989). The final goal is probably to activate the zygotic gap gene *tailless* (*tll*), and *huckerbein* (*hkb*) (Fig. 7.11). *tll* has a similar but less severe phenotype to the maternal effect terminal group and furthermore the gain of function phenotype of *torso* (defects in thorax and abdomen) are suppressed in *tll*⁻ embryos. The gain of function phenotype is ascribed to an altered *torso* product which remains on at all times without needing to be activated by its ligand. The terminal system has a mechanism remarkably similar to the dorsoventral system, and once again the activated receptor protein can be regarded as a cytoplasmic determinant.

The gap genes

The system of morphogens and determinants bequeathed by the mother becomes elaborated into more and more complex patterns of gene activity by successive levels of the developmental hierarchy. The first zygotic level is made up by the *gap genes*, so called because mutant embryos have patterns bearing gaps which lack the obvious periodicity of the pair-rule or segment polarity mutants. The best characterized gap genes are *hunchback* (*hb*), *Krüppel* (*Kr*) and *knirps* (*kni*).

hunchback

Embryos homozygous for null alleles of *hunchback* have a large anterior gap which removes the labium and thorax, and a small posterior gap comprising A7p and A8a (parasegment 13). It was mentioned above that transcription of *hunchback* occurs during oogenesis. Since *hunchback* is a zygotic lethal, it is not possible to grow up homozygous females to look for a maternal effect. However this can be done in females which have received

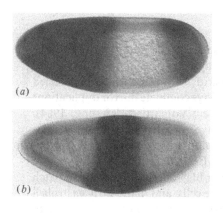
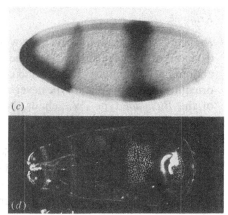

Fig. 7.12. Gap genes. (*a*) *hunchback* protein visualized by antibody staining, (*b*) *Krüppel* protein, (*c*) *in situ* hybridization of *knirps* mRNA, (*d*) cuticle preparation of *knirps⁻* embryo showing a large defect in the abdomen. All anterior to left. (*a*),(*b*) kindly provided by Drs U. Gaul and Dr H. Jäckle; (*c*) by Dr M. Rothe and H. Jäckle, (*d*) by Dr P. Ingham.)

hunchback⁻ pole cell grafts in early embryonic life. It turns out that there is no maternal effect if a good copy of the gene is introduced by the sperm, but if not then a different phenotype is obtained in which the telson, A8, the thorax and gnathal segments are all missing, and the abdomen is transformed into a double posterior duplication, reminiscent of *Bicaudal* (Lehmann & Nüsslein-Volhard, 1987*b*). As we shall see this is due to activation of *knirps* in the anterior.

Cloning and sequencing of *hunchback* revealed that it was a protein with the characteristic 'zinc finger' structure first discovered in the *Xenopus* transcription factor TFIIIA (Tautz *et al.*, 1987). This is, in fact, also true of *Krüppel* and *knirps*, and is consistent with the idea that all of the gap genes are regulators of other genes further down in the hierarchy. As mentioned above, transcription during oogenesis leaves mRNA uniformly distributed in the egg (Tautz, 1988). The translation of this is antagonized in the posterior half by the *nanos* protein and so an anterior to posterior gradient of protein is set up during cleavage. Also, the maternal mRNA decays more rapidly at the posterior end. Zygotic transcription commences in the syncytial blastoderm in a domain occupying the anterior half of the embryo, and at cellular blastoderm also in a posterior stripe about 20–10%EL (Fig. 7.12(*a*)). The anterior, but not the posterior domain is activated directly by the *bicoid* protein (Driever & Nüsslein-Volhard, 1989). In embryos from *bicoid⁻* mothers there is no anterior domain, and the posterior domain is duplicated in the anterior. The posterior domain depends in some way on the terminal system as it does not appear in *torso⁻* eggs.

Krüppel

Krüppel is an entirely zygotic gene whose amorphic mutants have a large deletion of the central part of the body, comprising the thorax and abdominal segments 1–5. A second copy of A6 is often found in inverted orientation, but the phenotype never displays complete mirror duplications of the *Bicaudal* type (Wieschaus, Nüsslein-Volhard & Kluding, 1984). Cloning and sequencing of the gene have revealed that it is a genetic regulator of the zinc finger class. Expression commences in the syncytial blastoderm as a band from about 60–50%EL, later extending somewhat to the posterior (Fig. 7.12(*b*)). By the cellular blastoderm there is also a patch at the posterior end, not involving the pole cells, and in the early gastrula a small patch anterior to the cephalic furrow (Knipple *et al.*, 1985). The gap phenotype corresponds to, but is wider than, the main central band. The regulation of *Krüppel* has been studied by looking at the mRNA in a variety of other mutants and at least the central domain appears to arise from a balance between activation by a low level of *bicoid* protein, and repression by a high level of *hunchback* protein.

knirps

Amorphic mutants of *knirps* are similar to the maternal posterior group, having abdominal segments 1–7 replaced by a single large abdominal segment of uncertain identity. Cloning and sequencing of the gene revealed that it is another finger-type DNA binding protein, this time belonging to the family of hormone binding proteins which in vertebrates are receptors for steroids, thyroid hormones or retinoic acid (Nauber *et al.*, 1988). The expression pattern shows a band between 45–30%EL, coming on at syncytial blastoderm and overlapping the main *Krüppel* domain. By cellular blastoderm this is joined by a patch in the ventral part of the head and another narrow stripe in the head (Rothe, Nauber & Jäckle 1989; Fig. 7.12(*c*)). As for *Krüppel*, the gap phenotype corresponds to, but is wider than, the principal band of expression (Fig. 7.12(*d*)). Examination of *knirps* expression in various other mutants leads to the conclusion that there is an element of constitutive expression and that the position of the main band in normal development is regulated by repression due to *hunchback*, activation due to *Krüppel* and repression due to *tailless*. There exists a second gene, *knirps-related*, which has a similar expression pattern to *knirps*.

Other gap genes

tailless has already been mentioned several times. It is a zygotic gene which is normally thought to be activated by the maternal terminal system since the mutant will suppress the gain of function phenotype of *torso*. However, its phenotype is not as extensive as that of the maternal group. Defects are found at the two termini of the body, but in the head they are confined to the cephalopharyngeal skeleton, the labrum being present, and in the posterior segments A8–10 and the Malpighian tubules are missing but the posterior midgut and proctodeum are present (Strecker *et al.*, 1986). The *tll* gene is now known to belong to the steroid receptor superfamily and to be expressed at the termini (Pignoni *et al.*, 1990). Another gene probably responding to the terminal system is *huckerbein*, whose mutants show those defects due to *torso* which are not shown in *tailless* mutants.

giant mutants have defects in the anterior thorax and in the abdomen at A5–A7. *giant* has been cloned and the expression pattern shows that transcription starts in the syncytial blastoderm in two zones, an anterior zone about 80–60%EL and a posterior zone about 33–0%EL. By the cellular blastoderm the posterior band is fading and the anterior one has resolved into three stripes (Mohler, Eldon & Pirrotta, 1989).

orthodenticle mutants have defects in the head. The gene contains a homeobox and is expressed in the anterior of the syncytial blastoderm and in a 90%–70% EL band in the cellular blastoderm (Finkelstein & Perrimon, 1990).

caudal

Although not normally counted as a gap gene, this is probably the most logical place to discuss *caudal* (*cad*). Unlike most of the other developmentally significant genes this was not identified initially from a mutant phenotype but was cloned because it contained a homeobox. When its expression pattern was studied it was found that the egg started with a uniform distribution of mRNA synthesized during oogenesis and at the syncytial blastoderm stage mRNA was differentially lost resulting in a posterior to anterior gradient. This was the first concentration gradient to be visualized in the *Drosophila* egg and so caused some excitement (Mlodzik, Fjose & Gehring, 1985). The gradient decays and zygotic expression leads to a posterior band of expression from 19–13%EL, later in the proctodeum and posterior midgut.

The formation of the gradient is probably regulated by *bicoid* since the protein distribution is uniform in *bicoid⁻* embryos. When it was possible to make mutants (Macdonald & Struhl, 1986) it was found that *caudal⁻* embryos were almost normal but with slight defects in structures derived from A10, around the level of the zygotic posterior band. *caudal⁻* embryos from *caudal⁻* mothers (actually germ cell chimaeras made by X-ray induced mitotic recombination) had severe disruption of abdominal segmentation and some head structures in A8. Examination of the expression pattern of the pair-rule gene *fushi tarazu* (*ftz*) in such mutants showed suppression of stripes 3, 5, 6 and 7, which is probably sufficient to account for the phenotype.

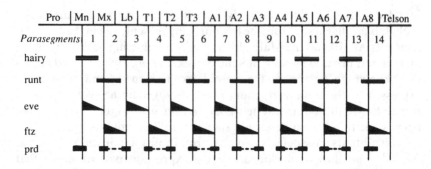

Fig. 7.13. Expression patterns of pair-rule genes during cellularization.

The pair-rule system

The pair-rule genes were discovered in 1980 as mutants which caused deletions in every other segment of the body (Nüsslein-Volhard & Wieschaus, 1980). Like the terminal genes, this class of functions had not previously been suspected from direct embryological studies. The majority of pair-rule genes have now been cloned and their expression patterns are known. It has become clear that they function as a layer of interpretation in the developmental hierarchy between the gap genes and the segment polarity genes. They also have a role, along with the gap genes, in controlling the expression of homeotic selector genes and aligning their domains with the segmental repeating pattern. The pair-rule genes cloned at the time of writing are described here, and a diagram of the principal expression patterns is shown in Fig. 7.13, although it should be borne in mind that all the expression patterns are dynamic. There are also three others: *odd skipped, odd paired* and *sloppy paired*, which are known and are listed in the gene table. The mutant phenotypes are not due to death of cells but to incorrect assignment of codings at lower levels of the genetic hierarchy, and it is because of this that the amorphic phenotypes are often more severe than deletion of every other segment.

hairy

The expression pattern for *hairy* (*h*) was published by Ingham *et al.*, (1985). Transcription starts with a uniform band of expression in the syncytial blastoderm, which has resolved into seven stripes plus a dorsal head patch by cellularization. The stripes later decay during germ band extension, with later expression in cells surrounding the tracheal pits. The protein expression pattern is very similar (Fig. 7.14(*b*)). The seven blastoderm stripes correspond well with the regions deleted by weak alleles (Fig. 7.14(*a*)) but the amorphic phenotype consists of more severe deletions resulting in a lawn of denticles. The gene structure shows a large regulatory region 5' to the coding sequence and by making many different deletions of DNA it has

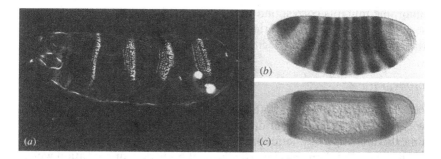

Fig. 7.14. Pair-rule genes. (*a*) cuticle preparation of *hairy⁻* larva showing defects with a 2-segment periodicity, (*b*) *hairy* protein visualized by antibody staining. (*c*) *eve* protein expression in the absence of maternal anterior and posterior systems (*vas⁻exu⁻*). All anterior to left. ((*a*) kindly provided by Dr P. Ingham, (*b*) by Dr M. Lardelli, (*c*) by Drs U. Gaul and H. Jäckle.)

been shown that individual stripes of expression depend on the presence of particular sequences (Howard, Ingham & Rushlow, 1988).

even-skipped

Transcription of *even-skipped* (*eve*) starts as a broad band in the anterior, this splits and further bands arise posteriorly to give seven stripes by cellularization. The stripes then narrow and sharpen anteriorly and during gastrulation split to give a single segmental repeat before fading. Later expression occurs in the posterior midgut and in some neurons (Harding *et al.*, 1986). The protein expression pattern is similar to the message. At the stage of the sharp anterior boundary to each stripe, the *eve* domains coincide exactly with the odd-numbered parasegment primordia. Weak alleles of *eve* show deletion of the odd numbered parasegments in the manner expected but the amorphic phenotype is more severe and consists of an abolition of all segmentation with the formation of a smooth lawn of denticles on the ventral surface. In case readers are wondering why *even-skipped* deletes *odd* parasegments, it should be explained that the original classification was based on visible segments in the abdomen. As for *hairy*, there is a large regulatory region 5' of the gene. This contains a site responsible for autoactivation and many other different sites which bind different gap gene products and could therefore control the expression of single stripes or pairs of stripes (eg, Stanojevic *et al.*, 1989).

runt

Transcription of *runt* starts as a uniform band then gaps develop to establish a seven-stripe pattern by nuclear elongation. At cellularization there is a change to a single segment periodicity which persists throughout germ band extension. There are also patches of expression in the dorsal head and the proctodeum (Gergen & Butler, 1988). The regions deleted in amorphic mutants correspond to the regions of expression at the earlier, seven-stripe, stage although there are, in addition, mirror duplications of the remaining parts of the segment.

fushi tarazu

fushi tarazu (ftz) is famous as the first gene whose stripy expression pattern was visualized by *in situ* hybridization (Hafen, Kuroiwa *et al.*, 1984). At the time this came as a dramatic confirmation of the inference from the mutant phenotypes that there was indeed a level of body plan organization which had a two-segment periodicity. Like the others an initially uniform expression pattern breaks up into seven stripes which fade during germ band extension. The mutant phenotype corresponds to this pattern. Later expression occurs in the central nervous system and hindgut.

Three control elements have been found to lie in the DNA 5' to the gene (Hiromi & Gehring, 1987). Nearest the gene is a 'zebra' element, which is required for periodic expression, then there is a neurogenic region, apparently used for the expression in the CNS, and lastly an 'upstream' element needed for autoregulation (see below).

paired

Transcription starts in the syncytial blastoderm with an anterior band. This divides and is joined by more posterior bands to give seven stripes each about 5–6 cells wide by cellularization. Then bands 2–7 split by loss of expression in the central two cells of each stripe, and another forms in the posterior to give a total of 14. The terminal bands remain but the rest fade during germ band extension. The mutant phenotype corresponds approximately to the deletion of those parasegment boundaries coincident with the posterior stripe of each pair (Kilchherr *et al.*, 1986).

Regulation of pair-rule gene expression

A large amount of work has been done in which the expression pattern of one gene is examined in a background of the absence of one or more others (Fig. 7.14(*c*)). As discussed above there are some problems in the interpretation of results, particularly that redundancy of function cannot be detected, and it is hard to tell whether effects are direct or indirect. However two broad generalizations have emerged which are unlikely to change with more detailed knowledge. Firstly the pair-rule genes themselves can be divided into 'primary' and 'secondary' groups. The primary genes, *hairy*, *eve* and *runt*, are required for the correct expression of the secondary genes but are themselves independently regulated. Secondly, the basic primary pair-rule patterns are set up in response to maternal and gap genes, and are only refined by interaction with other pair-rule genes. This means that the pair-rule stage of organization marks the point at which periodicity emerges from aperiodicity.

When it was first found that the pair-rule expression patterns often started uniform and then developed into periodic patterns it was thought by many that some sort of reaction–diffusion process must be operating which generated the stripes by autocatalysis and inhibition (see Figs 3.16 and 3.17). However, this has not been borne out by further investigation. As mentioned above, a study of the regulatory regions of the primary pair-rule genes *hairy* and *eve* has shown that they both possess a complex regulatory region in which there are many binding sites for positive and negative transcription factors. When parts of these regions are deleted and the genes re-introduced into embryos, then individual stripes, or pairs of stripes, are lost. This allows us to deduce that, in general, individual stripes in the normal expression patterns of pair-rule genes are independently regulated.

The gap genes certainly have a central role in control of the pair-rule patterns. There are suppressions, fusions or broadenings of stripes of any of the pair-rule genes in a mutant background of any of the gap genes, including *giant* and *tailless*. Some of these controls are now understood at the molecular level. For example, for *eve* it has been shown that *Krüppel* and *hunchback* proteins both bind to promoters within the regulatory region which are necessary for the formation of the second and third stripes (Stanojevic *et al.*, 1989). For *hairy* it has been shown that *Krüppel* and *knirps* proteins bind strongly to those control regions normally active at low levels and weakly to those control regions normally active at high levels (Pankratz *et al.*, 1990). There are also some effects from the maternal systems on the pair-rule genes which are probably not mediated via the gap genes.

In many cases it seems that a local gradient of a gap product can elicit a stripe rather than a step, in other words the same substance seems to be turning on the pair-rule gene at one concentration and turning it off at a higher concentration. This is also compatible with the observation that stripes are often lost in pairs in maternal or gap gene mutants since one would expect one stripe for each side of a bell shaped concentration profile of gap product (Fig. 7.14(*c*)). In molecular terms this may mean that at regulatory sites which can bind many molecules of an effector, binding of a certain number leads to activation while binding of more leads to repression.

To code for seven stripes the minimum requirement would be four gap genes each expressed in a single region, and this requirement is greatly exceeded in reality since there are actually at least eight zygotic gap and terminal genes all of which have expression patterns of more than one domain. The different phasing of different pair-rule genes may reflect rather similar regulation but with different thresholds of response for each effector, or of course different combinations of gap gene products may also be necessary, for example, *caudal* has a strong effect on *ftz* but probably does not have much effect on the others.

Interaction between pair-rule genes

Although periodic patterns of primary pair-rule gene expression are set up under the regulation of the gap and maternal genes they require further interactions to achieve the final sharpness, regularity and persistence of pattern. All of the evidence for this level of regulation has come from looking at the expression of one gene in a background mutant for another, so the directness of the effects is necessarily uncertain. Also the majority of the data so far concerns *hairy*, *runt*, *eve* and *ftz*, for which reagents have been available for some years (Ingham & Gergen, 1988). The story will remain incomplete until the others have been brought in as well. The important thing that these studies have established is the distinction between the primary and secondary pair-rule genes. The periodic pattern of the primaries arises from regulation in each region of the embryo by the overlapping combinations of gap proteins, while the periodic pattern of the secondaries is mainly due to periodic input from the primaries.

We shall consider just one example which is the regulation of *ftz*. *ftz* is thought to be repressed by *hairy* for three reasons: first, the normal expression patterns of the two genes are nearly out of phase. Secondly, the *ftz* stripes do not resolve properly in a *hairy⁻* background, wheras *hairy* is normal in a *ftz⁻* background (Howard & Ingham, 1986). Thirdly, uniform overexpression of *hairy* under the control of a heat shock promoter represses *ftz* all over (Ish Horowicz & Pinchin, 1987). *runt* is perhaps an activator of *ftz* since in its absence the *ftz* stripes decay earlier than usual. *eve* has complex effects, being apparently necessary for formation of the first *ftz* stripe, but acting as a repressor elsewhere. The *caudal* protein binds to a part of the zebra element required for expression of the posterior stripes (Dearolf *et al.*, 1989) and stripes 4–7 do not come on in *caudal⁻* embryos. In addition *ftz* is autocatalytic, as may be shown by the fact that reporter constructs can be activated by *ftz* protein via the upstream promoter (Hiromi & Gehring, 1987). So the seven stripes of *ftz* expression are achieved by activation by *caudal* and perhaps other gap genes, by *hairy* repression, *runt* activation, *eve* activation and repression and autoactivation. Probably something else is also required since the final phasing of the *ftz* stripes is such that the posterior cell of a *ftz* stripe is also the anterior cell of a *hairy* stripe. These putative relationships are shown in Fig. 7.15 and will undoubtedly become even more complex as more data is gathered.

The segment polarity genes

Unlike the pair-rule class, the discovery of the segment polarity genes was foreshadowed by embryological work. Studies of the regeneration of the cuticle in the larvae of hemimetabolous insects had led to the idea that there was a repeating gradient or repeating series of discrete codings which underlay the visible pattern of segmentation (Lawrence, 1973). Also work

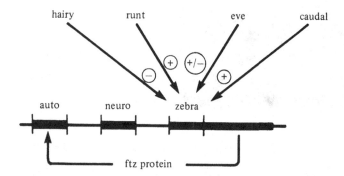

Fig. 7.15. Regulation of the *ftz* gene by the primary pair-rule genes and by *caudal*.

in the 1970s on the developmental genetics of the imaginal discs, visualized as changes in adult anatomy, had brought forth the compartment concept. As discussed above it had been established that clones did not cross between the anterior and posterior parts of each imaginal disc and that maintenance of this barrier depended on the activity of the *engrailed* gene in the posterior (Garcia Bellido *et al.*, 1975). When the segment polarity genes were discovered in the screens of Nüsslein-Volhard and Wieschaus it became clear that *engrailed* was also a critical gene for the organization of the embryonic segmental repeat.

Although work on the segment polarity genes has proceeded less far than that on the gap and pair-rule classes, it is clear that they represent the final stage in the determination of the repeating pattern. The parasegmental boundaries, which are the fundamental metameric subdivision of the early embryo, arise at the junction of the regions of expression of *engrailed* and *wingless*. Control of expression is initially by the pair-rule genes and subsequently by mutual interactions between each other. This switch is necessary because after cellularization of the blastoderm it is no longer possible for one nucleus to influence another simply by producing a genetic regulator. In a multicellular embryo communication necessarily involves secretion of inducing factors, cell surface receptors, and signal transduction mechanisms, as well as genetic regulation. A number of segment polarity mutants show considerable cell death in the affected regions and this led to speculation that the pattern duplications which are a frequent feature of the phenotypes, might be due to regeneration. However, it is now thought that the cell death is a consequence rather than a cause of the pattern alterations.

engrailed

The original *engrailed* (*en*) allele used in the compartment work was a hypomorphic allele compatible with viability and showing little or no embryonic phenotype. Stronger alleles are lethal and show a pair-rule effect

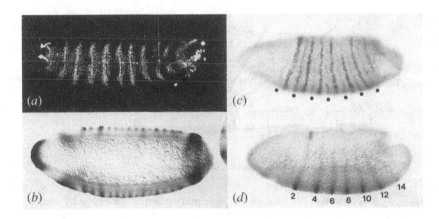

Fig. 7.16. Segment polarity genes. (*a*) cuticle preparation of *gooseberry* embryo showing transformation of anterior to posterior half parasegments. (*b*) expression of *wingless* mRNA visualized by *in situ* hybridization. (*c*) expression of *engrailed* mRNA, note that even stripes are more intense than odd ones. (*d*) expression of *engrailed* mRNA in *paired⁻* mutant, odd stripes have not formed. (Photographs kindly provided by Dr P. Ingham.)

with deletion of even numbered parasegments, and the amorphic phenotype shows a lawn of denticles with no segmentation at all. When the gene was cloned and the expression pattern studied it was found to come on at cellular blastoderm and to form a 14 stripe pattern by extended germ band (Kornberg *et al.*, 1985; Fig. 7.16*c*). Each band extends from the parasegmental groove to the tracheal pit, showing that *en* expression characterizes the anterior quarter of each parasegment. During development of the pattern the even numbered bands develop first, suggesting independent regulation from the odd numbered bands. There is later expression in the head and telson, in the mesoderm and in the CNS. In imaginal discs, as predicted, expression was found to be restricted to the posterior compartment.

wingless

Like *en*, *wingless* (*wg*) was originally known as a hypomorphic allele which caused proximal transformations of adult appendages, but the amorphic phenotype shows a complete loss of segmentation with the ventral cuticle transformed into a lawn of denticles. The expression pattern is quite similar to *en* in that 14 stripes are established by extended germ band and the even stripes grow faster than the odd ones (Baker, 1988 and Fig. 7.16(*b*)). However, the phasing is different: *wg* is expressed in the posterior quarter of each parasegment, just in front of each parasegmental groove hence adjacent to the *en* stripe in the anterior of the next parasegment. There are also early appearing patches in head and telson and transient expression in the mesoderm. The sequence of the gene shows a considerable homology to the murine proto-oncogene, *wnt-1*, which has all the hallmarks of a secreted factor, and *wg* product has been found outside *wg* producing cells by electron immunocytochemistry (van den Heuvel *et al.*, 1989).

gooseberry

The *gooseberry* (*gsb*) phenotype is somewhat similar to *wg* in that the ventral surface is covered with denticles, but it is more clearcut since the arrangement is of normal anterior half segments and replacement of the posterior halves, normally naked cuticle, with mirror duplications of the anterior half (Fig. 7.16(*a*)). The gene was cloned by means of some sequence homology with the *paired* gene. The expression pattern shows 14 bands initially about two cells wide and corresponding to the *en* + *wg* domain. As for *en*, the even bands appearing a little before the odd ones (Baumgartner *et al.*, 1987). There is also a closely related second gene whose expression is confined to neuroblasts from the extended germ band stage.

patched

patched (*ptc*) mutants show a rather different phasing of segment polarity phenotype from those above. The posterior part of each denticle belt, corresponding approximately to the second quarter of each parasegment, is deleted and replaced with a mirror duplication of the anterior part. Since the anterior part of each denticle belt includes the segment border it means that *ptc*$^-$ embryos have twice the normal number of segments. The expression pattern shows an initial uniform expression, except for the pole cells, at cellular blastoderm, then the formation of 14 stripes by extended germ band, each occupying the posterior $\frac{3}{4}$ of each parasegment (eg, Nakano *et al.*, 1989). Later each stripe splits into two, representing approximately the second and fourth quarter of each parasegment. There is also expression in the head and persistent expression in the mesoderm.

Other segment polarity genes

As will become clear below, the front running models for segmental organization involve a sequence of four states with E for *engrailed* as the first and W for *wingless* as the last. But each state will require several gene products for its definition. For example, it is believed that the *wg* protein is an inducing factor which can tell neighbouring cells that the source cell has the W coding. But there will also be one or more genetic regulators required to turn on *wg* and also a receptor through which it can exert its effect. Likewise *engrailed* probably is a genetic regulator but E cells will also need an associated inducing factor and receptor to inform neighbouring cells of

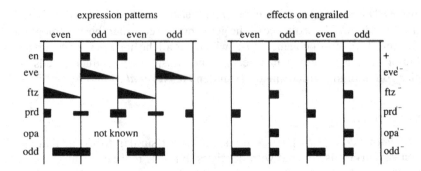

Fig. 7.17. Regulation of *engrailed* by pair-rule genes. On the left is shown the expression pattern of *en* together with five pair-rule genes. On the right is shown the effect on *en* of removing each of the pair-rule genes. It is thought that *en* can be activated by *ftz* + *opa*, normally giving the even stripes, and by *eve* + *prd*, normally giving the odd stripes. *odd* appears to act as a repressor and limits the size of the even stripes. Removal of *eve* causes loss of all the *en* stripes, probably because there are secondary effects on the phasing of *ftz*, which becomes out of register with *opa*.

their status. We therefore expect to find sets of genes which have similar null phenotypes corresponding to the removal of one cell state, but which code for products with different molecular functions. *armadillo*, *fused*, *hedgehog* and *cubitus interruptus* all have phenotypes similar to *gsb* and may all be involved in the W system. *armadillo* has been cloned and its expression pattern is uniform. The protein is rather abundant and its sequence suggests a cytoplasmic function, possibly a component of the cytoskeleton (Riggleman *et al.*, 1989). *naked* has a quite different phenotype in that the anterior (denticle bearing) half of each segment is deleted and presumably replaced by a posterior half consisting of more naked cuticle. As we shall see, *nkd* seems to be a repressor of *en* and is thus perhaps involved in a third, N state.

Control of segment polarity by pair-rule genes

As the expression patterns of the pair-rule genes become refined, *eve* and *ftz* form an alternating sawtooth pattern with the peaks of *eve* (confusingly) at the anterior margins of *odd* parasegments and the peaks of *ftz* at the anterior margins of *even* parasegments (Lawrence *et al.*, 1987). We might think that all that is happening is that these two products both activate *en*, perhaps with different time courses since we know that the even *en* stripes lead the odd ones. However, things are more complex, and it actually looks as though *combinations* of pair-rule genes are required to activate each *en* stripe. All the evidence to date comes from observing the expression patterns of *en* and *wg* in a pair-rule mutant background (DiNardo & O'Farrell, 1987; Ingham *et al.*, 1988; Fig. 7.17). The results suggest that *en* is activated by either (*ftz* + *opa* + *not odd*) or by (*eve* + *prd*), the former combination being responsible for the even stripes in normal development and the latter combination for the odd stripes. Complementary conclusions apply to *wg*, which is normally repressed by *eve* and *ftz* and comes on at the posterior edge of each parasegment in the absence of both.

This model will undoubtedly be extended as more expression patterns are examined in more mutant backgrounds. But the general conclusion seems clear: the initial regulation of the segment polarity genes is accomplished by pair-rule genes working in combination, and this means that the even and odd numbered parasegments are necessarily formed by a different routes although they may eventually become equivalent.

Interactions between segment polarity genes

Although the initial establishment of the segment polarity gene expression patterns are under the control of the pair-rule genes, their maintenance requires interaction within the group (Martinez-Arias *et al.*, 1988). In *wg*⁻ mutants the *en* pattern soon fades, and in *en*⁻ mutants the *wg* pattern soon fades, so the E and W states each seem to need the other in an adjacent cell file in order to persist. Martinez-Arias and coworkers have put forward a model in which each parasegmental repeating unit consists of four states, labelled E, N, P, W because the states require the genes *en*, *nkd*, *ptc* and *wg* respectively. Although these genes are each necessary they are unlikely to have equivalent functions. It is probable that several genes would underly each state since it is necessary both to encode it at the genetic level and to communicate it to neighbouring cells. In the cellular blastoderm, when the parasegmental primordium is about four cells wide, each state would represent a single file of cells. By the extended germ band when 1–2 more cell divisions have occurred, each normal stripe would represent 3–4 cells. It is postulated that the states are initially set up in response to different combinations of pair-rule gene activity, as discussed above, and that they can then change on cell division with each state depending for its maintenance on one or both of the states which are its normal neighbours (Fig. 7.18). Three states per segmental repeat is the minimum necessary to generate an anteroposterior polarity to the system, and four states is the minimum required to explain the intercalation of new states following the juxtaposition of states which are not normal neighbours.

Once again, it is obvious that this model is provisional and will be elaborated as more segment polarity genes are cloned and as expression of more genes is examined in more mutant backgrounds. But the general conclusion seems to be that the underlying cause of visible segmentation is a repeating pattern of states of activity of segment polarity genes. These are initially activated by appropriate combinations of pair-rule gene activity and are later sustained by mutual interactions depending on intercellular communication.

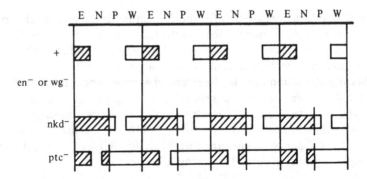

Fig. 7.18. Model of Martinez-Arias *et al.*, (1988) for interactions between segment polarity genes. Normally the *en* (hatched) and *wg* (open) domains occupy the anterior and posterior quarter respectively of each parasegment. Removal of either leads to loss of the other, suggesting mutual support. Loss of *nkd* prevents formation of the N state and leads to an initial sequence EEPW. On cell division this changes to EEEEWPWW because P is not stable next to E and so changes to W. A new parasegment boundary then arises between E and the ectopic W. A corresponding intercalation of E occurs in *ptc⁻*.

Homeotic selector genes

The homeotic genes of *Drosophila* were historically the first regional specification genes to be discovered, indeed a few have been known since the 1920s, and it has long been realized that homeotic mutations must represent errors in fundamental elements of the developmental program. Now it is clear that they are the coding factors for the anteroposterior axis of the body, not only in *Drosophila* but possibly in all animals. All of the genes which precede them in the developmental hierarchy have as their function to turn on the right homeotic gene at the appropriate body level, correctly aligned with the pattern of segmentation. As we have seen in Chapter 2, homeotic genes are defined as genes which, when mutated, cause transformations of one part of the body into another, and this is the hallmark of the genes to be considered in this section. On the whole, loss of function mutants of maternal, gap and pair-rule classes lead to phenotypes with gaps rather than to homeotic transformations. This is perhaps because they come earlier in the hierarchy than the selector genes and so mutations are more likely to produce 'nonsense' codings rather than codings appropriate to other parts of the body.

Most of the homeotic genes of *Drosophila* belong to two gene clusters called the *Antennapedia complex* (ANT-C) and the *Bithorax complex* (BX-C) which both lie on the right arm of the third chromosome. It seems probable that they were once a single complex since there is a clear homology between them and a single gene cluster in a beetle *Tribolium*, and a moderate homology between them and the *Hox-2* cluster in the mouse. All the genes in both complexes contain a homeobox, and, in addition to the selector genes considered here, the ANT-C contains two homeobox genes which have already been considered under separate headings: *bicoid* and *ftz*, and the zygotic dorsoventral gene: *zen*. There are also, as we shall see, many regulatory sequences in the intergenic regions and in the introns. Some of these produce non-translated RNAs and it is presumably the density of the regulatory connections that have kept the order of genes on the chromosome approximately constant through evolutionary time.

The term 'selector gene' was introduced by Garcia Bellido from work on imaginal discs which suggested that the function of homeotic genes was to select between different possible developmental pathways. It is usually supposed that the gene products activate batteries of other genes which make up the anatomy of the larva, but there is little concrete data to back this up and in fact the relationship between selector gene expression and terminal differentiation is now the least understood level of the developmental program. The names of the homeotic genes mainly derive from the adult phenotypes of the dominant or recessive viable alleles which were discovered first, but as with the other specification genes the phenotype of the null allele is the most significant. *labial* (*lab*), *Deformed* (*Dfd*), *Sex combs reduced* (*Scr*) and *Antennapedia* (*Antp*) belong to the ANT-C and *Ultrabithorax* (*Ubx*), *abdominal-A* (*abd-A*) and *Abdominal-B* (*Abd-B*) to the BX-C. The ANT-C also contains a gene, *proboscopedia*, which functions only in larval life. Except for *Deformed* where a substantial protein concentration is achieved by the cellular blastoderm, the selector genes are turned on slightly later than the gap and pair-rule classes, and their own protein products have generally built up to an effective level by the extended germ band stage. The order in which the genes are described is the order of their expression in the anteroposterior axis, and also their order of their arrangement on the chromosome. This fact is thought very significant by some, although the cause is more likely to lie in the evolutionary path by which the complexes were created rather than in any necessity of such an arrangement for a developmental program.

labial

The amorphic phenotype involves a loss of labial derivatives and defects in head involution. Homozygous clones induced during larval life by X-ray induced mitotic recombination show some transformation of dorsal posterior head to dorsal thorax. The gene is expressed from early germ band extension in the posterior midgut and the stomodeal region just anterior to the cephalic furrow (Diederich *et al.*, 1989). It is not expressed in the prospective labium and this mismatch is not currently understood.

Deformed

The original *Deformed* mutation was a viable dominant, but the amorphic phenotype approximates to a deletion of the mandibular and maxillary segments with a consequent failure of head involution. Homozygous clones in the adult show a transformation of head towards thoracic structures. Expression of the gene occurs somewhat earlier than the other selector genes, starting in the syncytial blastoderm as a band corresponding to parasegment 1, later transiently extending to parasegment 0 in the ectoderm and neurectoderm (Martinez-Arias *et al.*, 1987). In the larva it is found in the junctional region of the eye antennal disc. *Deformed* is not expressed in *bicoid⁻* eggs, suggesting activation by *bicoid⁺*. Either maternal of zygotic *hunchback* will support expression, but removal of both prevents it. Mutations of zygotic gap genes or other selector genes have very little effect on *Deformed* but all of the pair-rule mutants except *slp* have an effect, either extending or diminishing the domain of expression. So probably *Deformed* is activated by *bicoid* and *hunchback* and the domain is refined by the pair-rule genes (Jack & McGinnis, 1990). Once turned on, *Deformed* can maintain itself by autoactivation. This was shown by coupling the gene to a heat shock promoter and showing that not only the copy on the construct but also the endogenous copy could be turned on by heat shocking the transformed embryos (Kuziora & McGinnis, 1989). Such embryos, with ectopic expression of *Deformed*, show homeotic transformations of thoracic segments to maxillary character.

Sex combs reduced

The name of this gene derives from a dominant haploinsufficient phenotype in which prothoracic legs are transformed to mesothoracic. The amorphic phenotype involves a transformation of PS 3 to 4 and also the apparently opposite transformation of PS 2 to 1 resulting in defects in the mouthparts and incomplete head involution. The expression pattern shows transcription in a band corresponding to parasegment 2 in the cellular blastoderm, extending to parasegment 3 in the epidermis, but restricted to PS 2 in the neurogenic region and PS 3 in the mesoderm (Martinez-Arias *et al.*, 1987; Fig. 7.19(*b*)). Later there is also expression in the abdominal ganglia. Mutant studies show a possible activation by *hunchback* and *giant*, and possible repression by *Krüppel* and *Ubx*.

Antennapedia

Named after a dominant adult transformation of antenna to mesothoracic leg, *Antp* has an amorphic phenotype in which parasegments 4 and 5 are transformed towards PS 3. This is an interesting illustration of the fact that

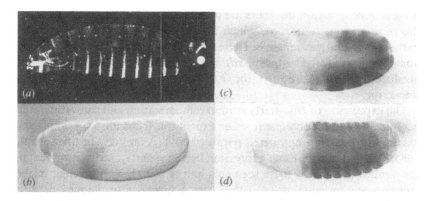

Fig. 7.19. Homeotic selector genes. (*a*) *abd-A-* larva, in which all abdominal segments have become transformed towards A1. (*b*) *Sex combs reduced* protein visualized by antibody staining at extended germ band stage. (*c*) *Ubx* protein at extended germ band stage. (*d*) *Ubx* protein after germ band shortening. (Photographs kindly provided by Dr P. Ingham.)

one may have a diametrically opposed phenotypes from loss of function and gain of function mutations, so it is important to know which is which.

Transcription commences in the cellular blastoderm with a peak in the prospective mesothorax (Levine *et al.*, 1983). In the extended germ band it is concentrated in parasegments 4 and 5, being present in all epidermal cells of PS 4 but not the tracheal pits of PS 5. In the mesoderm it is present only in PS 5. In addition there is expression in the neuromeres from PS 4-12 in the shortened germ band (Martinez-Arias, 1986). In the larva, the *Antp* protein is also found in the wing and leg discs but not the eye–antennal disc.

Antp contains two promoters which produce different mRNAs, but apparently the same protein. The P2 promoter is active only in parasegment 4, and the P1 promoter at the other levels as well. Mutant background experiments suggest that the P1 promoter may be be activated by *Krüppel*[+] and the P2 promoter may be activated by a combination of *hunchback*[+] and *ftz*[+] and probably something else as well (Harding & Levine, 1988). In *Ubx*[−] embryos, the expression of *Antp* is extended posteriorly to include PS 6, and if the entire bithorax complex is removed it is extended as far back as PS 12 (Hafen *et al.*, 1984), so it is probable that the posterior boundary of *Antp* expression depends on repression from the products of the BX-C. But this will be a later control than that by the gap and pair-rule genes because of the later appearance of BX-C proteins.

Ultrabithorax

Amorphic mutants of *Ubx* show a transformation of parasegments 5 and 6 to parasegment 4 (Sanchez-Herrero *et al.*, 1985). The *Ubx* gene itself is the protein coding region but there are also regulatory regions both flanking the gene and in its introns, which are responsible for its activation in particular spatial subdomains. These were formerly thought to be separate homeotic genes (*abx, bx, pbx, bxd*), but it is now known that they do not produce proteins, although the *bxd* region does produce an RNA transcript.

The expression of *Ubx* starts around cellularization and shows an initial peak in PS 6 and subsequent lesser expression from PS 5–13 with local maxima in the even numbered parasegments (Akam & Martinez-Arias, 1985; Fig. 7.19(*c*)(*d*)). Later expression becomes particularly prominent in the ventral ganglia of these segments and is found, as predicted, in the metathoracic imaginal discs of the larva. Expression in the mesoderm is more posterior, centring on A2–3. As for *Antp*, there is more than one transcript and a number of different protein forms for *Ubx*.

Initial control of *Ubx* expression is probably by *hunchback* which acts as a repressor (Irish *et al.*, 1989). The transient pair-rule character of the *Ubx* pattern also suggests control by one or more pair-rule genes. *ftz* is probably the most important here since the pair-rule character of *Ubx* disappears in *ftz⁻* mutants (Ingham & Martinez-Arias, 1986). This pair-rule control is important for aligning the parasegmental register of segment polarity and homeotic genes. Once it is established the *Ubx* pattern is sustained by repression from *abd-A* but it is not apparently affected by *Antp*, the selector gene expressed in the region immediately to the anterior. There is also autoactivation at least in the visceral mesoderm (Bienz & Tremml 1988).

abdominal-A and Abdominal-B

It was at one time thought that each abdominal segment would have its own selector gene and a number of recessive *infraabdominal* mutants were isolated by E. Lewis which seemed to support this view. However, it is now known that only two of these are protein coding genes and that they correspond to the two abdominal complementation groups of the BX-C: *iab-2* is the same as *abd-A* and *iab-7* is the same as *Abd-B*. The other *iab* genes, some of which have RNA products, turn out to be regulatory regions responsible for activation of the two structural genes in different parts of the body.

The amorphic phenotype of *abd-A* mutants is a transformation of PS 7–9 (ie, a zone overlapping abdominal segments 1–4) to PS 6 (Fig. 7.19(*a*)), with lesser effects on PS 10–13 (Sanchez-Herrero *et al.*, 1985). Expression of *abd-A* occurs at about the same time as *Ubx* in parasegments 7–12, and at the time of writing details of its regulation were not available.

There are two classes of *Abd-B* mutation causing anterior transformations either of PS 10–13 or PS 14, and these correspond to two groups of transcripts expressed in the predicted places (Sanchez-Herrero & Crosby,

1988). There is also later expression in the CNS, the mesoderm and the genital disc. The domain of expression is expanded to the anterior in *knirps⁻* mutants and even more so, reaching into the thorax, in *Krüppel⁻* mutants, providing some evidence for repression by the gap genes (Harding & Levine, 1988).

Terminal selector genes

Two further homeotic genes are *fork head* and *spalt*. Mutants of *fork head* have transformations of the stomodeum into bits of head skeleton and thorax, and transformations of the proctodeum into extra anal sense organs and abdominal denticles. Expression gives rise to small patches at the extreme termini of the cellular blastoderm (Weigel *et al.*, 1989). *spalt* mutants also have anterior and posterior defects but at less extreme positions. In the anterior there are defects in the maxillary and labial segments and some replacement by prothoracic structures. In the posterior there is a partial transformation of A9 and 10 towards A8, often with extra posterior spiracles. At cellular blastoderm *spalt* is expressed as two domains, from 70–60% EL in the anterior and 15–0% in the posterior (Frei *et al.*, 1988). At the time of writing the relationship of these genes to the terminal system and to the other gap genes was not known.

The anteroposterior body pattern

The long and complex series of interactions described above leads to a fully specified body plan by the extended germ band stage. This specification has two essential components. There is the metameric pattern composed of parasegments whose boundaries are defined by the juxtaposition of bands of cells in which the *engrailed* (anterior) and *wingless* (posterior) systems are active. There is also the non-repeating sequence of selector gene expression zones: *lb*, *Dfd*, *Scr*, *Antp*, *Ubx*, *abd-A* and *Abd-B*. Although these overlap, there is a clear sequence from anterior to posterior in which a single gene predominates (Fig. 7.20).

Each element of this pattern appears to be initiated locally by combinations of concentrations of the products of the maternal systems, the gap genes and the pair-rule genes. The pair-rule genes have a particularly important role in that they must control the register between the segment polarity and the selector genes so that each segment acquires the correct identity. By the extended germ band stage the products of the controlling systems have decayed or are decaying and so the maintenance of the pattern is ensured by separate means: such as autoactivation, for example of *Deformed*; mutual reinforcement of neighbouring states, for example, of *engrailed* and *wingless*; or inhibition between neighbouring states, for example of *Antp* by *Ubx*.

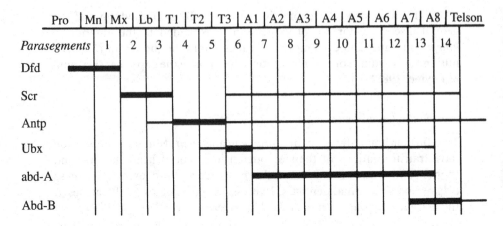

Fig. 7.20. Domains of activity at the extended germ band stage of the homeotic selector genes of the *Antennapedia* and *Bithorax* complexes.

Now the main mystery is how this cryptic pattern of gene activity is translated into the visible pattern of differentiation. Little is known about this and progress may be slow since the normal histology of *Drosophila* is very much less well understood that than of vertebrates. But one thing we can say is that the expression patterns of the selector genes themselves, although they commence fairly uniformly throughout a parasegment, later come to some extent to prefigure the pattern of cytodifferentiation. For example, expression of *Antp* in PS 5 excludes the tracheal pits, which are the only cells of PS 5 to express *Ubx*. So we may be able to approach the answer by more intensive study of genes that are already known.

RESULTS ON OTHER SPECIES

Descriptive embryology

Drosophila is not typical of other insects in several respects and in this final section a number of the differences will be pointed out in case they imply differences in the developmental program. Insects are divided into two main taxa: the Hemi and Holometabola (often called Exopterygota and Endopterygota by zoologists), and there is also a group of primitive wingless insects which does not concern us here. *Drosophila* is holometabolous and undergoes a complete metamorphosis. The Hemimetabola by contrast hatch from their eggs as larvae which are fairly similar in morphology to the adults and are called nymphs. As they grow the nymphs pass through a number of moults in each of which the old cuticle is shed and a slightly more adult-like nymph emerges. Most species take longer to develop than *Drosophila* and for Hemimetabola the time from fertilization to hatching is likely to be measured in weeks.

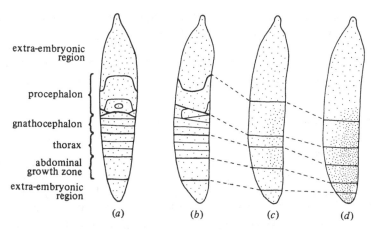

Fig. 7.21. Fate map of *Platycnemis* according to Seidel (1935). (*a*) and (*b*) blastoderm stage, (*c*) germ aggregation stage, (*d*) germ anlage stage.

Insect embryos can also be classified as: *long, intermediate,* or *short germ.* 'Germ' here refers to the *germ anlage* which is the condensation of cells giving rise to the embryo, as opposed to the extraembryonic parts. We have not referred to a germ anlage for *Drosophila* because almost the whole of the blastoderm is the germ anlage. But in most species this is not so and long, intermediate and short refer to the proportion of the larval body plan represented on the fate map of the germ anlage. On the whole, the Holometabola are long or intermediate germ species, and the Hemimetabola are intermediate or short germ species. The significance of this lies in the different sequences of segmentation which transform the germ anlage stage to the extended germ band.

In long germ insects the sequence of visible segmentation usually proceeds from anterior to posterior and rudiments for the whole length of the body are established quite quickly. The early germ band stretches itself so that the posterior end bends over to the dorsal side. *Drosophila* is a long germ insect, but development is so fast that there is little or no anteroposterior sequence to segmentation. In short germ insects only the rudiment for the head, sometimes just the procephalon, is present and the segments of thorax and abdomen are produced over a long period of time from a posterior growth zone. In insects with germ anlage of intermediate length, segmentation often commences in the thoracic region and proceeds from there both anteriorly and posteriorly, the abdomen being formed by a growth zone.

The differences from *Drosophila* are exemplified by the fate map of the dragonfly *Platycnemis*, an insect of intermediate length germ anlage, constructed by UV microbeam irradiation (Seidel, 1935; Fig. 7.21). The map starts off by covering most of the egg at the blastoderm stage and then contracts as the germ anlage is formed, presumably because of an aggregation of cells in the ventroposterior region. Although the prospective region for the abdominal growth zone is present on the fate maps, the individual abdominal segments are not. This implies that there is some indeterminancy between individuals regarding which particular cells in the growth zone will contribute to which segment.

Experimental studies

The gap phenomenon

In several types of insect, experiments have been performed in which the embryo is divided into two by constriction with a hair loop or by pinching with a blunt razor blade. The results follow a common pattern called the 'gap phenomenon'. This is the appearance of a gap in the normal sequence of segments at the constriction (review: Sander, 1976). There is no visible cell death in the region and it therefore seems as though the fate maps of both anterior and posterior fragments are shifted towards the constriction. The gap is large at early stages and becomes smaller at later stages, eventually disappearing altogether. Different insect species show the gap phenomenon to different extents, and in *Drosophila* it is quite limited. It is shown particularly clearly by the experiments of Jung (1966) on the bean weevil *Bruchidius*. In Fig. 7.22 is shown a graph representing the results. For each stage the upper line shows at which segment the embryo is truncated in the posterior fragment and the lower line shows at which segment the embryo is truncated in the anterior fragment following constriction at a given level. The gap in the sequence is large for early constrictions and has nearly vanished by the 'pre-germ anlage' (or 40 hour) stage. At the earlier stages of development in *Bruchidius* and several other species, most ligations only yield developing embryonic parts on one side of the constriction (see solid lines in Fig. 7.22). Since it is usually not possible to get structures on both sides of a constriction the idea grew up that *both* of the extreme regions, anterior and posterior, were required for the normal formation of the longitudinal pattern. These may perhaps be homologues of the *Drosophila* maternal anterior and posterior systems.

A posterior signalling centre

Seidel (1929) showed that cautery of the posterior 12% EL region at early stages of development of *Platycnemis* prevented any development of the germ band despite the fact that the prospective region for the germ band lies mainly anterior to this. The posterior pole region became known as the 'activation centre' and its role is shown clearly in experiments on the leaf hopper *Euscelis* (review: Sander, 1976). The eggs of this species contain a

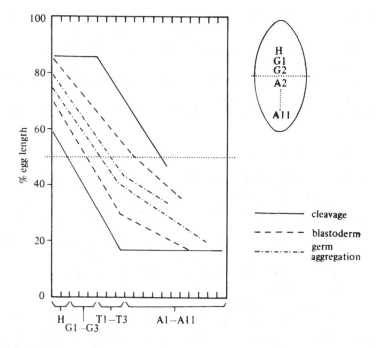

Fig. 7.22. The gap phenomenon. Embryos of the weevil *Bruchidius* were ligated at different levels along the egg axis. For each stage the top line shows at which structure a posterior fragment will terminate, and the bottom line shows at which structure an anterior fragment will terminate. The gap in the sequence of structures is smaller for late ligations than for early ones. An embryo ligated at 50% EL at blastoderm stage would accordingly lack G3,T1,T2,T3,A1 and so consist of H,G1,G2,A2–11 (inset). (After Jung, 1966.)

ball of symbiotic bacteria near the posterior pole and this can serve as a label for the position of the posterior cytoplasm. Sander combined displacement of the symbiont ball with constriction. *Euscelis* shows a gap phenomenon, so an anterior 60% fragment isolated during cleavage will typically produce just a procephalon. But when this fragment includes the material marked by the symbiont ball it will form more structures, often complete germ bands. The posterior 60% fragment will usually produce a sequence of structures from gnathocephalon to abdomen but when it includes the material marked by the symbiont ball at its anterior end it produces an inverted or a mirror duplicated set of posterior structures. Because of the continuity shown by the induced structures these results suggest that the posterior region is the source of a graded signal which would make it somewhat different from the maternal posterior system in *Drosophila*.

Embryos with double abdomens can also be produced by UV irradiation of the eggs of the midge *Smittia* and other Diptera (Kalthoff, 1971). It is worth noting that, like the *Bic-D* mutant in *Drosophila*, the double abdomens in *Smittia* do not have two sets of pole cells, but only those in the original abdomen. The double abdomens in a similar species *Chironomus* can be rescued by injection of poly A^+ RNA, which must be taken from the anterior end and must be injected into the anterior end (Elbetieha & Kalthoff, 1988). This might perhaps be due to a *hunchback* homologue.

Segment polarity genes

The *engrailed* gene, so fundamental to the organization of the segmental repeat in *Drosophila*, has been found to exist across the animal kingdom. In the grasshopper, *Schistocerca*, an intermediate germ species, it is also found to be expressed in the posterior part of each segment (Patel *et al.*, 1989*b*). However the sequence of expression is dissimilar. In *Drosophila en* comes on first in even then odd parasegments. In *Schistocerca* the sequence of expression starts in T1 then spreads to the anterior and the posterior, the posterior spread at least being sequential.

The similar expression pattern of *engrailed* in other insects is very important because it establishes in molecular terms the homology of segmentation which has been assumed for so long in classical zoology. As we shall see, *engrailed* also has a similar expression pattern in other arthropods, but not in other phyla.

At the time of writing, no data were available on expression of pair-rule genes in other insects although one might predict that the pair-rule level of control would be less important in creatures that form their segments sequentially.

Selector genes

Recently genetic and molecular studies have commenced on the HOM complex of the beetle *Tribolium* (Beeman *et al.*, 1989). This appears to be a gene cluster homologous to the ANT-C and BX-C together and consists of a series of homeobox containing genes arranged in the same order on the chromosome as their effects from anterior to posterior in the body. But the actual mutant phenotypes are somewhat different and do not suggest a one-to-one correspondence with the *Drosophila* genes at the level of developmental function.

Dorsoventral axis

As far as the dorsoventral axis is concerned (more appropriately described as mediolateral in embryos with a flat germ anlage), it is worth noting that some species can produce twins when cut into longitudinal halves. In the stone cricket *Tachycines* complete twins can be produced up to the gastrula stage (Krause, 1958) and in the silk moth *Bombyx mori* up to the germ anlage stage (Krause & Krause, 1965). As in other embryos the rule seems to be that in slowly developing species the determinative events occur later relative to morphological stage than in rapidly developing species. If these species have some homologue to the *dorsal* system then it must be able to reconstitute the pattern at a multicellular stage, and hence contain components capable of intercellular signalling which are not found in *Drosophila*.

WHAT *DROSOPHILA* HAS TAUGHT US

The amazing progress in the study of regional specification in *Drosophila* enables us to be quite definite about what mechanisms are and are not used. Yes, there are gradients and at least in the case of the *bicoid* and *dorsal* gradients they behave as real morphogens, evoking different cellular behaviours at different concentrations, and hence specifying both pattern and polarity simultaneously. Yes, there are determinants, any of the localized maternal gene products which control subsequent zygotic gene activity could count as a determinant, and in the case of *bicoid*, it is appropriate to regard the localized mRNA as a determinant and the gradient of protein product as a morphogen. Yes, there is a developmental hierarchy. Terminal cell types are not specified all at once but as the end products of a complex series of decisions. The necessary corollary of this is that there are such things as selector genes, which code for particular parts of the body rather than for terminal cell types. The intermediate cell states are controlled in a combinatorial way: each domain of gene activity usually depends on a specific combination of presences and absences of preceding gene products in the region in question. Furthermore these controls often include autoactivation, and as we saw in Chapter 3 under 'thresholds' this is one way of forming sharp boundaries and maintaining stability of expression. All these elements of the developmental program were predicted by theorists and by experimental embryologists, but it has taken the technology of the molecular era to give us the concrete evidence required to convince the general scientific public of their reality.

Various other mechanisms have not turned up so far. For example, periodic patterns in the *Drosophila* egg are not made by reaction diffusion mechanisms, but instead every stripe is independently controlled. There is also, so far, no evidence of directed chromosome modification as a mechanism of regional specification.

Although most of the predicted mechanisms are there, the biochemical realization is very varied. Some determinants are mRNA (*bicoid, nanos*), while others are localized states of receptor activation (*torso, Toll*). Some genes are controlled by transcriptional activation, others by mRNA degradation. It seems that all that is required of any particular component is the ability to modify the activity of another component. What is important is not the biochemistry but the logical sequence of the program, and there is absolutely no way one can predict which particular sorts of molecular nuts and bolts will be used to build it.

Drosophila has shown us that a complete explanation of early development can be achieved. The number of components involved is large, but not astronomical, and it is even possible for one individual to remember for a few days the names of all the genes and their principal characteristics. *Drosophila* has also clearly pointed the finger at genetics as being the road to the breakthrough, and whether a similar course can be followed for the vertebrates we shall consider in the concluding Chapter.

GENERAL REFERENCES

Ingham, P.W. (1988). The molecular genetics of embryonic pattern formation in *Drosophila*. *Nature, London*, **335**, 25–34.

Lawrence, P.A. ed. (1976). *Insect Development. Royal Entomological Society Symposium* **8**. Oxford: Blackwell.

Sander, K. (1976). Specification of the basic body pattern in insect embryogenesis. *Adv. Insect Physiol.* **12**, 125–238.

Drosophila *gene tables*

Notes on the table. Expression patterns are usually given for transcription. If protein is mentioned explicitly then its pattern differs from that of the RNA. If not it can be assumed to follow the RNA after a short time interval. The mutant effects are usually null phenotypes from homozygous loss of function alleles. Weak alleles or gain of function effects are mentioned explicitly. Molecular natures are usually inferred from the sequence by comparison with data bases but in some cases are supported by direct experimentation. Developmental functions are mainly inferred from looking at one expression pattern in embryos mutant for another gene, so directness is always uncertain. The word 'regulate' here signifies probable activation at one concentration and repression at another, thus giving rise to a stripe of the regulated product. References will be found in the text.

Gene	Expression pattern	Mutant phenotype	Molecular nature	Developmental function
Maternal dorsoventral system				
dorsal (*dl*)	Transcription nurse cells, uniform mRNA in oocyte and egg, V-D gradient of protein in syncytial blastoderm nuclei	Extreme dorsalization. Embryos become tubes of dorsal epidermis. Some haploinsufficient alleles	Nuclear protein. (*c-rel* homologue)	Activates *twi* Represses *zen*
tube (*tub*)	Maternal, action in embryo	Similar to *dl*		
pelle (*pll*)	Maternal, action in embryo	Similar to *dl*		
cactus (*cac*)	Maternal, action in embryo	Ventralizing		Inhibits entry of *dl* to nuclei
Toll (*Tl*)	Maternal, uniform mRNA in oocyte and egg	Similar to *dl* Gain of function alleles ventralize producing denticle belts all round.	Cell surface protein	Activation of dl protein
snake (*snk*)	Maternal, needed in oocyte	Similar to *dl*	Serine protease	Activation of *Tl* ligand
easter (*ea*)	Maternal, needed in oocyte	Similar to *dl* Also weak ventralizing gain of function alleles	Serine protease	Activation of *Tl* ligand

Gene	Expression pattern	Mutant phenotype	Molecular nature	Developmental function
pipe (*pip*)	Maternal, needed in follicle cells	Similar to *dl*		Possible ligand for *Tl*
nudel (*ndl*)	Maternal, needed in follicle cells	Similar to *dl*		
windbeutel (*wbl*)	Maternal, needed in follicle cells	Similar to *dl*		
gastrulation defective (*gd*)	Maternal, needed in oocyte	Similar to *dl*		
spätzle (*spz*)	Maternal, needed in oocyte	Similar to *dl*		
torpedo (*tor*)	Maternal, needed in follicle cells	Ventralizing, affects egg shell and embryo	EGF receptor	
gurken (*grk*)	Maternal, needed in oocyte	Similar to *tor*		
K10	Maternal, needed in oocyte	Dorsalizing, affects egg shell and embryo		
Zygotic dorsoventral system				
twist (*twi*)	Transcription from syncytial blastoderm in midventral strip. Later in mesoderm, fades by maximum extension.	Loss of mesoderm	Nuclear protein (helix–loop–helix)	
snail (*sna*)	Zygotic	Loss of mesoderm		

Gene	Expression/transcription	Mutant phenotype	Product	Function/regulation
decapentaplegic (*dpp*)	Transcription syncytial blastoderm on dorsal side, curling round to ventral at poles. Later two stripes in lateral epidermis. Later visceral mesoderm, fore and hind gut. Imaginal discs	Loss of amnioserosa and reduction of dorsal epidermis. Viable alleles produce multiple defects in imaginal disc derivatives	Signalling molecule (TGFß homologue)	
zerknüllt (*zen*)	Transcription syncytial blastoderm on dorsal side, concentrated in prospective amnioserosa by beginning gastrulation	Loss of amnioserosa and optic lobe	Genetic regulator (homeobox)	
twisted gastrulation (*twg*)	Zygotic	Similar to *zen*		
tolloid (*tld*)	Zygotic	Loss of amnioserosa, reduction of dorsal epidermis	Bone morphogenetic protein homologue	
shrew (*srw*)	Zygotic	Similar to *tld*		
single minded (*sim*)	Transcription longitudinal strips just above prospective mesoderm	Loss of certain neurons	Nuclear protein	

Maternal anterior posterior axis

Anterior system

Gene	Expression/transcription	Mutant phenotype	Product	Function/regulation
bicoid (*bic*)	Transcription in oocyte & nurse cells. mRNA localized in anterior of oocyte and egg. Translation cleavage-blastoderm forms A–P gradient	Deletion of head and thorax. Substitution second telson for head	Genetic regulator (homeobox PRD repeat)	Activates *hb*, *otd* transcription. Activates *Kr* at low level and represses *Kr* at higher level Represses *cad* translation. Switches telson to head. Regulates *eve*, *ftz*, *h* Activates *Dfd*

Gene	Expression pattern	Mutant phenotype	Molecular nature	Developmental function
exuperantia (*exu*)	Maternal	Reduction of head, replacement by telson	*bicoid* binding protein?	Needed to transport *bcd* mRNA from nurse cells
swallow (*swa*)	Maternal	Reduction of head, cellularization defects	cytoskeleton?	Needed to localize *bcd* mRNA to anterior
Posterior system				
nanos (*nos*)	Transcription in oocyte and nurse cells. Localized to posterior in late oocyte and egg, then to pole cells	Absence of abdomen. Pole cells normal		Inhibits translation of maternal *hb* mRNA, hence allows activation of *kni* in posterior. Regulates *h*, *eve*, *ftz*
pumilio (*pum*)	Transcription maternal. Protein probably localized to posterior (20–0% EL)	Most abdomen missing Pole cells normal		Needed for transport of *nos* product from pole cells to prospective abdomen
vasa (*vas*)	Transcription in nurse cells and oocyte, mRNA uniform in early zygotic embryo. Later zygotic transcription in pole cells. Protein localized in posterior of oocyte, egg, pole cells	A2-7 missing. No pole cells	Translation factor?	
tudor (*tud*)	Maternal	Similar to *vasa*		Establishment of pole plasm
oskar (*osk*)	Maternal	Similar to *vasa*		Establishment of pole plasm
valois (*vls*)	Maternal	Similar to *vasa* with cellularization defect		Establishment of pole plasm

Gene				
staufen (*stau*)	Maternal	Similar to *vasa* but with head defect. Allows posterior structures in anterior if combined with *Bic*		Establishment of pole plasm
Bicaudal (*Bic-D*)	Maternal, uniform	Dominant. Symmetrical double abdomen with mirror plane A3-A8. No duplication of pole cells	Fibrous protein, maybe cytoskeletal	Transport of *nos* mRNA even stripes develop first.
Terminal system				
torso (*tor*)	Transcription in nurse cells. Transport to oocyte where uniform	Absence of labrum, reduction head skeleton, absence A8 and telson. Gain of function: suppression of segmentation in thorax and abdomen	Cell surface receptor (Tyr kinase)	Activates *tll*, *hkb* Activates zygotic transcription of *cad*
torsolike (*tsl*)	Maternal, required in follicle cells not germ line	Similar to *tor* (loss of function only)		Probable ligand for *tor*
polehole (*ph*)	Maternal and zygotic. Uniform mRNA in egg	Similar to *tsl* if good copy introduced by sperm. Maternal and zygotic absence suppresses most pattern	SerThr kinase (*raf* homologue)	Downstream of *tor*
trunk (*trk*)	Maternal	Similar to *tsl*		
Nasrat (*N*)	Maternal	Similar to *tsl*		

Gene	Expression pattern	Mutant phenotype	Molecular nature	Developmental function
Zygotic anteroposterior gap class				
hunchback (*hb*)	Maternal mRNA initially uniform, protein A–P gradient. Zygotic transcription 100–55% EL and 20–10% EL. Posterior domain long lasting	Zygotic: deletion labium, thorax A7a/8p (PS13). No pure maternal effect, but if mother and sperm hb⁻ then deletion all gnathal and thoracic segments and A1–3, with mirror duplication of abdomen	Genetic regulator (finger)	Represses *kni* Represses *Kr* at high level and may activate at low level Regulates *eve* Regulates *h, runt, ftz* Activates *Antp P2* (with *ftz*) Represses *Ubx* Activates *Dfd* Activates *Scr*
Krüppel (*Kr*)	Transcription starts syncytial blastoderm in central zone, eventually expanding to 60–30% EL. At cellular blastoderm also posterior patch, later posterior plate, and patch in head	Purely zygotic. Head normal, deletion thorax and A1–5. Duplicated inverted A6	Genetic regulator (finger)	Activates *kni* Represses *gt* Regulates *eve* Regulates *h, runt, ftz* Represses *Scr* Activates *Antp P1* Represses *Abd B*
knirps (*kni*)	Transcription starts in syncytial blastoderm forming zone 45–30% EL. Ventral patch and narrow ring form in anterior by cellular blastoderm	Replacement A1–7 by single A type segment	Genetic regulator (finger)	Represses *Kr* Regulates *h, runt, eve, ftz* Represses *Abd-B*

Gene	Expression	Phenotype	Type	Interactions
giant (*gt*)	Transcription syncytial blastoderm 82–60 and 33–0% EL, then ant. band narrows, divides into 2, another ant. band appears at 97–91% EL and post. band fades	Defects in or absence of labrum, labium, A5–7	Genetic regulator (Leu zipper)	Activates *Scr*
orthodenticle (*otd*)	Transcription in anterior of syncytial blastoderm 90–70% band in cellular blastoderm	Defects in head	Genetic regulator (homeobox)	
tailless (*tll*)	Transcription in termini of syncytial blastoderm	Head skeleton reduced but labrum present. Hindgut and Malpighian tubules absent but post. midgut present	steroid receptor family	Activates *hb* Represses *Kr* Represses *kni* Activates *h*
huckerbein (*hkb*)		Defects in termini, *tor* effects not attributable to *tll*		
caudal (*cad*)	Maternal mRNA initially uniform in egg then forms P-A gradient, as does protein, by syncytial blastoderm. Zygotic transcription contributes to gradient but most pronounced in zone 19–13% EL	Little maternal effect if *cad*$^+$ introduced by sperm. Zygotic phenotype defects A10. Maternal and zygotic absence causes severe defects in abdomen with mouthhooks in A8	Genetic regulator (homeobox)	Activates *ftz*

Zygotic pair-rule class

Gene	Expression	Phenotype	Type	Interactions
hairy (*h*)	Transcription starts syncytial blastoderm, 7 stripes + dorsal head patch by cellularization. Fades during gastrulation. Later expressed in cells around tracheal pits	Range from deletion of odd parasegments to formation of lawn of denticles. Also loss of labral tooth. Other alleles give extra hairs in adult	Genetic regulator (helix–loop–helix)	Represses *runt, ftz* Activates *eve*

Gene	Expression pattern	Mutant phenotype	Molecular nature	Developmental function
even skipped (*eve*)	Transcription starts syncytial blastoderm, 7 stripes corresponding to parasegments 1,3,5 etc by cellularization. Transiently 14 stripes during gastrulation then fades. Later in proctodeum and CNS	Abolishes segmentation giving lawn of denticles. Weak alleles delete odd parasegments	Genetic regulator (homeobox)	Activates *h*, *ftz*, *runt* (early) Represses *ftz*, *runt* (later) Activates itself Represses *odd* Represses *wg* Activates *Dfd*
runt (*run*)	Transcription starts syncytial blastoderm, 7 stripes out of phase with *h*. After cellularization becomes single segment stripes which persist through extension, also dorsal head patch	Deletions more than one segment wide centred on T2, A1,3,5,7, with mirror duplication of what is left	Nuclear protein	Represses *h*, *eve* Activates *ftz*
fushi tarazu (*ftz*)	Transcription starts syncytial blastoderm, 7 stripes corresponding to parasegments 2,4,6 etc by cellularization. Fades during extension, later expression in CNS	Deletes even parasegments	Genetic regulator (homeobox)	Activates itself Activates *en* Activates *Antp P2* Activates *Ubx* Represses *wg*
paired (*prd*)	Transcription starts syncytial blastoderm, 8 stripes with dorsal head patch by cellularization. Stripes 2–7 split then fade during extension	Deletions about 1 segment wide starting in middle of denticle band affecting mainly odd abdominal segments	Genetic regulator (homeobox)	Activates *wg*, *en*
odd paired (*opa*)		Opposite to *prd*		
sloppy paired (*slp*)		Like *prd* but weaker		Activates *wg*, *en*

Gene	Expression	Function	Interactions
odd skipped (*odd*)	Transcription starts syncytial blastoderm, stripes 1 cell posterior to *ftz*	Deletions of less than 1 segment width including posterior part of denticle bands in odd abdominal segments, with mirror duplications of anterior part. Also defects in head and anal tufts.	Genetic regulator (finger)	Represses *en*

Zygotic segment polarity class

Gene	Expression	Function	Interactions
engrailed (*en*)	Transcription starts at cellular blastoderm and produces 14 narrow stripes by extended germ band, marking the anterior of each parasegment. Even stripes develop before odd ones. Later stripes on head and telson, transient mesoderm expression. Expression in posterior compartments of larval imaginal discs	Ventral cuticle a continuous lawn of denticles. Less severe alleles delete even parasegments	Genetic regulator (homeobox)	Maintains *wg* Defines posterior compartment
wingless (*wg*)	Transcription starts at cellular blastoderm and produces 14 narrow stripes by extended germ band, marking the posterior of each parasegment. As for *en*, even stripes develop first. Also patches on head and telson	Ventral cuticle a continuous lawn of denticles, indicating deletion of ¾ of each segmental repeating unit	Signalling molecule (*wnt-1* homologue)	Maintains *en*
gooseberry (*gsb*)	Transcription starts late syncytial blastoderm. 14 bands form in anteroposterior sequence but with even bands leading. Bands correspond to *en + wg*. Transient expression in mesoderm	Ventral cuticle a lawn of denticles formed by the deletion of each posterior half segment and replacement by mirror copy of anterior half	Genetic regulator (homeobox, paired box)	

Gene	Expression pattern	Mutant phenotype	Molecular nature	Developmental function
armadillo (*arm*)	Uniform maternal and zygotic	Similar to *gsb*	Cytoskeletal	
fused (*fus*)	Maternal and zygotic	Similar to *gsb*. Both mother and embryo must be *fus*⁻	SerThr kinase	Maintains *en*
hedgehog (*hh*)		Similar to *gsb* but more severe		Maintains *en*
cubitus interruptus (*ci^D*)		Similar to *gsb*		
patched (*ptc*)	Uniform zygotic expression at cellular blastoderm, resolves into 14 stripes, 2 cells wide, plus dorsal head patch. 28 stripes by extended germ band, occupying second and fourth quarter of each parasegment. Also in mesoderm	Deletes posterior half of each denticle belt and replaces it with a mirror copy of the anterior half	Cell surface receptor?	Represses *wg*
naked (*nkd*)		Deletes most of denticle bands		Represses *en*
Homeotic selector genes				
labial (*lb*)	Transcription extended germ band anterior to cephalic furrow and posterior midgut	Deletes labial derivitives	Genetic regulator (homeobox)	Selector gene
Deformed (*Dfd*)	Transcription syncytial blastoderm in PS 1, later also PS 0. Eye-antennal disc	Deletion of mandibular and maxillary segments	Genetic regulator (homeobox)	Activates itself Selector gene

Gene	Transcription	Transformation	Protein type	Function
Sex combs reduced (*Scr*)	Transcription PS 2 in cellular blastoderm, later PS 3 epidermis and mesoderm, and abdominal ganglia	Transforms PS 3 to PS 4 and PS 2 to PS 1	Genetic regulator (homeobox)	Selector gene
Antennapedia (*Antp*)	Transcription starts cellular blastoderm, mainly PS 4 and 5 in extended germ band (separate promoters), later neuromeres in PS 4–12. Thoracic imaginal discs	Transforms PS 4,5 to PS 3	Genetic regulator (homeobox)	Selector gene
Ultrabithorax (*Ubx*)	Transcription starts cellular blastoderm, mainly in PS 6, lower levels PS 5–13, more in even parasegments. Neuromeres. Metathoracic discs	Transforms PS 5,6 to PS 4	Genetic regulator (homeobox)	Represses *Scr* Represses *Antp* Activates itself Selector gene
abdominal A (*abd-A*)	Transcribed PS 7–12 by extended germ band	Transforms PS 7–9 to PS 6	Genetic regulator (homeobox)	Represses *Ubx* Selector gene
Abdominal B (*Abd-B*)	Transcribed PS 10–13 and PS 14 (separate promoters) by extended germ band	Transforms PS 10–14 to PS 9.	Genetic regulator (homeobox)	Represses *Ubx* Selector gene
fork head (*fkh*)	Transcription cellular blastoderm, 100–95 and 15–0% EL	Stomodeum and proctodeum replaced by ectopic head structures	Nuclear protein	
spalt (*sal*)	Transcription cellular blastoderm, 70–60 and 15–0% EL	Transformation *mx* and *lb* toward T1. Transformation A9,10 toward A8	Nuclear protein	Regulates *eve*

8

What does it all mean?

Embryos used to be very mysterious but our knowledge has now advanced so far that most of the mystery has gone. We can now imagine building an embryo ourselves if we were provided with the correct components, and in the Appendix that follows such a developmental program is presented to illustrate the principles involved. Regional specification was until a few years ago a neglected problem persued only by a dedicated few but has now become the best understood aspect of early development. By contrast, the poorly understood aspects are how the epigenetic codings are converted into shape changes (morphogenesis) and into visible pattern (terminal differentiation), and it is in these areas that the proposed program is most deficient.

The fantastic progress of the past few years has vindicated the idea that if you want to understand embryos you should study the embryo itself. Studying other developmental problems such as virus assembly or gene regulation in terminally differentiating cells is no substitute. In this work the heritage of classical experimental embryology has been of crucial importance and many of the concepts which arose from it are now again on the centre stage. The power of molecular biology has also begun to show itself, and when appropriately applied, as to *Drosophila*, it has decisively transformed the kinds of experiments we can do.

A book such as this which deals with a series of animal types obviously invites comparison and generalization, and we do now have a reasonable idea of the sequence of processes through which a typical animal embryo will pass. There is often some significant regional localization in the egg, in at least one of the principal axes. There is often a period of cytoplasmic rearrangement following fertilization at which further localizations occur, conferring bilateral symmetry on the organism. Usually, early development is marked by rapid synchronous cleavage giving a ball or sheet of cells. Despite its featureless appearance this is the stage at which morphogen gradients specify major subdivisions of the body plan. Because of the global or long range character of the signals it is only at these early stages that significant embryonic regulation can occur. There is then a phase of complex morphogenetic movements to create a multilayered anatomy.

This may be accompanied by further regional specification within each of the original domains, and by appositional inductions between the different cell sheets. Determinations which are achieved early on must be stabilized in some way and it seems that this may often be done by mechanisms involving autoactivation.

The recent progress has not given us any wholly new developmental mechanisms. Cytoplasmic localization and induction remain the only two strategies known to bring about regional specification. The alternative possibility of directed sequential chromosome modification has yet to be discovered in an animal embryo. But the actual molecular realization of these two types of process shows great variability. Determinants can, it seems, be localized RNAs, localized proteins, or localized states of protein modification. Inducing factors can be protein cytokines, small molecules like retinoic acid, or even in the special case of the insect syncytium, nuclear proteins. After the early stages when persistent maternal components are important, changes in cell states will usually involve activation or repression of genes, but a protein can be caused to appear in active form in many ways: by transcriptional activation, reduction of mRNA instability, differential splicing, translational control or protein phosphorylation. Which particular biochemical mechanism is used in a particular case does not really matter. What does matter are the logical connections between different elements, in other words the developmental program. There seems no reason from what we know about the mechanisms at present why any body plan whatever cannot be made from living matter. An explanation of why we have on Earth a particular set of body plans is presumably historical and this remains a important problem for evolutionary theorists to solve.

Learning from *Drosophila*

There is no doubt that the success of work on *Drosophila* has had a great impact on our thinking. What we have to ask now is: can a similar genetic approach be made to other types of embryo, particularly to vertebrates? and: is the developmental program of other types, particularly vertebrates, the same as *Drosophila*?

A genetic approach to early vertebrate development has been difficult because the most experimentally tractable organisms, the amphibians, are non-starters when it comes to genetics. Amphibians are notorious for their large genome sizes and long generation times. The mouse is much better, but the numbers of offspring and accessibility of early stages falls far short of *Drosophila*. For these reasons the zebrafish has recently been introduced as a vertebrate on which large-scale mutagenesis screens could be carried out (review: Kimmel, 1989) and we await the results of this programme with interest.

There are two substantial differences from *Drosophila* which may cause problems. Firstly, *Drosophila* forms many visible regional specializations on its cuticle which means that a lot of information was obtained from looking at the phenotypes of embryo–lethal mutations. If the existing range of defects obtained from mutation or chemical teratogenesis is any guide about what to expect, then vertebrates show a much smaller range of defects, tending to concentrate in a few structures at a few critical developmental stages. In fact, most of the defects appear to be failures of morphogenesis: cell movements, tissue sheet extensions and fusions, rather than of regional specification. Secondly, the larger genome size of vertebrates may mean that there is more redundancy, each function being carried out by several genes rather than one. This would mean that mutating any particular gene to inactivity would have little effect, and this has been the experience with some early gene ablation experiments in the mouse using homologous recombination.

How serious these problems will be has yet to be established. But if a genetic approach to vertebrates is not going to be fruitful then some other way must be found of assessing biological function. Any organism can be subjected to *in situs* to reveal expression patterns of potentially interesting genes but the work with *Drosophila* has shown that it is not possible to deduce the developmental program from expression patterns alone. When phenotypes and expression patterns are put together, and when one expression pattern is looked at in a mutant for another gene, then useful information is obtained.

But perhaps we do not need to do all this work on vertebrates? Perhaps the problem is already solved because at a deep level the developmental program for *Drosophila* is also the program for all other animals? This is a popular view among *Drosophila* workers and is worth examining carefully. Undoubtedly *Drosophila* has yielded a number of genes, particularly the homeobox genes, which look like they are important in the early development of other animals as well. Also some *Drosophila* genes have turned out to code for components already identified in vertebrates and suspected of being important in development, chiefly cytokines and their receptors. But it is important to remember that a primary sequence resemblance is no guarantee of conserved biochemical function. For example, many genes belong to the TGFβ superfamily but do not have TGFβ activity. Moreover, even when the biochemical activity is the same, is the position in the developmental program the same? The programs must obviously differ somewhere otherwise all animals would look alike, and an instructive set of data comes from examination of the expression patterns of the *engrailed* gene in a variety of animals (Patel *et al.*, 1989*a*). In insects other than *Drosophila* and also in crustacea, *engrailed* is expressed in a stripe at the posterior of each segment, consistent with a similar role in the program to that which it plays in *Drosophila*. But in the leech and earthworm the patterns, although reiterated, are quite different; and in chick and *Xenopus*, expression is confined to a patch in the hindbrain. Although the biochemical function is probably conserved at least to the extent of being a transcription factor, the *engrailed* product must almost certainly play a different developmental role outside the arthropods.

So where could *Drosophila* and vertebrate developmental programs be the same? All animals have some sort of anteroposterior pattern, and almost all have a dorsal and ventral side. The discovery of the homology between the *Antennapedia–Bithorax* complexes in *Drosophila* with the Hox clusters in the mouse, where both the gene order along the chromosome and the order of expression along the anteroposterior axis of the organism have been conserved, is a powerful argument in favour of the evolutionary persistence of a primordial epigenetic coding system for head, trunk and tail (Graham *et al.*, 1989; Duboule & Dollé, 1989). As far as the dorsoventral axis is concerned, vertebrates do all the interesting things in the dorsal midline and insects in the ventral midline, so this seems on the face of it very different. But it is perhaps possible that the same coding factors are present but are interpreted differently. The sole piece of evidence at the time of writing concerns the gene *twist*, whose expression patterns in *Drosophila* and *Xenopus* do not, in fact, seem very similar.

Moving to earlier stages, it seems unlikely that the maternal systems in *Drosophila* could operate in vertebrates as they depend to a large extent on the anatomy of the egg chamber and on interactions between oocyte, nurse cells and follicle cells. Also in the syncytial stages it is possible for the transcription factors made by one nucleus to influence neighbouring nuclei by simple diffusion whereas in a multicellular embryo any corresponding processes would have to involve intercellular signalling.

In the end the similarity or difference of developmental programs can only be decided by experiment, but whatever the outcome, no one can deny the crucial role which the study of model systems has already had both in the discovery of important molecular components and in informing our modes of thought about early development. Those medical research organizations which have spent some of their money on *Drosophila* or *Xenopus* instead of spending it all on direct clinical applications have undoubtedly spent it wisely and will reap the benefits in due course.

Appendix: How to write a program for development

In this appendix we shall write a *developmental program* for making a simple animal. The animal in question, called the Crendonian snapper, belongs to no known phylum but uses the same processes of regional specification that have been uncovered in *Drosophila* and elsewhere. The developmental program is a set of rules for establishing the cell states in each part of the embryo. It is a self-contained dynamical system but one whose language is not differential equations but simple logic. Although the Crendonian snapper itself is an entirely new and original creation, the thinking behind such a representation will be found in some important theoretical articles by Kauffman (1971), Thomas (1973) and Wolpert & Lewis (1975).

The main gaps in the completeness of the program lie in the acquisition of an executive function by a group of cells with a particular coding. In other words, we still do not know exactly which set of genes we need to make one cell sheet move in a particular way, or adhere to another sheet, or differentiate into a particular terminal cell type. As discussed above, these are problems of morphogenesis and cell differentiation, rather than of regional specification.

Nomenclature

In what follows, the state of a gene will be represented by a small letter:

$$g \text{ means gene g is } \textit{on}$$

$$\bar{g} \text{ means gene g is } \textit{off}$$

Combinations of gene activity will be represented by capital letters or names:

$$A = abc \text{ means the state 'A' is defined by a and b and c } \textit{on}.$$
$$\text{Leg} = a\bar{b}\bar{c} \text{ means that the state 'Leg' is defined by a } \textit{on} \text{ and}$$
$$b \textit{ off} \text{ and c } \textit{off}.$$

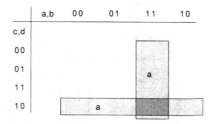

Fig. A.1 Veitch matrix for the function $a' = ab + c\bar{d}$.

Sometimes a state can be defined by alternative combinations of gene activity:

Liver = $a\bar{b} + c\bar{d}$ means that the state 'Liver' is defined by
(a *on* and b *off*) or (c *on* and d *off*).

(The '+' sign here denotes the logical *or* operation rather than arithmetic addition.)

The states of activity of individual genes are controlled by the presence and absence of other gene products:

$a' = ab + c\bar{d}$ means that a is *on* at time t if
either (a and b were on at t-1) or
(c was on and d was off at time t-1).

We shall have many occasions to express this type of relationship graphically as a so-called Veitch matrix in which all possible combinations are written out. In Fig. A.1 is shown the matrix for $a' = ab + c\bar{d}$, with the values satisfying the equation shown in grey. It will be noted that each term in the equation corresponds to a connected block of the possible combinations and that the overall solution is a simple superimposition of the two blocks.

A high proportion of the program will be written in such language, where components are either on or off. Although chemical concentrations are inherently continuous, the binary representation is often adequate because in molecular terms *on* means *enough to do the job*, and *off* means *not enough*. But there will also be some occasions where quantity is important, and in each of these cases we will have two threshold responses occurring at different concentrations of the effector.

The symbols used in the flow charts are the standard ones used in computing science: a rectangle for an autonomous event and a diamond for a decision involving a choice of pathway. But unlike a computer program, the developmental program embodies a high degree of *parallel processing* because every cell is making decisions simultaneously and because several genes may be regulated in the same cell at the same time. Parallel processes are shown in circles. Time is assumed to be divided into discrete steps, so in sequential processes one event has finished before the next one begins. If the

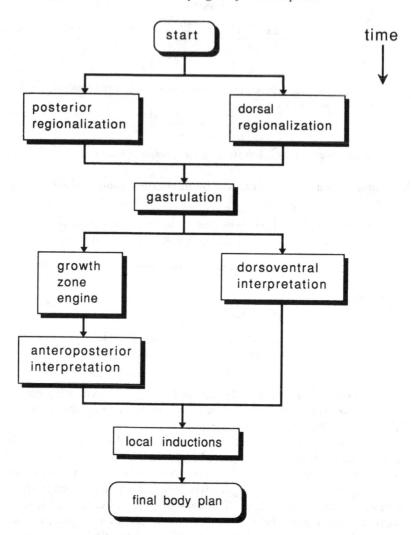

Fig. A.2 Overall program for making the Crendonian snapper.

snapper were ever built in reality it would probably incorporate a clock of some sort, for instance a series of genes each turning on the next in sequence, which would serve to break time into discrete steps and to synchronize events in different regions of the embryo.

The program

The overall program is shown in Fig. A.2, the embryonic development of the snapper in Fig. A.3 and the final adult form in Fig. A. 4. In the program, each box represents a subprogram started by a condition generated in the

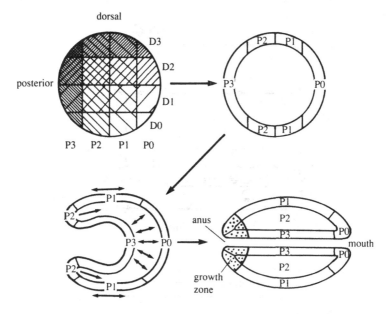

Fig. A.3 Embryonic development of the Crendonian snapper.

previous subprogram. Each subprogram occupies several discrete time steps. Initially the egg cleaves and secretes fluid into an internal cavity to create a spherical hollow blastula. The egg is equipped with determinants localized to the dorsal and posterior extremes and the cells inheriting these emit gradients across the blastula and activate a pair of genes for each axis. The posterior and dorsal regionalization divides an egg initially consisting of three cytoplasmic zones into an embryo consisting of 16 differently specified territories. Gastrulation consists of a set of cell behaviours, arising from the mutual relationships of the territories and from their different adhesive properties, in the course of which the embryo becomes three layered. As a result of these tissue movements a growth zone is established at the posterior end and this produces nine differently specified states along the anteroposterior axis of the organism. The processes of anteroposterior (AP) interpretation refer to the conversion of the codings generated by the growth zone into the principal structures of the AP axis. Dorsoventral (DV) interpretation is a similar process but based on the codings established by the original DV gradient. The final subprogram involves formation of structures on the surface of the body by localized inductive interactions with the underlying tissues.

As shown in Fig. A.4, the snapper is a rather simple animal and differs in several respects form all naturally occurring animals. In particular it has been found convenient to form the nervous system, lungs and liver from the mesoderm, and to form the sense organ and nostril as single midline organs.

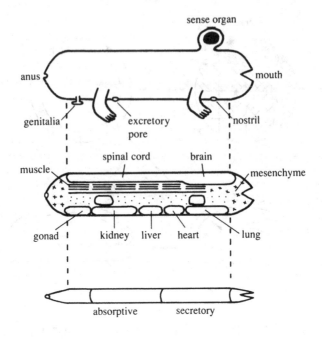

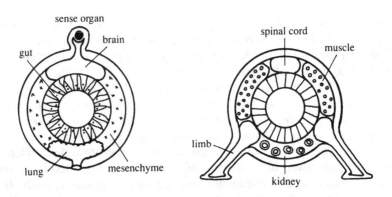

Fig. A.4 Anatomy of the adult snapper.

Posterior and dorsal regionalization

The snapper commences its development in the usual way by cleavage into a ball of cells and secretion of fluid to the interior to create a blastocoelic cavity. The creature is equipped with three dorsal genes: d_3, d_2, d_1. The first, d_3, is expressed only in the dorsal quarter of the oocyte, and its product causes a graded signal to be emitted at the blastula stage which turns on d_2 at a particular threshold and d_1 at a lower threshold (Fig. A.5). So the

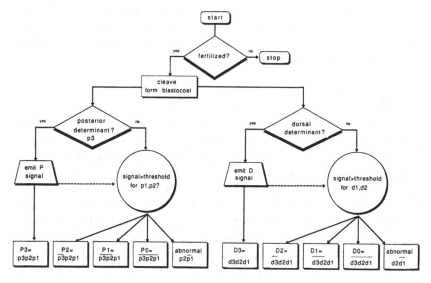

Fig. A.5 Subprogram for dorsal and posterior specification.

embryo becomes divided into four slices, from dorsal to ventral, for which the codings are:

$$D3 = d_3d_2d_1$$

$$D2 = \overline{d_3}d_2d_1$$

$$D1 = \overline{d_3d_2}d_1$$

$$D0 = \overline{d_3d_2d_1}$$

An exactly similar process involving an independent set of posterior genes occurs in the posterior to anterior axis leading to the subdivision of the embryo into four slices from posterior to anterior (Fig. A.5):

$$P3 = p_3p_2p_1$$

$$P2 = \overline{p_3}p_2p_1$$

$$P1 = \overline{p_3p_2}p_1$$

$$P0 = \overline{p_3p_2p_1}$$

The whole embryo is now a hollow ball of cells transected by two sets of three planes and is hence divided into 16 differently coded territories involving the differential activity of six genes (Fig. A.3). The codings used here are serial, in that they represent the simplest readout from the two gradients. It would obviously be possible to label 16 regions using only four genes if all the possible combinations were used, but then a more complex signalling system would be required to set up the spatial pattern.

At this early stage, a graft of dorsal tissue to ventral would reprogram the surroundings to give a double-dorsal duplication (D3, D2, D2, D3) without affecting the anteroposterior axis. A graft of posterior tissue to the anterior would likewise give a double posterior duplication.

Gastrulation

This most important stage of the snapper's life involves different morphogenetic behaviour in each of the territories P3, P2, P1, P0, and the action is radially symmetrical around the posterior-anterior axis (Fig. A.3).

The territory P3 emits processes which are long enough to explore the whole internal surface of the blastula and adhere to the territory P0. The processes contract, drawing the territory P3 into an internal sausage, later forming the gut. Where it meets P0 cell death takes place in both layers to create the mouth. While this is going on the remainder of P0 and P1 expand autonomously, for example by the action of microfilament bundles drawing the apical and basal surfaces of the cells closer together, and eventually cover most of the external surface. The territory P2 has adhesive properties which cause it to slip between the layers P3 and P1 to become a middle layer. The layers P0/1, P2 and P3 now correspond to the classical germ layers ectoderm, mesoderm and endoderm. The juxtaposition of the three territories P1, P2 and P3 in a ring at the posterior end leads to the activation of genes f and g. This region of three layers with g on is the growth zone which will make the anteroposterior body levels.

Dorsoventral interpretation

The dorsoventral codings established in the blastula are now examined and recorded in the mesoderm by the activation of four genes, one for each level.

d_3	d_2	d_1	turns on
0	0	0	v
0	0	1	1
0	1	1	m
1	1	1	n

There are several mappings which would accomplish this, they differ in the consequences of the abnormal codings which are not generated during normal development but which might be created by mutation or ectopic overexpression. One possible mapping which covers all the eight possible gene combinations is:

$$v' = \overline{d_3 d_2 d_1}$$
$$l' = \overline{d_3} d_2 d_1$$
$$m' = \overline{d_3} d_2$$
$$n' = d_3 \qquad \text{(Fig. A.6)}.$$

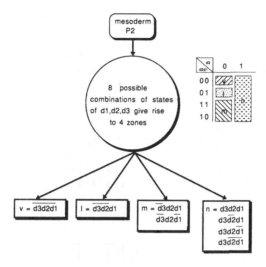

Fig. A.6 Dorsoventral interpretation

With this mapping, mutations of any of the d genes to inactivity would yield gaps in the pattern, rather like the *Drosophila* gap genes:

wild type has n m l v from dorsal to ventral
$d_3{}^-$ gives m m l v
$d_2{}^-$ gives n l l v
$d_1{}^-$ gives n m v v

and if adjacent territories with the same coding fuse into single territories then this means gaps.

The growth zone engine

We have thoughtfully provided the snapper with a mouth at an early stage of its development, so it is able to feed itself to obtain material to fuel the growth zone. The growth zone elongates the body by sequentially adding regions of different specification to the posterior end. At each time step the growth zone doubles in size and the part that is adjacent to the non-growing stock ceases to grow. The growth zone itself undergoes a sequence of changes in the state of two genes rather reminiscent of the activator and the inhibitor in the Gierer-Meinhardt model of Chapter 3 (GM1 model). The changes are initiated by the juxtaposition of the tissues P1, P2 and P3 at the posterior end. This turns on a gene g which, when on, both causes growth and allows the transformations of the activator, a, and the inhibitor, h, to occur. The juxtaposition also turns on a gene f, which activates the gene a after one time step and is then turned off autonomously. The activator, a, both turns on h and turns on itself in the absence of h. The gene g is

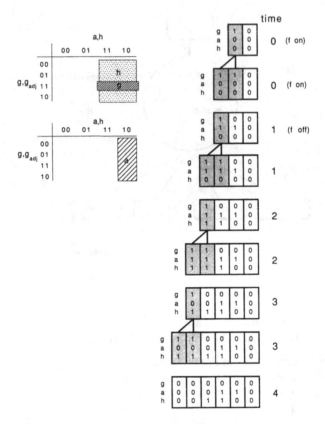

Fig. A.7 Laying down of a series of states from anterior to posterior by the growth zone.

maintained by itself, by a and by the presence of g in the adjacent internal territory, g_{adj}. In short:

$$a' = f + a\bar{h}$$
$$h' = a$$
$$g' = fg_{adj} + gg_{adj}a$$

The effect of this is turn on first a and then h in the growth zone, then a goes off, then h goes off. Because the growth zone is growing, it deposits regions of tissue with each of these states, $\bar{a}h$, $a\bar{h}$, ah, $\bar{a}h$, $a\bar{h}$ as it grows. The changes are defined as being dependent on the presence of g, and since g itself is turned off away from the tip the states in the interior do not evolve further. The sequence of events is shown in Fig. A.7. This produces a creature with five territories along the anteroposterior axis, of which the termini are equivalent codings $\bar{a}h$.

Five is not enough to make an animal so in *Drosophila* style we increase

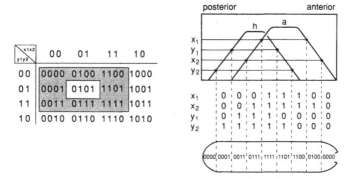

Fig. A.8 Subdivision of the territories formed by the growth zone.

the number by using the quantitative changes in a and h to turn on a second layer of genes: x_1, x_2, y_1, y_2. We assume that a and h do not rise and fall instantaneously, as implied by the logical equations, and that the true situation is as depicted in Fig. A.8. x_1 and x_2 are turned on at different threshold concentrations by a, and y_1 and y_2 are turned on at different threshold concentrations by h. Since the domains of a and h overlap in the manner shown, the effect of this is to divide the central part of the animal into seven distinctly coded zones instead of three. The two termini still have equivalent null codings so the number of zones from head to tail is nine, having eight distinct codings. On the Veitch matrix the codings form a simple ring, whose members are traversed in a clockwise direction proceeding from anterior to posterior (Fig. A.8). Because the growth zone included parts from the three germ layers and their states (P1-3) are clonally propagated, all three layers are subdivided in the same way with respect to x_1, x_2, y_1, y_2. The later differentiation will depend on the entire coding, which is composed of the states of the posterior genes, the dorsal genes and the anteroposterior genes.

Anteroposterior specification

Starting with the endoderm (P3), this becomes divided up into zones of squamous epithelium at the termini (mouth and anus), secretory epithelium in the anterior region where the food is being digested, and absorptive epithelium in the posterior part where it is being absorbed. In this case the mapping is simple and there are no abnormal codings (Fig. A.9).

In the mesoderm the most dorsal part is the future nerve cord, defined at this stage as P2n. This becomes divided into an anterior brain, a posterior spinal cord, and loose mesenchyme at the termini. The mapping differs from that in the endoderm, because we want the brain/spinal cord

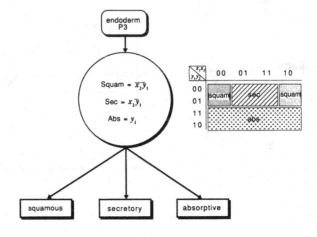

Fig. A.9 Subdivision of the gut.

boundary to be more anterior than the secretory/absorptive boundary in the gut (see Fig. A.4).

$$Ms = \overline{x_1 x_2 y_2}$$

$$Br = x_2 \bar{y}_2$$

$$Sc = y_2$$

Ventral to the nerve cord lies the strip P2m. This becomes striated muscle at all levels except the termini (see Fig. A.4). It later becomes segmented into equivalent blocks with septa anchored to the outer body wall, so as to provide the organism with a simple means of propulsion in water.

$$Mu = x_1 + y_2$$

$$Ms = \overline{x_1 y_2}$$

The l belt

The next most ventral layer is the l belt, which forms the limbs. The snapper is an amphibious creature and needs to be able to walk as well as swim. The parts which do not become limbs will become haemolymph, an arthropod type of blood which is propelled about the body cavity by the heart. It is convenient to make the limb equation of two terms:

$$limb = x_1 \bar{y}_2 + y_1 \bar{x}_2$$

since this makes the forelimb and the hindlimb non-equivalent and would enable us to introduce differences between them if this was required in a later version. The forelimb will develop from the third territory of the l belt

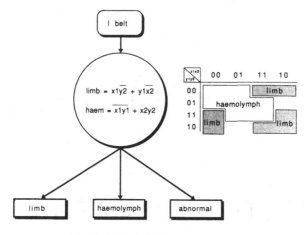

Fig. A.10 Subdivision of the l belt.

and the hindlimb in the seventh. If we wished we could easily program in a third pair of limbs in the middle by adding another term:

$$\text{limb} = x_1\overline{y}_2 + y_1\overline{x}_2 + x_2 y_1$$

This would give us pairs of limbs from the third, fifth and seventh levels, as may easily be confirmed by inspection of Figs A.10 and A.8.

Since the limb arrangement is a little more complex than that of the gut, n or m belts, it is worth considering what would happen to the snapper if one of the x or y genes were mutated. Consider first the loss of function mutations. If the gene x_1 were mutated to inactivity then any coding with $x_1 = 1$ would be changed such that $x_1 = 0$. Such changes do not necessarily lead to changes in anatomy because each tissue maps onto several genetic states. However, the state 1100 would be converted to 0100 and this represents a conversion from limb to haemolymph. So a loss of function mutation of x_1 would cause loss of the forelimb. If we write the normal anatomy of the l belt from posterior to anterior as HHLHHHLHH, then the x_1^- mutant would have a phenotype HHLHHHHHH.

The corresponding gain of function mutation would turn x_1 on everywhere and so any coding with x_1 off would be converted to the corresponding coding with x_1 on. This will have quite a profound effect on the pattern because the terminal regions will be recoded to limb. The pattern from posterior to anterior will be L-LHHHLLL. The - refers to the new coding 1001 which is not assigned in the program and which would therefore lead to no further change in state. We assume that repeated territories such as LLL would form a single organ, perhaps larger than normal, while an interrupted rudiment such as L-L would form two organs. So the most obvious result of overexpression of x_1 is the creation of an extra pair of limbs in the posterior.

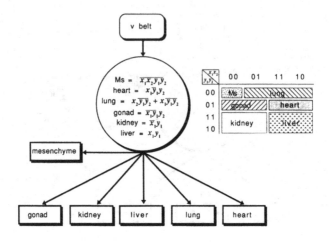

Fig. A.11 Subdivision of the v belt.

We can list all of the predicted phenotypes arising from loss of function and gain of function mutations as follows:

wildtype   HHLHHHLHH (posterior to anterior)

mutant	*phenotype*
x_1^-	HHLHHHHHH
x_2^-	HHLLL-LHH
y_1^-	HHHHHHLHH
y_2^-	HHL-LLLHH
x_1 ectopic	L-LHHHLLL
x_2 ectopic	HHHHHHLHH
y_1 ectopic	LLLHHHL-L
y_2 ectopic	HHLHHHHHH

If we had obtained these results from a real organism whose developmental program we did not know, then we might interpret them as follows. x_1 is obviously necessary for formation of the forelimb and y_1 for the hindlimb. Since overexpression in both cases gives extra limbs we can confidently regard these as limb forming genes. Loss of function mutations in x_2 and y_2 have little effect but the gain of function phenotypes are similar to loss of function of y_1 and x_1 respectively, so they would be regarded as repressors of hind and forelimb respectively. Priority would undoubtedly be given to cloning x_1 and y_1 but of course we really need the whole system to be exposed to understand what is going on.

The v belt

The most ventral part of the mesoderm is the P2v strip which is destined to form a variety of internal organs, namely the lung (Lu), heart(He), liver(Li), kidney(Ki) and gonad(Go) in anterior to posterior sequence. The mapping is shown in Fig. A.11. Once again we can enquire about the effects of mutations in the four selector genes:

wildtype   MsGoKiKiLiHeLuLuMs (posterior to anterior)

mutant	*phenotype*
x_1^-	MsGoKiKiKiGoLuLuMs
x_2^-	MsGoKiKiLiHeLuMsMs
y_1^-	MsGoGoGoHeHeLuLuMs
y_2^-	MsMsKiKiLiLuLuLuMs
x_1 ectopic	LuHeLiLiLiHeLuLuLu
x_2 ectopic	LuGoKiKiLiHeLuLuLu
y_1 ectopic	KiKiKiKiLiLiLiKiKi
y_2 ectopic	GoGoKiKiLiHeHeGoGo

So loss of function of x_1 produces a homeotic transformation of heart to gonad and suppresses the liver. Loss of function of x_2 has little effect, loss of function of y_1 causes loss of the kidney and liver, and of y_2 causes loss of the heart and gonad. Overexpression of x_1 causes a nearly mirror-symmetrical double anterior duplication, and overexpression of y_1 causes an extreme type of double posterior duplication in which only two sorts of structure remain. Overexpression of x_2 and y_2 cause rather striking homeotic transformations of the termini, in the case of x_2 we get extra lung at the posterior and in the case of y_2 we get extra gonad in the anterior. Given this genetic data alone we would doubtless know that we had identified some important genes but would probably not be able to reconstruct the subroutine for regionalization of the P2v strip. However, if we had the mutant data *and* the normal expression patterns we probably would be able to work it out. After all, that was how it was done in *Drosophila*.

The ectoderm

The ectoderm, P1/P0, is not directly subdivided in relation to the antero-posterior genes, although they are expressed in this layer. Instead, a series of local inductive interactions creates specializations and the remainder of the ectoderm becomes epidermis. The inductions required to provide the bare minimum of relationship with the external world are shown in Fig. A.12.

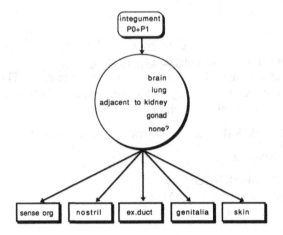

Fig. A.12 Local inductions leading to specializations in the ectoderm.

The body plan complete

The organism as shown in Fig. A.4 is built using only 14 genes which have explicit informational functions (not counting v, l, m, n here used only as labels), plus six local interactions making a total of 19 variables. The number of distinctly coded parts in the final organism is 48 onto which are mapped 19 tissue types, so the same tissue in different parts of the body is often non-equivalent. 48 parts could be uniquely labelled with as few as 6 genes, but the signalling systems for activating the genes in such a tightly combinatorial way would have to be rather complex.

Some readers may feel that the design of the Crendonian snapper is so arbitrary as to be pointless and that it does not contribute anything to our knowledge of how real animals develop. But in contrast to the experimental observations which are the main topic of this book it is not meant as an explanation of any particular problem. It is rather an exercise in thinking about body plan formation and is intended to give an idea of what a complete explanation would look like. The justification for this is that, in the final analysis, you only really understand something when you can make one yourself.

References

Akam, M.E. & Martinez-Arias, A. (1985). The distribution of *Ultrabithorax* transcripts in *Drosophila* embryos. *EMBO J.*, **4**, 1689–700.

Albers, B. (1987). Competence as the main factor determining the size of the neural plate. *Dev. Growth Diffn.* **29**, 535–45.

Allen, N.D. and others (1988). Transgenes as probes for active chromosomal domains in mouse development. *Nature, London*, **333**, 852–5.

Ambrosio, L., Mahowald, A.P. & Perrimon, N. (1989). Requirement of the *Drosophila raf* homologue for *torso* function. *Nature, London*, **342**, 288–91.

Anderson, K.V. (1987). Dorso-ventral embryonic pattern genes of *Drosophila*. *Trends in Genetics*, **3**, 91–7.

Anderson, K.V., Bokla, L. & Nüsslein-Volhard, C. (1985). Establishment of dorsoventral polarity in the *Drosophila* embryo: the induction of polarity by the *Toll* gene product. *Cell*, **42**, 791–8.

Anderson, K.V., Jürgens, G. & Nüsslein-Volhard, C. (1985). Establishment of dorsoventral polarity in the *Drosophila* embryo: genetic studies on the role of the *Toll* gene product. *Cell*, **42**, 779–89.

Anderson, K.V. & Nüsslein-Volhard, C. (1984). Information for the dorsoventral pattern of the *Drosophila* embryo is stored as maternal mRNA. *Nature, London*, **311**, 223–7.

Astrow, S., Holton, B. & Weisblat, D.A. (1987). Centrifugation redistributes factors determining cleavage patterns in leech embryos. *Dev. Biol.*, **120**, 270–83.

Astrow, S.H., Holton, B. & Weisblat, D.A. (1989). Teloplasm formation in a leech, *Helobdella triserialis*, is a microtubule dependent process. *Dev. Biol.*, **135**, 306–19.

Atkinson, J.W. (1971). Organogenesis in normal and lobeless embryos of the marine prosobranch gastropod *Ilyanassa obsoleta*. *J. Morphol.*, **133**, 339–52.

Azar, Y. & Eyal-Giladi, H. (1979). Marginal zone cells – the primitive streak inducing component of the primary hypoblast in the chick. *J. Embryol. Exp. Morph.*, **52**, 79–88.

Bäckstrom, S. (1954). Morphogenetic effects of lithium on the embryonic development of *Xenopus*. *Arkiv for Zoologi*, **6**, 527–36.

Baker, N.E. (1988). Localization of transcripts from the *wingless* gene in whole *Drosophila* embryos. *Development*, **103**, 289–98.

Balinsky, B.I. (1947). Kinematik des endodermalen Materials bei der Gestaltung der wichstigsten Teile des Darmkanals bei den Amphibien. *Wilhelm Roux Arch. für EntwMech. Org.*, **143**, 126–66.

(1948). Korrelation in der Entwicklung der Mund und Kiemenregion und des Darmkanals bei Amphibien. *Wilhelm Roux Arch. für EntwMech. Orgs.*, **143**, 365–95.

297

Barton, M.C. & Shapiro, D.J. (1988). Transient administration of estradiol 17 beta establishes an autoregulatory loop permanently inducing oestrogen receptor mRNA. *Proc. Natl Acad. Sci USA*, **85**, 7119–23.

Barton, S.C., Surani, M.A.H. & Norris, M.L. (1984). Role of paternal and maternal genomes in mouse development. *Nature, London*, **311**, 374–6.

Bates, W.R. (1988). Development of myoplasm enriched ascidian embryos. *Dev. Biol.*, **129**, 241–52.

Bates, W.R. & Jeffrey, W.R. (1987*a*). Alkaline phosphatase expression in ascidian egg fragments and andromerogones. *Dev. Biol.*, **119**, 382–9.

(1987*b*). Localization of axial determinants in the vegetal pole region of ascidian eggs. *Dev. Biol.*, **124**, 65–76.

Bateson, W. (1894). *Materials for the Study of Variation*. London, Macmillan.

Baumgartner, S., Bopp, D., Burri, M. & Noll, M. (1987). Structure of two genes at the *gooseberry* locus related to the *paired* gene, and their spatial expression during *Drosophila* embryogenesis. *Genes and Dev.*, **1**, 1247–67.

Bautzmann, H., Holtfreter, J., Spemann, H. & Mangold, O. (1932). Versuche zür Analyse der Induktionsmittel in der Embryonalentwicklung. *Naturwissenschaften*, **20**, 971–4.

Beddington, R.S.P. (1987). Isolation, culture and manipulation of post-implantation mouse embryos. In *Mammalian Development: a Practical Approach*, ed. Monk, M., pp. 43–69, IRL Press, Oxford.

Beddington, R.S.P., Morgernstern, J., Land, H. & Hogan, A. (1989). An *in situ* transgenic enzyme marker for the midgestation mouse embryo and the visualisation of inner cell mass clones during early organogenesis. *Development*, **106**, 37–46.

Beddington, R.S.P. & Robertson, E.J. (1989). An assessment of the developmental potential of embryonic stem cells in the midgestation mouse embryo. *Development*, **105**, 733–7.

Beeman, R.W., Stuart, J.J., Haas, M. & Denell, R.E. (1989). Genetic analysis of the homeotic gene complex (*HOM-C*) in the beetle *Tribolium castaneum*. *Dev. Biol.*, **133**, 196–209.

Bellairs, R. (1963). The development of somites in the chick embryo. *J. Embryol. Exp. Morph.*, **11**, 697–714.

Benson, S., Sucov, H., Stephens, L., Davidson, E. & Wilt, F. (1987). A lineage specific gene encoding a major matrix protein of the sea urchin embryo spicule. *Dev. Biol.*, **120**, 499–506.

Berleth, T., Burri, M., Thoma, G., Bopp, D., Richstein, S., Frigero, G. *et al.* (1988). The role of localization of *bicoid* RNA in organizing the anterior pattern of the *Drosophila* embryo. *EMBO J.*, **7**, 1749–56.

Bienz, M. & Tremml, G. (1988). Domain of *Ultrabithorax* expression in *Drosophila* visceral mesoderm from autoregulation and exclusion. *Nature, London*, **333**, 576–578.

Black, S.D. & Gerhart J.C. (1986). High frequency twinning of *Xenopus laevis* embryos from eggs centrifuged before first cleavage. *Dev. Biol.*, **116**, 228–40.

Blair, S.S. & Weisblat, D.A. (1984). Cell interactions in the developing epidermis of the leech *Helobdella triserialis*. *Dev. Biol.*, **101**, 318–25.

Boucaut, J.C.,Darribère, T., Boulekbache, H. & Thiery, J.P. (1984). Prevention of gastrulation but not neurulation by antibodies to fibronectin in amphibian embryos. *Nature, London*, **307**, 364–7.

Boycott, A.E., Diver, C., Garstang, S.L. & Turner, F.M. (1930). The inheritance of sinestrality in *Lymnaea peregra*. *Phil. Trans. Roy. Soc. B*, **219**, 51–131.

Bradley, A., Evans, M., Kaufman, M.H. & Robertson, E. (1984). Formation of germ line chimeras from embryo derived teratocarcinoma cell lines. *Nature, London*, **309**, 255–6.

Brandriff, B. & Hinegardner, R. (1975). Centrifugation and alteration of cleavage pattern. In *The Sea Urchin Embryo*, ed. Czihak, G. pp. 333–344, Springer-Verlag, Berlin and Heidelberg.

Briggs, R. & King, T.J. (1952). Transplantation of living nuclei from blastula cells into enucleated frogs' eggs. *Proc. Natl. Acad. Sci. USA*, **38**, 455–63.

Brinster, L. (1974). The effect of cells transferred into the mouse blastocyst on subsequent development. *J. Exp. Med.*, **140**, 1049–56.

Brinster, R.L., Chon, H.Y., Trumbauer, M.E., Yagle, M.K. & Palmiter, R.D. (1985). Factors affecting the efficiency of introducing foreign DNA into mice by microinjecting cells. *Proc. Natl. Acad. Sci. USA*, **82**, 4438–42.

Busa, W.B. & Gimlich, R.L. (1989). Lithium induced teratogenesis in frog embryos prevented by a polyphosphoinositide cycle intermediate or a diacylglycerol analog. *Dev. Biol.*, **132**, 315–24.

Cameron, R.A., Fraser, S.E., Britten, R.J. & Davidson, E.H. (1989). The oral-aboral axis of a sea urchin embryo is specified by first cleavage. *Development*, **106**, 641–7.

Cameron, R.A., Hough-Evans, B.R., Britten, R.J. & Davidson, E.H. (1987). Lineage and fate of each blastomere of the eight-cell sea urchin embryo. *Genes and Development*, **1**, 75–84.

Carpenter, C.D., Bruskin, A.M., Hardin, P.E., Keast, M.J., Anstrom, J., Tyner, A.L., *et al.* (1984). Novel proteins belonging to the troponin C superfamily are encoded by a set of mRNAs in sea urchin embryos. *Cell*, **36**, 663–71.

Cather, J.N. (1967). Cellular interactions in the development of the shell gland of the gastropod *Ilyanassa*. *J. Exp. Zool.*, **166**, 205–23.

Cather, J.N. & Verdonk, N.H. (1979). Development of *Dentalium* following removal of the D-quadrant blastomeres at successive cleavage stages. *Wilhelm Roux Arch. Dev. Biol.*, **187**, 355–66.

Cattanach, B.M. (1986). Parental origin effects in mice. *J. Embryol. Exp. Morph.*, **97** (Suppl), 137–50.

Chabry, L. (1887). Contribution à l'embryologie normale et teratologie des ascidies simples. *J. Anat. Physiol. (Paris)*, **23**, 167–319.

Child, C.M. (1936). Differential reduction of vital dyes in the early development of echinoderms. *Wilhelm Roux Arch. EntwMech. Orgs.*, **135**, 426–56.

Clement, A.C. (1952). Experimental studies on germinal localisation in *Ilyanassa*. I The role of the polar lobe in determination of the cleavage pattern and its influence in later development. *J. Exp. Zool.*, **121**, 593–625

(1956). Experimental studies on germinal localisation in *Ilyanassa*. II The development of isolated blastomeres. *J. Exp. Zool.*, **132**, 427–45.

(1962). Development of *Ilyanassa* following removal of the D macromere at successive cleavage stages. *J. Exp. Zool.*, **149**, 193–215.

(1967). The embryonic values of the micromeres in *Ilyanassa obsoleta* as determined by deletion experiments. I. The first quartet cells. *J. Exp. Zool.*, **166**, 77–88.

(1968). Development of the vegetal half of the *Ilyanassa* egg after removal of most of the yolk by centrifugal force compared with the development of animal halves of similar visible composition. *Dev. Biol.*, **17**, 165–86.

Cockroft, D.L. & Gardner, R.L. (1987). Clonal analysis of the developmental potential of 6th and 7th day visceral endoderm cells in the mouse. *Development*, **101**, 143–55.

Conklin, E.G. (1897). The embryology of *Crepidula*. *J. Morphol.*, **13**, 1–226.

(1905a). The organization and cell lineage of the ascidian egg. *J. Acad. Nat. Sci., Philadelphia*, **13**, 1–119.

(1905b). Mosaic development in ascidian eggs. *J. Exp. Zool.*, **2**, 145–223.

Cooke, J.C. (1988). A note on segmentation and the scale of pattern formation in insects and in vertebrates. *Development*, **104**, 245–8.

Cooke, J. (1989a). *Xenopus* mesoderm induction: evidence for early size control and partial autonomy for pattern development by onset of gastrulation. *Development*, **106**, 519–29.

(1989b). Mesoderm inducing factors and Spemann's organiser phenomenon in amphibian development. *Development*, **107**, 229–41.

Cooke J. & Smith, E.J. (1988). The restrictive effect of early exposure to lithium upon body pattern in *Xenopus* development studied by quantitative anatomy and immunofluorescence. *Development*, **102**, 85–99.

Cooke J. & Smith, J.C. (1987). The mid-blastula cell cycle transition and the character of mesoderm in UV induced non-axial *Xenopus* development. *Development*, **99**, 197–210.

Cooke, J. & Webber, J.A. (1985). Dynamics of the control of body pattern in the development of *Xenopus laevis*. II Timing and pattern in the development of single blastomeres (presumptive lateral halves), isolated at the 2 cell stage. *J. Embryol. Exp. Morph.*, **88**, 113–33.

Cooke, J. & Zeeman, E.C. (1976). A clock and wave-front model for control of the number of repeated structures during animal morphogenesis. *J. Theor. Biol.* **58**, 455–76.

Costello, D.P. (1945). Experimental studies of germinal localization in *Nereis*. I. The development of isolated blastomeres. *J. Exp. Zool.*, **100**, 19–66.

Cowan, A.E. & McIntosh (1985). Mapping the distribution of differentiation potential for intestine, muscle and hypodermis during early development in *Caenorhabditis elegans*. *Cell*, **41**, 923–32.

Cox, K.H., Angerer, L.M., Lee, J.J., Davidson, E.H. & Angerer, R.C. (1986). Cell lineage specific programs of expression of multiple actin genes during sea urchin embryogenesis. *J. Mol. Biol.*, **188**, 159–72.

Crampton, H.E. (1896). Experimental studies on gasteropod development. *Arch. f. Entw. Mech. Orgs.*, **3**, 1–26.

Crick, F. (1970). Diffusion in embryogenesis. *Nature, London*, **225**, 420–2.

Crick, F.H.C. & Lawrence, P.A. (1975). Compartments and polyclones in insect development. *Science*, **189**, 340–7.

Crowther, R.J. & Whittaker, J.R. (1986). Differentiation without cleavage. Multiple cytospecific ultrastructural expressions in individual one celled ascidian embryos. *Dev. Biol.*, **117**, 114–26.

Cruz, Y.P. & Pedersen, R.A. (1985). Cell fate in the polar trophectoderm of mouse blastocysts as studied by microinjection of cell lineage tracers. *Dev. Biol.*, **112**, 73–83.

Czihak, G. (1963). Entwicklungsphysiologische Untersuchungen an Echidnen. Verteilung und Bedeutung der Cytochromoxydase. *Wilhelm Roux Arch. EntwMech. Orgs.*, **154**, 272–95.

Dalcq, A. & Pasteels, J. (1937). Une conception nouvelle des bases physiologiques de la morphogénèse. *Arch. Biol.*, **48**, 669–710.

Dale, L. & Slack, J.M.W. (1987a). Fate map for the 32 cell stage of *Xenopus laevis*. *Development*, **99**, 527–51.

(1987b). Regional specification within the mesoderm of early embryos of *Xenopus laevis*. *Development*, **100**, 279–95.

Dale, L., Smith, J.C. & Slack, J.M.W. (1985). Mesoderm induction in *Xenopus laevis*. *J. Embryol. Exp. Morph.*, **89**, 289–313.

Dan-Sohkawa, M. & Satoh, N. (1978). Studies on dwarf larvae developed from isolated blastomeres of the starfish *Asterina pectinifera*. *J. Embryol. Exp. Morph.* **46**, 171–85.

Davidson, E.H. (1986). *Gene Activity in Early Development*, third edn. Orlando, Academic Press.

(1989). Lineage specific gene expression and the regulative capacities of the sea urchin embryo: a proposed mechanism. *Development*, **105**, 421–45.

Dearolf, C.R., Topol, J. & Parker, C.S. (1989). The *caudal* gene product is a direct activator of *fushi tarazu* transcription during *Drosophila* embryogenesis. *Nature, London*, **341**, 340–3.

Deno, T. & Satoh, N. (1984). Studies on the cytoplasmic determinant for muscle cell differentiation in ascidian embryos: an attempt at transplantation of the myoplasm. *Dev. Growth Diffn.*, **26**, 43–8.

Deol, M.S. & Whitten, W.K. (1972). Time of X chromosome inactivation in retinal melanocytes of the mouse. *Nature New Biol.*, **238**, 159–60.

Deuchar, E.M. & Burgess, A.M.C. (1967). Somite segmentation in amphibian embryos: is there a transmitted control mechanism? *J. Embryol. Exp. Morph.*, **17**, 349–58

DiBeradino, M.A., Orr, N.H. & McKinnell, R.G. (1986). Feeding tadpoles cloned from *Rana* erythrocyte nuclei. *Proc. Natl Acad. Sci. USA*, **83**, 8231–4.

Diederich, R.J., Merrill, V.K.L., Pultz, M.A. & Kaufman, T.C. (1989). Isolation, structure and expression of *labial*, a homeotic gene of the Antennapedia complex involved in *Drosophila* head development. *Genes and Dev.*, **3**, 399–414.

DiNardo, S. & O'Farrell, P.H. (1987). Establishment and refinement of segmental pattern in the *Drosophila* embryo: spatial control of *engrailed* expression by pair rule genes. *Genes and Dev.*, **1**, 1212–25.

Dohmen, M.R. & Verdonk, N.H. (1979). The ultrastructure and role of the polar lobe in development of molluscs. In *Determinants of Spatial Organization* ed. Subtelny, S. & Konigsberg, I.R. pp. 3–27. Academic Press.

Dorresteijn A.W.C., Wagemaker, H.A., de Laat, S.W., & van den Biggelaar, J.A.M. (1983). Dye coupling between blastomeres in early embryos of *Patella vulgata* (Mollusca, Gastropoda). Its relevance for cell determination. *Wilhelm Roux's Arch. Dev. Biol.*, **192**, 262–9.

Doyle, H.J., Harding, K., Hoey, T. & Levine , M. (1986). Transcripts encoded by a homeobox gene are restricted to dorsal tissues of *Drosophila* embryos. *Nature, London*, **323**, 76–9.

Driever, W. & Nüsslein-Volhard, C. (1988a). A gradient of *bicoid* protein in *Drosophila* embryos. *Cell*, **54**, 83–93.

(1988b). The *bicoid* protein determines position in the *Drosophila* embryo in a concentration-dependent manner. *Cell*, **54**, 95–104.

(1989). The *bicoid* protein is a positive regulator of *hunchback* transcription in the early *Drosophila* embryo. *Nature, London*, **337**, 138–43.

Duboule, D. & Dollé, P. (1989). The structural and functional organization of the murine *Hox* gene family resembles that of *Drosophila* homeotic genes. *EMBO J.*, **8**, 1497–505.

Durston, A.J., Timmermans, J.P.M., Hage, W.J., Hendriks, H.F.J., de Vries, N.J. Heideveld, M. & Nieuwkoop, P.D. (1989). Retinoic acid causes an anteroposterior transformation in the developing central nervous system. *Nature, London*, **340**, 140–4.

Dyce, J., George, M., Goodall, H. & Fleming, T.P. (1987). Do trophectoderm and inner cell mass cells in the mouse blastocyst maintain discrete lineages? *Development*, **100**, 685–98.

Edgar, L.G. & McGhee, J.D. (1986). Embryonic expression of a gut specific esterase in *Caenorhabditis elegans. Dev. Biol.* **114**, 109–18.

Eichele, G. & Thaller, C. (1987). Characterization of concentration gradients of a morphogenetically active retinoic acid in the chick limb bud. *J. Cell Biol.*, **105**, 1917–23.

Elbetieha, A. & Kalthoff, K. (1988). Anterior determinants in embryos of *Chironomus samoensis*: characterization by rescue bioassay. *Development*, **104**, 61–75.

Elinson, R.P. & Pasceri, P. (1989). Two UV sensitive targets in dorsoanterior specification of frog embryos. *Development*, **106**, 501–8.

Elinson, R.P. & Rowning, B. (1988). A transient array of parallel microtubules in frog eggs: potential tracks for a cytoplasmic rotation that specifies the dorsoventral axis. *Dev. Biol.*, **128**, 185–97.

Ettensohn, C.A. & McClay, D.R. (1988). Cell lineage conversion in the sea urchin embryo. *Dev. Biol.*, **125**, 396–409.

Evans, M.J. & Kaufman, M.H. (1981). Establishment in culture of pluripotential cells from mouse embryos. *Nature, London*, **292**, 154–6.

Eyal-Giladi, H. (1984). The gradual establishment of cell commitments during the early stages of chick development. *Cell Diffn.*, **14**, 245–55.

Eyal-Giladi, H. & Khaner, O. (1989). The chick's marginal zone and primitive streak formation. II. Quantification of the marginal zone's potencies–temporal and spatial aspects. *Dev. Biol.*, **134**, 215–21.

Eyal-Giladi, H. & Kochav, S. (1976). From cleavage to primitive streak formation: a complementary normal table and a new look at the first stages of the development of the chick. *Dev. Biol.*, **49**, 321–37.

Fankhauser, G. (1945). The effects of changes in chromosome number on amphibian development. *Quart. Rev. Biol.*, **20**, 20–78.

Fenderson, B.A., Eddy, E.M. & Hakomori, S.I. (1988). The blood group I antigen defined by monoclonal antibody C6 is a marker of early mesoderm during murine embryogenesis. *Differentiation*, **38**, 124–33.

Finkelstein, R. & Perrimon, N. (1990). The *orthodenticle* gene is regulated by *bicoid* and *torso* and specifies *Drosophila* head development. *Nature, London*, **346**, 485–8.

Forman, D. & Slack, J.M.W. (1980). Determination and cellular commitment in the embryonic amphibian mesoderm. *Nature, London*, **286**, 492–4.

Fraser, R.C. (1960). Somite genesis in the chick. III. The role of induction. *J. Exp. Zool.*, **145**, 151–67.

Fraser, S., Keynes, R. & Lumsden, A. (1990). Segmentation in the chick hindbrain is defined by cell lineage restrictions. *Nature, London*, **344**, 431–5.

Freeman, G. (1988). The factors that promote the development of symmetry properties in aggregates from dissociated echinoid embryos. *Wilhelm Roux Arch. Dev. Biol.*, **197**, 394–405.

Freeman, G. & Lundelius, J.W. (1982). The developmental genetics of dextrality and sinestrality in the gastropod *Lymnaea peregra. Wilhelm Roux Arch. Dev. Biol.*, **191**, 69–83.

Frei, E., Schuh, R., Baumgartner, S., Burri, M., Noll, M., Jürgens, G. *et al.* (1988). Molecular characterization of *spalt*, a homeotic gene required for head and tail development in the *Drosophila* embryo. *EMBO J.* 7, 197–204.

Frohnhöfer, H.G. & Nüsslein-Volhard, C. (1986). Organization of anterior pattern in the *Drosophila* embryo by the maternal gene *bicoid. Nature, London*, **324**, 120–5.

(1987). Maternal genes required for anterior localization of *bicoid* activity in the embryo of *Drosophila. Genes and Development*, **1**, 880–90.

Gallera, J. (1971). Primary induction in birds. *Adv. Morphogen.*, **9**, 149–80.

Garcia Bellido, A., Lawrence, P.A. & Morata, G. (1979). Compartments in animal development. *Sci. Am.*, **241**, 102–11.

Gardner, R.L. (1978). The relationship between cell lineage and differentiation in the early mouse embryo. In *Genetic Mosaics and Cell Differentiation.* ed. Gehring, W.J. pp. 205–41, Springer Verlag : Berlin and Heidelberg.

(1984). An *in situ* marker for clonal analysis of development of the extraembryonic endoderm in the mouse. *J. Embryol. Exp. Morph.*, **80**, 251–88.

(1985). Regeneration of endoderm from primitive ectoderm in the mouse embryo: fact or artefact? *J. Embryol. Exp. Morph.*, **88**, 303–26.

Gardner, R.L. & Johnson, M.H. (1973). Investigation of early mammalian development using interspecific chimeras between rat and mouse. *Nature New Biology*, **246**, 86–9.

Gardner, R.L., Papaioannou, V.E. & Barton, S.C. (1973). Origin of the ectoplacental cone and secondary giant cells in mouse blastocysts reconstituted from isolated trophoblast and inner cell mass. *J. Embryol. Exp. Morph.*, **30**, 561–72.

Gardner, R.L. & Rossant, J. (1979). Investigation of the fate of 4.5 day post coitum mouse inner cell mass cells by blastocyst injection. *J. Embryol. Exp. Morph.*, **52**, 141–52.

Garner, W. & McLaren, A. (1974). Cell distribution in chimeric mouse embryos before implantation. *J. Embryol. Exp. Morph.*, **32**, 495–503.

Gergen, J.P. & Butler, B.A. (1988). Isolation of the *Drosophila* segmentation gene *runt* and analysis of its expression during embryogenesis. *Genes and Dev.*, **2**, 1179–93.

Gerhart, J., Danilchik, M., Doniach, T., Roberts, S., Rowning, B., & Stewart, R. (1989). Cortical rotation of the *Xenopus* egg: consequences for the anteroposterior pattern of embryonic dorsal development. *Development*, (Suppl.) 37–51.

Gerhart, J., Ubbels, G., Black, S., Hara, K. & Kirschner, M. (1981). A reinvestigation of the role of the grey crescent in axis formation in *Xenopus laevis*. *Nature, London*, **292**, 511–16

Gierer, A. (1981). Some physical, mathematical and evolutionary aspects of biological pattern formation. *Phil. Trans. Roy. Soc. Lond. B*, **295**, 429–40.

Gierer, A. & Meinhardt, H. (1972). A theory of biological pattern formation. *Kybernetik*, **12**, 30–9.

Gillespie, L.L., Paterno, G.D., & Slack, J.M.W. (1989). Analysis of competence: receptors for fibroblast growth factor in early *Xenopus* embryos. *Development*, **106**, 203–8.

Gimlich, R.L. (1986). Acquisition of developmental autonomy in the equatorial region of the *Xenopus* embryo. *Dev. Biol.*, **115**, 340–52.

Gimlich. R.L. & Braun, J. (1985). Improved fluorescent compounds for tracing cell lineage. *Dev. Biol.*, **109**, 509–14.

Giudice, G. (1986). *The Sea Urchin Embryo.* pp. 61–68. Springer Verlag: Berlin and Heidelberg.

Godsave, S.F. & Slack, J.M.W. (1989). Clonal analysis of mesoderm induction in *Xenopus laevis*. *Dev. Biol.* **134**, 486–90.

Gospodarowicz, D., Ferrara, N., Schweigerer, L. & Neufeld, G. (1987). Structural characterization and biological functions of fibroblast growth factor. *Endocrine Rev.*, **8**, 95–114.

Gossler A., Joyner, A.L., Rossant, J. & Skarnes, W.C. (1989). Mouse embryonic stem cells and reporter constructs to detect developmentally regulated genes. *Science*, **244**, 463–5.

Grabowski, C.T. (1956). The effects of the excision of Hensen's node on the early development of the chick embryo. *J. Exp. Zool.*, **133**, 301–44.

(1957). The induction of secondary embryos in the early chick blastoderm by grafts of Hensen's node. *Am. J. Anat.*, **101**, 101–27.

Graham, A., Papalopulu, N. & Krumlauf, R. (1989). The murine and *Drosophila* homeobox gene complexes have common features of organization and expression. *Cell*, **57**, 367–78.

Green, J.B.A., Howes, G., Symes, K., Cooke, J. & Smith, J.C. (1990). The biological effects of XTC-MIF: quantitative comparison with *Xenopus* bFGF. *Development*, **108**, 173–83.

Greenwald, I.S., Sternberg, P.W. & Horvitz, H.R. (1983). The *lin-12* locus specifies cell fates in *Caenorhabditis elegans*. *Cell*, **34**, 435–44.

Grunz, H. & Tacke, L. (1986). The inducing capacity of the presumptive endoderm of *Xenopus laevis* studied by transfilter experiments. *Wilhelm Roux's Arch. Dev. Biol.*, **195**, 467–73.

Guerrier, P. (1970). Les caractères de la segmentation et la détermination de la polarité dorsoventrale dans le développment de quelques *Spiralia*. II *Sabellaria alveolata* (Annelide polychete). *J. Embryol. Exp. Morph.*, **23**, 639–65.

Guerrier, P., van den Biggelaar, J.A.M., van Dongen, C.A.M. & Verdonk, N.H. (1978). Significance of the polar lobe for the determination of dorsoventral polarity in *Dentalium vulgare* (da Costa). *Dev. Biol.*, **63**, 233–42.

Gurdon, J.B. (1974). *The Control of Gene Expression in Animal Development*. Oxford University Press.

(1987). Embryonic induction - molecular prospects. *Development*, **99**, 285–306.

(1989). The localization of an inductive response. *Development*, **105**, 27–33.

Gurdon, J.B., Fairman, S., Mohun, T.J. & Brennan, S. (1985). The activation of muscle specific actin genes in *Xenopus* development by an induction between animal and vegetal cells of a blastula. *Cell*, **41**, 913–22.

Hafen, E., Kuroiwa, A. & Gehring, W.J. (1984). Spatial distribution of transcripts from the segmentation gene *fushi tarazu* during *Drosophila* embryonic development. *Cell*, **37**, 833–41

Hafen, E., Levine, M. & Gehring, W.J. (1984). Regulation of *Antennapedia* transcript distribution by the Bithorax complex in *Drosophila*. *Nature, London*, **307**, 287–9.

Hamilton, L. (1969). The formation of somites in *Xenopus*. *J. Embryol. Exp. Morph.*, 22, 253–64.

Hamilton, W.J. & Mossman, H.W. (1976). *Human Embryology*. Macmillan: London.

Handyside, A.H. (1978). Time of commitment of inside cells isolated from preimplantation mouse embryos. *J. Embryol. Exp. Morph.*, **45**, 37–53

Harding, K. & Levine, M. (1988). Gap genes define the limits of *Antennapedia* and *bithorax* gene expression during early development in *Drosophila*. *EMBO J.* 7, 205–14.

Harding, K., Rushlow, C., Doyle, H.J., Hoey, T. & Levine, M. (1986). Cross regulatory interactions among pair rule genes in *Drosophila*. *Science*, **233**, 953–9.

Hartenstein, V., Technau, G.M. & Campos Ortega, J.A. (1985). Fate mapping in wild type *Drosophila melanogaster*. III A fate map of the blastoderm. *Wilhelm Roux Arch. Dev. Biol.*, **194**, 213–16.

Harvey, R.P. & Melton, D.A. (1988). Microinjection of synthetic *Xhox-1A* homeobox mRNA disrupts somite formation in developing *Xenopus* embryos. *Cell*, **53**, 687–97.

Hashimoto, C., Hudson, K.L. & Anderson, K.V. (1988). The *Toll* gene of *Drosophila*, required for dorsoventral embryonic polarity, appears to encode a transmembrane protein. *Cell*, **52**, 269–79.

Hay, B., Ackerman, L., Barbel, S., Jan, L.Y. & Jan, Y.N. (1988). Identification of a component of *Drosophila* polar granules. *Development*, **103**, 625–40.

Heath, J.K. & Rees, A.R. (1985). Growth factors in mammalian embryogenesis. In *Growth Factors in Biology and Medicine*. CIBA Symposium 116. pp. 3–22, Pitman: London.

Henry, J.J., Amemiya, S., Wray, G.A. & Raff, R.A. (1989). Early inductive interactions are involved in restricting cell fates of mesomeres in sea urchin embryos. *Dev. Biol.*, **136**, 140–53.

Henry, J.J. & Martindale, M.Q. (1987). The organizing role of the D quadrant as revealed through the phenomenon of twinning in the polychaete *Chaetopterus variopedatus*. *Wilhelm Roux's Arch. Dev. Biol.*, **196**, 499–510.

Hill, D.P. & Strome, S. (1990). Brief cytochalasin induced disruption of microfilaments during a critical interval in 1 cell *C. elegans* embryos alters the partitioning of developmental instructions to the 2 cell embryo. *Development*, **108**, 159–72.

Hillman, N., Sherman, M.I. & Graham, C. (1972). The effect of spatial arrangement on cell determination during mouse development. *J. Embryol. Exp. Morph.*, **28**, 263–78

Hiromi, Y. & Gehring, W. (1987). Regulation and function of the *Drosophila* segmentation gene *fushi tarazu*. *Cell*, 963–74.

Hogan, B.L.M., Taylor, A. & Adamson, E. (1981). Cell interactions modulate embryonal carcinoma cell differentiation into parietal or visceral endoderm. *Nature, London*, **291**, 235–37.

Hogan, B. & Tilly, R. (1978). *In vitro* development of inner cell masses isolated immunosurgically from mouse blastocysts. II Inner cell masses from 3.5 to 4.0 day p.c. blastocysts. *J. Embryol. Exp. Morph.*, **45**, 107–21.

Hogan, B.L.M. & Tilly, R. (1981). Cell interactions and endoderm differentiation in cultured mouse embryos. *J. Embryol. Exp. Morph.*, **62**, 379–91.

Holland, P.H. & Hogan B.L.M. (1988). Expression of homeobox genes during mouse development: a review. *Genes and Development*, **2**, 773–82.

Holliday, R. (1987). The inheritance of epigenetic defects. *Science*, **238**, 163–70.

Holtfreter, J. (1933). Organisierungsstufen nach regionaler Kombination von Entomesoderm mit Ektoderm. *Biologisches Zentralblatt*, **53**, 404–31.

(1938a). Differenzierungspotenzen isolierter Teile der Urodelengastrula. *Wilhelm Roux Arch. EntwMech. Orgs.*, **138**, 522–656.

(1938b). Differenzierungspotenzen isolierter Teile der Anurengastrula. *Wilhelm Roux Arch. EntwMech. Orgs.*, **138**, 657–738.

Hopwood, N.D., Pluck, A. & Gurdon, J.B. (1989a). A *Xenopus* mRNA related to *Drosophila twist* is expressed in response to induction in the mesoderm and in the neural crest. *Cell*, **59**, 893–903.

(1989b). *Myo D* expression in the forming somites is an early response to mesoderm induction in *Xenopus* embryos. *EMBO J.*, **8**, 3409–17.

Hornbruch, A., Summerbell, D. & Wolpert, L. (1979). Somite formation in the early chick embryo following grafts of Hensen's node. *J. Embryol. Exp. Morph.*, **51**, 51–62.

Horst, J. Ter (1948). Differenzierungs und Induktions-leistungen verschiedener Abschnitte der Medullarplatte und der Urdarmdaches von *Triton* im Kombinat. *Wilhelm Roux Arch. EntwMech. Orgs.*, **143**, 275–303.

Hörstadius, S. (1935). Uber die Determination im Verlaufe der Eiachse bei Seeigeln. *Publ. Staz. Zool. Napoli.*, **14**, 251–429.

(1939). The mechanics of sea urchin development studied by operative methods. *Biol. Rev.*, **14**, 132–79.

(1952). Induction & inhibition of reduction gradients by the micromeres in the sea urchin egg. *J. Exp. Zool.*, **120**, 421–36.

(1957). On the regulation of bilateral symmetry in plutei with exchanged meridional halves and in giant plutei. *J. Embryol. Exp. Morph.*, **5**, 60–73.

(1973). *The Experimental Embryology of Echinoderms*. Clarendon Press: Oxford.

Hörstadius, S. & Wolsky, A. (1936). Studien Über die Determination der Bilateralsymmetrie der jungen Seeigelkeimes. *Wilhelm Roux Arch. EntwMech. Orgs.*, **135**, 69–113.

Howard, K. & Ingham, P. (1986). Regulatory interactions between the segmentation genes *fushi tarazu*, *hairy* and *engrailed* in the *Drosophila* blastoderm. *Cell*, **44**, 949–57.

Howard, K., Ingham, P. & Rushlow, C. (1988). Region specific alleles of the *Drosophila* segmentation gene *hairy*. *Genes and Dev.*, **2**, 1037–46.

Hülskamp, M., Schröder, C., Pfeifle, C. & Jäckle, H. (1989). Posterior segmentation of the *Drosophila* embryo in the absence of a maternal posterior organizer gene. *Nature, London*, **338**, 629–32.

Hurley, D.L., Angerer, L.M. & Angerer, R.C. (1989). Altered expression of spatially regulated embryonic genes in the progeny of separated sea urchin blastomeres. *Development*, **106**, 567–79.

Illmensee, K. & Mahowald, A.P. (1974). Transplantation of posterior polar plasm in *Drosophila*. Induction of germ cells at the anterior pole of the egg. *Proc. Natl. Acad. Sci. USA*, **71**, 1016–20.

Ingham, P.W., Baker, N.E. & Martinez-Arias, A. (1988). Regulation of segment polarity genes in the *Drosophila* blastoderm by *fushi tarazu* and *even skipped*. *Nature, London*, **331**, 73–5.

Ingham, P. & Gergen, P. (1988). Interactions between the pair rule genes *runt*, *hairy*, *even-skipped* & *fushi tarazu* and the establishment of pattern in the *Drosophila* embryo. *Development*, **104 (Suppl.)**, 51–60.

Ingham, P.W., Howard, K.R. & Ish Horowicz, D. (1985). Transcription pattern of the *Drosophila* segmentation gene *hairy*. *Nature, London*, **318**, 439–45.

Ingham, P.W. & Martinez-Arias, M. (1986). The correct activation of Antennapedia and Bithorax complex genes requires the *fushi tarazu* gene. *Nature, London*, **324**, 592–7.

Irish, V., Martinez-Arias, M. & Akam, M. (1989). Spatial regulation of the *Antennapedia* and *Ultrabithorax* homeotic genes during *Drosophila* early development. *EMBO J.*, **8**, 1527–37.

Ish Horowicz, D. & Pinchin, S.M. (1987). Pattern abnormalities induced by ectopic expression of the *Drosophila* gene *hairy* are associated with repression of *ftz* transcription. *Cell*, **51**, 405–15.

Ito, K. & McGhee, J.D. (1987). Parental DNA strands segregate randomly during embryonic development of *Caenorhabditis elegans*. *Cell*, **49**, 329–36.

Jack, T., & McGinnis, W. (1990). Establishment of the *Deformed* expression stripe requires the combinatorial action of coordinate, gap and pair rule proteins. *EMBO J.*, **9**, 1187–98.

Jacobson, A.G. (1988). Somitomeres: mesodermal segments of vertebrate embryos. *Development*, **104 suppl**, 209–20.

Jacobson, M. (1983). Clonal organization of the central nervous system of the frog. III Clones stemming from individual blastomeres of the 128, 256 and 512 cell stages. *J. Neurosci.*, **3**, 1019–38.

Jacobson, M. & Hirose, G. (1978). Origin of the retina from both sides of the embryonic brain: A contribution to the problem of crossing over at the optic chiasma. *Science*, **202**, 637–9.

Jacobson, M. & Hirose, G. (1981). Clonal organization of the central nervous system of the frog II. Clones stemming from individual blastomeres of the 32 and 64 cell stages. *J. Neurosci.*, **1**, 271–84.

Jacobson, M. & Rutishauser, U. (1986). Induction of neural cell adhesion molecule (N-CAM). in *Xenopus* embryos. *Dev. Biol.*, **116**, 524–31.

Jacobson, M. & Xu, W. (1989). States of determination of single cells transplanted between 512 cell *Xenopus* embryos. *Dev. Biol.*, **131**, 119–25.

Jaenisch, R., Jähner, D., Nobis, P., Simon, I., Löhler, J., Harbers, K., & Crotkopp, D. (1981). Chromosomal position and activation of retroviral genomes inserted into the germ line of mice. *Cell*, **24**, 519–29.

Janning, W. (1978). Gynandromorph fate maps in *Drosophila*. In *Genetic Mosaics & Cell Differentiation*. ed. Gehring, W.J. pp. 1–28 Springer Verlag: Berlin and Heidelberg.

Jeffery, W.R. & Meier S. (1983). A yellow crescent cytoskeletal domain in ascidian eggs & its role in early development. *Dev. Biol.*, **96**, 125–143.

(1984). Ooplasmic segregation of the myoplasmic actin network in stratified ascidian eggs. *Wilhelm Roux Arch. Dev. Biol.*, **193**, 257–62.

Jeffery, W.R. (1990). Ultraviolet irradiation during ooplasmic segregation prevents gastrulation, sensory cell induction and axis formation in the ascidian embryo. *Dev. Biol.*, **140**, 388–400.

Johnson, M.H. & Ziomek, C.A. (1983). Cell interactions influence the fate of mouse blastomeres undergoing the transition from the 16 to the 32 cell stage. *Dev. Biol.*, **95**, 211–18.

Jones, E. & Woodland, H. (1987). The development of animal caps in *Xenopus*; a measurement of the start of animal competence to form mesoderm. *Development*, **101**, 557–63.

Jones, E.A. & Woodland, H.R. (1989). Spatial aspects of neural induction in *Xenopus laevis*. *Development*, **107**, 785–791.

Jung, E. (1966). Untersuchungen am Ei des Speisebohnenkäfers *Brachidius obtectus* SAY (Coleoptera). II Entwick lungsphysiologische Ergebnisse der Schnurungsexperimente. *Wilhelm Roux Arch. EntwMech. Orgs.*, **157**, 320–92.

Jürgens, G., Wieschaus, E., Nüsslein-Volhard, C. & Kluding, H. (1984). Mutations affecting the pattern of the larval cuticle in *Drosophila melanogaster*. II. Zygotic loci on the third chromosome. *Wilhelm Roux Arch. Dev. Biol.*, **193**, 283–95.

Kageura, H. & Yamana, K. (1983). Pattern regulation in isolated halves and blastomeres of early *Xenopus laevis*. *J. Embryol. Exp. Morph.*, **74**, 221–34.

(1984). Pattern regulation in defect embryos of *Xenopus laevis*. *Dev. Biol.*, **101**, 410–15.

(1986). Pattern formation in 8 cell composite embryos of *Xenopus laevis*. *J. Embryol. Exp. Morph.*, **91**, 79–100.

Kalthoff, J. (1971). Position of targets and period of competence for UV induction of the malformation 'double abdomen' in the egg of *Smittia* sp. *Wilhelm Roux Arch. EntwMech Orgs.*, **168**, 63–84.

Kao, K.R., Masui, Y. & Elinson, R.P. (1986). Lithium induced respecification of pattern in *Xenopus laevis* embryos. *Nature, London*, **322**, 371–3.

Kaneda, T. & Hama, T. (1979). Studies on the formation and state of determination of the trunk organiser in the newt *C. pyrrhogaster*. *Wilhelm Roux Arch. of Dev. Biol.*, **187**, 25–34.

Kauffman, S.A. (1971). Gene regulation networks: a theory for their global structure and behaviours. *Curr. Top. Dev. Biol.*, **6**, 145–82.

Kaufman, M.H., Barton, S.C. & Surani, M.A.H. (1977). Normal postimplantation development of mouse parthenogenetic embryos to the forelimb bud stage. *Nature, London*, **265**, 53–5.

Kaufman, M.H. & O'Shea, K.S. (1978). Induction of monozygotic twinning in the mouse. *Nature, London*, **276**, 707–8.

Keller, R.E. (1975). Vital dye mapping of the gastrula and neurula of *Xenopus laevis* I. Prospective areas and morphogenetic movements of the superficial layer. *Dev. Biol.*, **42**, 222–41.

(1976). Vital dye mapping of the gastrula and neurula of *Xenopus laevis* II Prospective areas and morphogenetic movements of the deep layer. *Dev. Biol.*, **51**, 118–37.

Keller, R.E. & Danilchik, M. (1988). Regional expression, pattern and timing of convergence and extension during gastrulation of *Xenopus laevis*. *Development*, **103**, 193–209.

Keller, R.E., Danilchik, M., Gimlich, R. & Shih, J. (1985). The function and mechanism of convergent extension during gastrulation of *Xenopus laevis*. *J. Embryol. Exp. Morph.*, **89 Suppl**. 185–209.

Kemphues, K.J., Priess, J.R., Morton, D.G. & Cheng, N. (1988). Identification of genes required for cytoplasmic localization in early *C.elegans* embryos. *Cell*, **52**, 311–20.

Kessel, M., Balling, R. & Gruss, P. (1990). Variations of cervical vertebrae after expression of a Hox 1.1 transgene in mice. *Cell*, **61**, 301–8.

Kilchherr, F., Baumgartner, S., Bopp, D., Frei E. & Noll, M. (1986). Isolation of the *paired* gene of *Drosophila* and its spatial expression during early embryogenesis. *Nature, London*, **321**, 493–9.

Kimble, J.E. (1981). Strategies for control of pattern formation in *Caenorhabditis elegans*. *Proc. Roy. Soc. B*, **295**, 539–51.

Kimmel, C.B. (1989). Genetics and early development of zebrafish. *Trends in Genetics*, **5**, 283–8.

Kimmel, C.B. & Warga, R.M. (1986). Tissue specific cell lineages originate in the gastrula of the zebrafish. *Science*, **231**, 365–8.

King, R.C. (1970). *Ovarian development in Drosophila melanogaster*. Academic Press: New York.

Kintner, C.R. & Brockes, J.P. (1984). Monoclonal antibodies identify cells derived from dedifferentiating muscle in newt limb regeneration. *Nature, London*, **308**, 67–9.

Kintner, C.R. & Melton, D.A. (1987). Expression of *Xenopus* N-CAM RNA in ectoderm is an early response to neural induction. *Development*, **99**, 311–25.

Klar, A.J.S. (1990). The developmental fate of fission yeast cells is determined by the pattern of inheritance of parental and grandparental DNA strands. *EMBO J.*, **9**, 1407–15.

Klinger, M., Erdelyi, M., Szabad, J. & Nüsslein-Volhard, C. (1988). Function of *torso* in determining the terminal anlagen of the *Drosophila* embryo. *Nature, London*, **335**, 275–7.

Knipple, D.C., Seifert, E., Rosenberg, W.B., Preiss, A. & Jäckle, H. (1985). Spatial and temporal patterns of *Krüppel* gene expression in early *Drosophila* embryos. *Nature, London*, **317**, 40–4.

Kobayakawa, Y. & Kubota, H.Y. (1981). Temporal pattern of cleavage and the onset of gastrulation in amphibian embryos developed from eggs with reduced cytoplasm. *J. Embryol. Exp. Morph.*, **62**, 83–94.

Kochav, S. & Eyal-Giladi, H. (1971). Bilateral symmetry in the chick embryo. Determination by gravity. *Science*, **171**, 1027–9.

Komamini, T. (1988). Determination of dorsoventral axis in early embryos of the sea urchin *Hemicentrotus pulcherrimus*. *Dev. Biol.*, **127**, 187–96.

Kornberg, T., Siden, I., O'Farrell, P. & Simon, M. (1985). The *engrailed* locus of *Drosophila*: *in situ* localization of transcripts reveals compartment specific expression, *Cell*, **40**, 45–53.

Krause, G. (1958). Induktionssysteme in der Embryonalentwicklung von Insekten. *Ergebn. Biol.*, **20**, 159–98.

Krause, G. & Krause, J. (1965). Uber das Vermogen median durchschnittener Keimanlagen von *Bombyx mori* L. Sich *in ovo* und sich ohne Dottersystem *in vitro* zwillingsartig zu entwickeln. *Z. Naturf.*, **20b** 334–9.

Krumlauf, R., Hammer, R., Tilghman, S. & Brinster R.L. (1985). Developmental regulation of alpha-foetoprotein genes in transgenic mice. *Mol. Cell. Biol.*, **5**, 1639–48.

Kühltreiber, W.M., Serras, F., and van den Biggelaar, J.A.M. (1987). Spreading of microinjected horseradish peroxidase to nondescendent cells in embryos of *Patella* (Mollusca, Gastropoda). *Development*, **100**, 713–22.

Kumé, M. & Dan, K. (1957). *Invertebrate Embryology*. Reprinted Garland, New York 1988.

Kuziora, M.A. & McGinnis, W. (1989). Autoregulation of a *Drosophila* homeotic selector gene. *Cell*, **55**, 477–85.

Langelan, R.E. & Whiteley, A.H. (1985). Unequal cleavage and the differentiation of echinoid primary mesenchyme. *Dev. Biol.*, **109**, 464–75.

Laufer, J.S. & von Ehrenstein, G. (1981). Nematode development after removal of egg cytoplasm: absence of localized unbound determinants. *Science*, **211**, 402–5.

Lawrence, P.A. (1973). A clonal analysis of segment development in *Oncopeltus* (Hemiptera). *J. Embryol. Exp. Morph.*, **30**, 681–99.

Lawrence, P.A., Johnston, P., Macdonald, P. & Struhl, G. (1987). Borders of parasegments in *Drosophila* embryos are delimited by the *fushi tarazu* and *even skipped* genes. *Nature, London*, **328**, 440–3.

Lawson, K.A. & Pedersen, R.A. (1987). Cell fate, morphogenetic movement and population kinetics of embryonic endoderm at the time of germ layer formation in the mouse. *Development*, **101**, 627–52.

Le Douarin, N. (1982). *The Neural Crest*. Cambridge University Press: Cambridge.

Lehmann, F.E. (1937). Mesodermisierung des praesumptiven Chordamaterials durch Einwirkung von Lithiumchlorid auf die Gastrula von *Triturus alpestris*. *Wilhelm Roux Arch. EntwMech. Orgs.*, **136**, 112–46.

Lehmann, R. & Nüsslein-Volhard, C. (1986). Abdominal segmentation, pole cell formation, and embryonic polarity require the localized activity of *oskar*, a maternal gene in *Drosophila*. *Cell*, **47**, 141–52.

(1987*a*). Involvement of the *pumilio* gene in the transport of an abdominal signal in the *Drosophila* embryo. *Nature, London*, **329**, 167–70.

(1987*b*). *hunchback-* a gene required for segmentation of an anterior and posterior region of the *Drosophila* embryo. *Dev. Biol.*, **119**, 402–17.

Levine, M., Hafen, E. , Garber, R.L. & Gehring, W.J. (1983). Spatial distribution of *Antennapedia* transcripts during *Drosophila* development. *EMBO J.*, **2**, 2037–46.

Levy, J.B., Johnson, M.H., Goodall, H. & Maro, B. (1986). The timing of compaction: control of a major developmental transition in mouse early embryogenesis. *J. Embryol. Exp. Morph.*, **95**, 213–37.

Lewis, J.H. & Wolpert, L. (1976). The principle of non-equivalence in development. *J. Theor. Biol.*, **62**, 479–90.

Lewis, J., Slack, J.M.W. & Wolpert, L. (1977). Thresholds in development. *J. Theor. Biol.*, **65**, 579–90.

Lewis, N.E. & Rossant, J. (1982). Mechanism of size regulation in mouse embryo aggregates. *J. Embryol. Exp. Morph.*, **72**, 169–81.

Ling, N., Ueno, N., Ying, S.Y., Esch, F., Shimasaki, S., Hotta, M. *et al.* (1988).

Inhibins and activins. *Vitamins and Hormones*, **44**, 1–46.

Livingston, B.T. & Wilt, F.H. (1989). Lithium evokes expression of vegetal specific molecules in the animal blastomeres of sea urchin embryos. *Proc. Natl Acad. Sci. USA*, **86**, 3669–73.

—— (1990). Range and stability of cell fate determination in isolated sea urchin blastomeres. *Development*, **108**, 403–10.

Lohs-Schardin, M., Cremer, Ch. & Nusslein-Volhard, Ch. (1979). A fate map for the larval epidermis of *Drosophila melanogaster*. Localized cuticular defects following irradiation of the blastoderm with an ultra-violet laser microbeam. *Dev. Biol.*, **73**, 239–55.

Lovell-Badge, R.H., Evans, M.J. & Bellairs, R. (1975). Protein synthetic patterns of tissues in the early chick embryo. *J. Embryol. Exp. Morph.*, **85**, 65–80.

Lumsden, A. & Keynes, R. (1989). Segmental patterns of neuronal development in the chick hindbrain. *Nature, London*, **337**, 424–8.

Lutz, H. (1949). Sur la production expérimentale de la polyembryonie et de la monstruosité double chez les oiseaux. *Arch. d'Anat. Micro. et de Morph.*, **38**, 79–144.

Macdonald, P.M. & Struhl, G. (1986). A molecular gradient in early *Drosophila* embryos and its role in specifying the body pattern. *Nature, London*, **324**, 537–45.

McGrath, J. & Solter, D. (1984a). Inability of mouse blastomere nuclei transferred to enucleated zygotes to support development *in vitro*. *Science*, **226**, 1317–19.

—— (1984b). Completion of mouse embryogenesis requires both the maternal and paternal genomes. *Cell*, **37**, 179–83.

McKinnell, R.G. (1978). *Cloning. Nuclear Transplantation in Amphibia*. University of Minnesota Press: Minneapolis.

McLaren, A. (1976). *Mammalian Chimaeras*. Cambridge University Press: London.

McMahon, A.P. & Bradley, A. (1990). The *wnt-1 (int-1)* proto-oncogene is required for development of a large region of the mouse brain. *Cell*, **62**, 1073–85.

Malacinski, G.M., Benford, H. & Chung, H.M. (1975). Association of an ultraviolet irradiation sensitive cytoplasmic localization with the future dorsal side of the amphibian egg. *J. Exp. Zool.*, **191**, 97–110.

Mangold, O. (1933). Uber die Induktionsfähigkeit der verscheidener Bezirke der Neurula von Urodelen. *Naturwissenschaften*, **21**, 761–6.

Mangold, O. & Seidel, F. (1927). Homoplastiche und heteroplastiche Verschmelzung ganzer Tritonkeime. *Wilhelm Roux Arch. EntwMech. Orgs.*, **111**, 593–665.

Mansour, S.L., Thomas, K.R. & Capecchi, M.R. (1988). Disruption of the proto-oncogene *int-2* in mouse embryo-derived stem cells: a general strategy for targeting mutations to non-selectable genes. *Nature, London*, **336**, 348–52.

Martin, G.R. (1981). Isolation of a pluripotent cell line from early mouse embryos cultured in a medium conditioned by teratocarcinoma stem cells. *Proc. Natl Acad. Sci. USA*, **78**, 7634–8.

Martindale, M.Q., Doe, C.Q. & Morrill, J.B. (1985). The role of animal–vegetal interaction with respect to the determination of dorsoventral polarity in the equal cleaving spiralian, *Lymnaea palustris*. *Wilhelm Roux Arch. Dev. Biol.*, **194**, 281–95.

Martinez-Arias, A. (1986). The *Antennapedia* gene is required and expressed in parasegments 4 and 5 of the *Drosophila* embryo. *EMBO J.*, **5**, 135–41.

Martinez-Arias, A., Baker, N.E. & Ingham, P.W. (1988). Role of segment polarity genes in the definition & maintenance of cell states in the *Drosophila* embryo. *Development*, **103**, 157–70.

Martinez-Arias, A., Ingham, P.W., Scott, M.P. & Akam, M.E. (1987). The spatial

& temporal deployment of *Dfd* and *Scr* transcripts throughout development of *Drosophila. Development*, **100**, 673–83.

Martinez-Arias, A. & Lawrence, P.A. (1985). Parasegments and compartments in the *Drosophila* embryo. *Nature, London*, **313**, 639–42.

Maruyama, Y.K., Nakaseko, Y. & Yagi, S. (1985). Localization of cytoplasmic determinants responsible for primary mesenchyme formation and gastrulation in the unfertilized egg of the sea urchin *Hemicentrotus pulcherrimus. J. Exp. Zool.*, **236**, 155–63.

Massagué, J. (1987). The TFG-ß family of growth and differentiation factors. *Cell*, **49**, 437–8.

Meedel, T.H., Crowther, R.J. & Whittaker, J.R. (1987). Determinative properties of muscle lineages in ascidian embryos. *Development*, **100**, 245–60.

Meedel, T.H. & Whittaker, J.R. (1983). Development of translationally active mRNA for larval muscle acetylcholinesterase during ascidian embryogenesis. *Proc. Natl. Acad. Sci. USA*, **80**, 4761–5.

Meinhardt, H. & Gierer, A. (1980). Generation and regeneration of sequences of structures during morphogenesis. *J. Theor. Biol.*, **85**, 429–50.

Melton, D.A. & Whitman, M. (1989). Growth factors in early embryogenesis. *Ann. Rev. Cell Biol.*, **5**, 93–117.

Menkes, B. & Sandor, S. (1977). Somitogenesis: regulation, potencies, sequence determination & primordial interactions. In *Vertebrate Limb and Somite Development* 3rd Symposium of British Soc. Dev. Biol., ed. Ede, D.A., Hinchliffe, J.R. & Balls, M., pp. 403–419, Cambridge University Press, Cambridge.

Mintz, B. (1965). Experimental genetic mosaicism in the mouse. In *Preimplantation Stages of Pregnancy*, pp. 194–216, CIBA Foundation Symposium. J & A Churchill Ltd: London.

Mita-Miyazawa, I., Nishikata, T. & Satoh, N. (1987). Cell and tissue specific monoclonal antibodies in eggs and embryos of the ascidian *Halocynthia roretzi. Development*, **99**, 155–62.

Mitrani, E. & Eyal-Giladi, H. (1981). Hypoblastic cells can form a disk inducing an embryonic axis in chick epiblast. *Nature, London*, **289**, 800–2.

Mitrani, E. & Shimoni, Y. (1990). Induction by soluble factors of organized axial structures in chick epiblasts. *Science*, **247**, 1092–4.

Mlodzik, M., Fjose, A. & Gehring, W.J. (1985). Isolation of *caudal*, a *Drosophila* homeobox containing gene with maternal expression, whose transcripts form a concentration gradient at the preblastoderm stage. *EMBO J.*, **4**, 2961–69.

Mohler, J., Eldon, E.D. & Pirrotta, V. (1989). A novel spatial transcription pattern associated with the segmentation gene, *giant*, of *Drosophila. EMBO J.*, **8**, 1539–48.

Mohler, J. & Wieschaus, E.F. (1986). Dominant maternal effect mutations of *Drosophila melanogaster* causing the production of double abdomen embryos. *Genetics*, **112**, 803–22.

Mohun, T.J., Brennan, S., Dathan, N., Fairman, S. & Gurdon, J.B. (1984). Cell type specific activation of actin genes in the early amphibian embryo. *Nature, London*, **311**, 716–21.

Monk, M. (1986). Methylation and the X chromosome. *Bioessays*, **4**, 204–8.

Monk, M. & Harper, M.I. (1979). Sequential X chromosome inctivation coupled with cellular differentiation in early mouse embryos. *Nature, London*, **281**, 311–13.

Moody, S.A. (1987). Fates of the blastomeres of the 32 cell *Xenopus* embryo. *Dev. Biol.*, **122**, 300–19.

Morgan, T.H. (1901). *Regeneration*. Macmillan: London.

(1927). *Experimental Embryology.* Columbia University Press: New York.

Nakano, Y., Guerrero, I., Hidalgo, A., Taylor, A., Whittle, J.R.S. & Ingham, P.W. (1989). A protein with several possible membrane spanning domains encoded by the *Drosophila* segment polarity gene *patched. Nature, London*, **341**, 508–13.

Nakauchi, M. & Takashita, T. (1983). Ascidian one half embryos can develop into functional adult ascidians. *J. Exp. Zool.*, **227**, 155–8.

Nauber, U., Pankratz, M.J., Kienlin, A., Siefert, E., Klemm, U. & Jäckle, H. (1988). Abdominal segmentation of the *Drosophila* embryo requires a hormone receptor like protein encoded by the gap gene *knirps. Nature, London*, **336**, 489–92.

Needham, A.E. (1965). Regeneration in the arthropods and its endocrine control. In *Regeneration in Animals and Related Problems.* ed. Kiortsis, V. & Trampusch, H.A.L. pp. 283–323, Amsterdam: N. Holland.

Nelson, S.H. & McClay, D.R. (1988). Cell polarity in sea urchin embryos; reorientation of cells occurs quickly in aggregates. *Dev. Biol.*, **127**, 235–47.

Nemer, M. (1986). An altered series of ectodermal gene expressions accompanying the reversible suspension of differentiation in the zinc animalized sea urchin embryo. *Dev. Biol.*, **114**, 214–24.

Newport, J. & Kirschner, M. (1982*a*). A major developmental transition in early *Xenopus* embryos: I. Characterization and timing of cellular changes at the midblastula stage. *Cell*, **30**, 675–86.

(1982*b*). A major developmental transition in early *Xenopus* embryos II. Control of the onset of transcription. *Cell*, **30**, 687–96.

Newrock, K.M. & Raff, R.A. (1975). Polar lobe specific regulation of translation in embryos of *Ilyanassa obsoleta. Dev. Biol.*, **42**, 242–61.

Nicolet, G. (1970). Is the presumptive notochord responsible for somite genesis in the chick? *J. Embryol. Exp. Morph.*, **24**, 467–78.

(1971). Avian gastrulation. *Adv. Morphogen.*, **9**, 231–62.

Nicolis, G. & Prigogine, I. (1977). *Self Organization in Nonequilibrium Systems.* John Wiley: New York.

Nieuwkoop, P.D. (1952*a*). Activation and organization of the central nervous system in amphibians. I. Induction and activation. *J. Exp. Zool.*, **120**, 1–31.

(1952*b*). Activation and organization of the central nervous system in amphibians. II. Differentiation and organization. *J. Exp. Zool.*, **120**, 33–81.

(1952*c*). Activation and organization of the central nervous system in amphibians. III. Synthesis of a new working hypothesis. *J. Exp. Zool.*, **120**, 83–108.

(1969). The formation of the mesoderm in urodelean amphibians. I. The induction by the endoderm. *Wilhelm Roux Arch. Dev. Biol.*, **162**, 341–73.

Nieuwkoop, P.D. & Faber, J. (1967). Normal Table of *Xenopus laevis* (Daudin). 2nd edn. N. Holland; Amsterdam.

Nishida, H. (1987). Cell lineage analysis in ascidian embryos by intracellular injection of a tracer enzyme. III. Up to the tissue restricted stage. *Dev. Biol.*, **121**, 526–41.

Nishida, H. & Satoh, N. (1983). Cell lineage analysis in ascidian embryos by intracellular injection of a tracer enzyme. I. Up to the 8 cell stage. *Dev. Biol.*, **99**, 382–94.

(1985). Cell lineage analysis in ascidian embryos by intracellular injection of a tracer enzyme. II. The 16 and 32 cell stages. *Dev. Biol.*, **110**, 440–54.

Nishikata, T., Mita-Miyazawa, I., Deno, T. & Satoh, N. (1987*a*). Muscle cell differentiation in ascidian embryos analysed with a tissue specific monoclonal antibody. *Development*, **99**, 163–71.

(1987*b*). Monoclonal antibodies against components of the myoplasm of eggs of the ascidian *Ciona intestinalis* partially block the development of muscle specific acetylcholinesterase. *Development*, **100**, 577–86.

Nishikata, T., Mita-Miyazawa, I., Deno, T., Takamura, K. & Satoh, N. (1987). Expression of epidermis specific antigens during embryogenesis of the ascidian *Halocynthia roretzi*. *Dev. Biol.*, **121**, 408–16.

Nishikata, T., Mita-Miyazawa, I. & Satoh, N. (1988). Differentiation expression in blastomeres of cleavage arrested embryos of the ascidian *Halocynthia roretzi*. *Dev. Growth Diffn.*, **30**, 371–81.

Nüsslein-Volhard, C., Frohnhöfer, H.,G. & Lehmann, R. (1987). Determination of anteroposterior polarity in *Drosophila*. *Science*, **238**, 1675–81.

Nüsslein-Volhard, C. & Wieschaus, E. (1980). Mutations affecting segment number & polarity in *Drosophila*. *Nature, London*, **287**, 795–801.

Nüsslein-Volhard, C., Wieschaus, E. & Kluding, H. (1984). Mutations affecting the pattern of the larval cuticle in *Drosophila melanogaster*. I. Zygotic loci on the second chromosome. *Wilhelm Roux Arch. Dev. Biol.*, **193**, 267–82.

Odell, G., Oster, G., Burnside, B. & Alberch, P. (1981). The mechanical basis of morphogenesis I: Epithelial folding and invagination. *Dev. Biol.*, **85**, 446–62.

Okada, T.S. (1953). Role of the mesoderm in the differentiation of endodermal organs. *Memoirs of the College of Science, University of Kyoto*. **20**, 157–62.

— (1957). The pluripotency of the pharyngeal primordium in urodelan neurulae. *J. Embryol. Exp. Morph.*, **5**, 438–48.

— (1960). Epitheliomesenchymal relationships in the regional differentiation of the digestive tract in the amphibian embryo. *Wilhelm Roux Arch. EntwMech. Orgs.*, **152**, 1–21.

Okada, Y.K. & Hama, T. (1945). Prospective fate & inductive capacity of the dorsal lip of the blastopore of the *Triturus* gastrula. *Proc. Imperial Acad. (Tokyo).*, **21**, 342–8.

O'Kane, C. & Gehring, W. (1987). Detection *in situ* of genomic regulatory elements in *Drosophila*. *Proc. Natl Acad. Sci. USA*, **84**, 9123–7.

Okazaki, K. (1975). Normal development to metamorphosis. In *The Sea Urchin Embryo*. ed G. Czihak, pp. 177–232, Springer Verlag: Berlin and Heidelberg.

Oster, G.F., Murray, J.D. & Harris, A.K. (1983). Mechanical aspects of mesenchymal morphogenesis. *J. Embryol. Exp. Morph.*, **78**, 83–125.

Otte, A.P., Van Run, P., Heideveld, M., van Driel, R. & Durston, A.J. (1989). Neural induction is mediated by cross talk between the protein kinase C and cyclic AMP pathways. *Cell*, **58**, 641–8.

Palmiter, R.D., Behringer, R.R., Quaife, C.J., Maxwell, I.H. & Brinster, R.L. (1987). Cell, lineage ablation in transgenic mice by by cell specific expression of a toxic gene. *Cell*, **50**, 435–43.

Pankratz, M.J., Seifert, E., Gerwin, N., Billi, B., Nauber, U. & Jäckle, H. (1990). Gradients of *Krüppel* and *knirps* gene products direct pair-rule gene stripe patterning in the posterior region of the *Drosophila* embryo. *Cell*, **61**, 309–17.

Papaiannou, V.E. (1982). Lineage analysis of inner cell mass & trophectoderm using microsurgically reconstituted mouse blastocysts. *J. Embryol. Exp. Morph.*, **68**, 199–209.

Pasteels, J. (1942). New observations concerning the maps of presumptive areas of the young amphibian gastrula (*Ambystoma* and *Discoglossus*). *J. Exp. Zool.*, **89**, 255–281.

Patel, N.H., Kornberg, T.B. & Goodman, G.S. (1989*b*). Expression of *engrailed* during segmentation in grasshopper and crayfish. *Development*, **107**, 201–12.

Patel, N.H., Martin-Blanco, E., Coleman, K.G., Poole, S.J., Ellis, M.C., Kernberg, T.B. & Goodman, C.S. (1989*a*). Expression of *engrailed* proteins in arthropods, annelids and chordates. *Cell*, **58**, 955–68.

Paterno, G.D., Gillespie, L.L., Dixon, M.. Slack, J.M.W. & Heath, J.K. (1989). Mesoderm inducing properties of *INT-2* and *kFGF*: two oncogene encoded growth factors related to FGF. *Development*, **106**, 79–83.

Pearson, M. & Elsdale, T. (1979). Somitogenesis in amphibian embryos. I. Experimental evidence for an interaction between two temporal factors in the specification of somite pattern. *J. Embryol. Exp. Morph.*, **51**, 27–50.

Pedersen, R.A., Wu, K. & Balakier,H. (1986). Origin of the inner cell mass in mouse embryos: cell lineage analysis by microinjection. *Dev. Biol.*, **117**, 581–95.

Penners, A. (1926). Experimentelle Untersuchungen zum Determinations problem am Keim von *Tubifex rivulorum* Lam. II. Die Entwicklung teilwize abgetoteter Keime. *Z. Wiss. Zool.*, **127**, 1–140.

Perrimon, N., Mohler, J.D., Engstrom, L. & Mahowald, A.P. (1986). X-linked female sterile loci in *Drosophila melanogaster*. *Genetics*, **113**, 695–712.

Pignoni, F., Baldarelli, R.M., Steingrimsson, E., Diaz, R.J., Patapoutian, A., Merriam, J.R. & Lengyel, J. (1990). The *Drosophila* gene *tailless* is expressed at the embryonic termini and is a member of the steroid receptor superfamily. *Cell*, **62**, 151–63.

Priess, J.R., Schnaebel H. & Schnaebel R. (1987). The *glp-1* locus and cellular interactions in early *C.elegans* embryos. *Cell*, **51**, 601–11.

Priess, J.R. & Thomson, J.N. (1987). Cellular interactions in early *C.elegans* embryos. *Cell*, **48**, 241–50.

Primmett, D.R.N., Norris, W.E., Carlson, G.J., Keynes, R.J. & Stern, C.D. (1989). Periodic segmental anomalies induced by heat shock in the chick embryo are associated with the cell cycle. *Development*, **105**, 119–30.

Prost, E., Deryckere, F., Roos, C., Haenlin, M., Pantesco, V. & Mohier, E. (1988). Role of the oocyte nucleus in determination of the dorsoventral polarity of *Drosophila* as revealed by molecular analysis of the *K10* gene. *Genes and Dev.*, **2**, 891–900.

Raff, R.A. (1987). Constraint, flexibility and phylogenetic change in the evolution of direct development in sea urchins. *Dev. Biol.*, **119**, 6–19.

Rands, G.F. (1985). Cell allocation in half and quadruple sized preimplantation mouse embryos. *J. Exp. Zool.*, **236**, 67–70.

Raven, Ch.P. (1966). *Morphogenesis: The Analysis of Molluscan Development*. 2nd Edn Pergamon Press: New York.

Rawles, M.E. (1936). A study in the localization of the organ forming areas in the chick blastoderm of the head process stage. *J. Exp. Zool.*, **32**, 271–315.

Razin, A. & Riggs, A.D. (1980). DNA methylation and gene function. *Science*, **210**, 604–10.

Render, J.A. (1983). The second polar lobe of the *Sabellaria cementarium* embryo plays an inhibitory role in apical tuft formation. *Wilhelm Roux Arch. Dev. Biol.*, **192**, 120–9.

Render, J.A. & Guerrier P. (1984). Size regulation & morphogenetic localization in the *Dentalium* polar lobe. *J. Exp. Zool.*, **232**, 79–86.

Represa, J. & Slack, J.M.W. (1989). Mesoderm induction by the mesoderm of *Xenopus* neurulae. *Int. J. Dev. Biol.*, **33**, 397–401.

Reverberi, G. & Minganti, A. (1946). Fenomeni di evocazione nello sviluppo dell'uovo di Ascidie. *Publ. Staz. Zool. Napoli*, **20**, 199–252.

Reverberi, G. & Ortolani, G. (1962). Twin larvae from halves of the same egg in ascidians. *Dev. Biol.*, **5**, 84–100.

Riggleman, B., Wieschaus, E. & Schedl, P. (1989). Molecular analysis of the *armadillo* locus: uniformly distributed transcripts and a protein with novel internal repeats are associated with a *Drosophila* segment polarity gene. *Genes and Dev.*, **3**, 96–113.

Robertson, E., Bradley, A., Kuehn, M. & Evans, M. (1986). Germ line transmission of genes introduced into cultured pluripotential cells by retroviral vector. *Nature, London*, **323**, 445–8.

Rosa, F.M. (1989). *Mix-1*, a homeobox mRNA inducible by mesoderm inducers, is expressed mostly in the presumptive endodermal cells of *Xenopus* embryos. *Cell*, **57**, 965–74.

Roth, S., Stein, D. & Nüsslein-Volhard, C. (1989). A gradient of nuclear localization of the *dorsal* protein determined dorsoventral pattern in the *Drosophila* embryo. *Cell*, **59**, 1189–202.

Rothe, M., Nauber, U. & Jäckle, H. (1989). Three hormone receptor like *Drosophila* genes encode an identical DNA binding finger. *EMBO J.*, **8**, 3087–94.

Ruiz i Altaba, A. & Melton, D.A. (1989*a*). Bimodal and graded expression of the *Xenopus* homeobox gene *Xhox3* during embryonic development. *Development*, **106**, 173–83.

(1989*b*). Interaction between peptide growth factors and homeobox genes in the establishment of anteroposterior polarity in frog embryos. *Nature, London*, **341**, 33–8.

Runnström, J. (1975). Integrating factors. In *The Sea Urchin Embryo*, ed. G. Czihak. pp 646–670, Springer Verlag:

Rutishauser, U. & Jessell, T.M. (1988). Cell adhesion molecules in vertebrate neural development. *Physiol. Rev.*, **68**, 819–57.

St Johnston, R. & Gelbart, W.M. (1987). *Dekapentaplegic* transcripts are localized along the dorsoventral axis of the *Drosophila* embryo. *EMBO J.*, **6**, 2785–91.

Sanchez-Herrero, E. & Crosby, M.A. (1988). The Abdominal B gene of *Drosophila melanogaster*. Overlapping transcripts exhibit two different spatial distributions. *EMBO J.*, **7**, 2163–73.

Sanchez-Herrero, E., Vernos, I., Marco, R. & Morata, G. (1985). Genetic organization of the *Drosophila* bithorax complex. *Nature, London*, **313**, 108–13.

Sander, K. (1976). Specification of the basic body pattern in insect embryogenesis. *Adv. Insect. Physiol.*, **12**, 125–238.

Sanes, J.R., Rubenstein, J.L.R. & Nicholas, J.F. (1986). Use of a recombinant retrovirus to study postimplantation cell lineage in mouse embryos. *EMBO J.*, **5**, 3133–42.

Sardet, C., Speksnijder, J., Inoue, S. & Jaffe, L. (1989). Fertilization and ooplasmic movements in the ascidian egg. *Development*, **105**, 237–49.

Satoh, N. (1979). On the 'clock' mechanism determining the time of tissue specific enzyme development during ascidian embryogenesis. I. Acetylcholinesterase development in cleavage arrested embryos. *J. Embryol. Exp. Morph.*, **54**, 131–39.

Satoh, N. & Ikegami, S. (1981). On the 'clock' mechanism determining the time of tissue specific enzyme development during ascidian embryogenesis. II. Evidence for association of the clock with the cycle of DNA replication. *J. Embryol. Exp. Morph.*, **64**, 61–71.

Sawada, T.O. & Schatten G. (1989). Effects of cytoskeletal inhibitors on ooplasmic segregation and microtubule organization during fertilization and early development in the ascidian *Molgula occidentalis*. *Dev. Biol.*, **132**, 331–42.

Saxen, L., Karkinen-Jaaskelainen, Lehtonen, M., Nordling, S. & Wartiovaara, J. (1976). Inductive tissue interactions. In Cell Surface Interactions in Embryogenesis ed. Poste, G. & Nicholson, G.L., pp. 331–407, Amsterdam; N.Holland.

Scharf, S.R. & Gerhart, J.C. (1980). Determination of the dorsoventral axis in eggs of *Xenopus laevis*: Complete rescue of UV impaired eggs by oblique orientation before first cleavage. *Dev. Biol.*, **79**, 181–98.

Scharf, S.R., Rowning, B., Wu, M. & Gerhart, J.C. (1989). Hyperdorsoanterior embryos from *Xenopus* eggs treated with D_2O. *Dev. Biol.*, **134**, 175–88.

Schierenberg, E. (1985). Cell determination during early embryogenesis of the nematode *Caenorhabditis elegans*. *Cold Spring Harbor Symp. Quant. Biol.*, **50**, 59–68.

(1987). Reversal of cellular polarity and early cell–cell interaction in the embryo of *Caenorhabditis elegans*. *Dev. Biol.*, **122**, 452–63.

Schüpbach, T. (1987). Germ line and soma cooperate during oogenesis to establish the dorsoventral pattern of egg shell and embryo in *Drosophila melanogaster*. *Cell*, **49**, 699–707.

Schüpbach, T. & Wieschaus, E. (1986). Maternal effect mutations altering the anterior posterior pattern of the *Drosophila* embryo. *Wilhelm Roux Arch. Dev. Biol.*, **195**, 302–17.

Seidel, F. (1929). Untersuchungen über das Bildungsprinzip der Keimanlage im Ei der Libelle *Platycnemis pennipes*. *Wilhelm Roux Arch. EntwMech. Orgs.*, **119**, 322–440.

(1935). Der Anlageplan im Libellenei. *Wilhelm Roux Arch. EntwMech. Orgs.*, **132**, 671–751.

Sengel, P. (1976). *The Morphogenesis of Skin*. Cambridge University Press.

Serras, F. & van den Biggelaar, J.A.M. (1987). Is a mosaic embryo also a mosaic of communication compartments? *Dev. Biol.*, **120**, 132–8.

Seydoux, G. & Greenwald, L. (1989). Cell autonomy of *lin-12* function in a cell fate decision in *C.elegans*. *Cell*, **57**, 1237–45.

Shankland, M. (1984). Positional determination of supernumerary blast cell death in the leech embryo. *Nature, London*, **307**, 541–3.

Sharpe, C.R. (1988). Developmental expression of a neurofilament M and two vimentin-like genes in *Xenopus laevis*. *Development*, **103**, 269–77.

Sharpe, C.R., Fritz, A., de Robertis, E.M. & Gurdon, J.B. (1987). A homeobox containing marker of posterior neural differentiation shows the importance of predetermination in neural induction. *Cell*, **50**, 749–58.

Shen, G. (1937). Experimente zür Analyse der Regulationsfähigkeit der frühen gastrula von *Triton*, zugleich ein Betrag zum problem der Cyclopie. *Wilhelm Roux Arch. EntwMech. Orgs.*, **137**, 271–316.

Shimizu, T. (1982). Ooplasmic segregation in the *Tubifex* egg: mode of pole plasm formation and possible involvement of microfilaments. *Wilhelm Roux Arch. Dev. Biol.*, **191**, 246–56.

Simcox, A.A. & Sang, J.H. (1983). When does determination occur in *Drosophila* embryos? *Dev. Biol.*, **97**, 212–21.

Simeone, A., Acampora, D., Arcioni, L., Andrews, P.W., Boncinelli, E. & Mavilio, F. (1990). Sequential activation of HOX2 homeobox genes by retinoic acid in human embryonal carcinoma cells. *Nature, London*, **346**, 763–66.

Sive, H.L., Hattori, K. & Weintraub, H. (1989). Progressive determination during formation of the anteroposterior axis in *Xenopus laevis*. *Cell*, **58**, 171–80.

Slack, J.M.W. (1980). A serial threshold theory of regeneration. *J. Theor. Biol.*, **82**, 105–40.

(1985). Homoeotic transformations in man: implications for the mechanism of embryonic development and for the organization of epithelia. *J. Theor. Biol.*, **114**, 463–90.

(1987*a*). We have a morphogen! *Nature, London*, **327**, 553–4.

(1987*b*). Morphogenetic gradients, past and present. *Trends Biochem. Sci.*, **12**, 200–4.

Slack, J.M.W., Darlington, B.G., Gillespie, L.L., Godsave, S.F., Isaacs, H.V. & Paterno, G.D. (1989). The role of fibroblast growth factor in early *Xenopus* development. *Development Suppl.*, 141–8.

Smith, A.G., and others (1988). Inhibition of pluripotential embryonic stem cell differentiation by purified polypeptides. *Nature, London*, **336**, 688–90.

Smith, J.C. (1989). Mesoderm induction and mesoderm inducing factors in early amphibian development. *Development*, **105**, 665–77.

Smith, J.C., Price, B.M.J., Van Nimmen, K. & Huylebroek, D. (1990). XTC-MIF: a potent *Xenopus* mesoderm inducing factor, is a homologue of activin A. *Nature, London*, 345, 729–31.

Smith, J.C. & Slack, J.M.W. (1983). Dorsalization and neural induction: properties of the organizer in *Xenopus laevis. J. Embryol. Exp. Morph.*, **78**, 299–317.

Smith, L.D. (1966). The role of a 'germinal plasm' in the formation of primordial germ cells in *Rana pipiens. Dev. Biol.*, **14**, 330–47.

Snape, A., Wylie, C.C., Smith, J.C. & Heasman, J. (1987). Changes in states of commitment of single animal cell blastomeres of *Xenopus laevis. Dev. Biol.*, **119**, 503–10.

Snell, G.D. & Stevens, L.C. (1966). Early Embryology. In *Biology of the Laboratory Mouse* ed. Green E.L., pp. 205–245, Dover Publications Inc.: New York.

Spemann, H. (1931). Uber den Anteil von Implantat und Wirtskeim an der Orientierung und Beschaffenheit der induzierten Embryonalanlage. *Wilhelm Roux Arch. EntwMech. Orgs.*, **123**, 389–517.

(1938). *Embryonic Development and Induction*. Reprinted by Garland Publishing 1988.

Spemann, H. & Mangold, H. (1924). Uber Induktion von Embryonenanlagen dürch Implantation artfremder Organisatoren. *Arch. für microscopische Anat. und Entwicklungsmechanik*, **100**, 599–638.

Spratt, N.T. (1955). Analysis of the organizer center in the early chick embryo. I. Localization of prospective notochord and somite cells. *J. Exp. Zool.*, **128**, 121–63.

Spratt, N.T. & Haas, H. (1960). Integrative mechanisms in the development of the early chick blastoderm. I. Regulative potentiality of separated parts. *J. Exp. Zool.*, **145**, 97–137.

(1961). Integrative mechanisms in the development of the early chick blastoderm. III. Role of cell population size and growth potentiality in synthetic systems larger than normal. *J. Exp. Zool.*, **147**, 271–93.

Sprengler, F., Stevens, L.M. & Nüsslein-Volhard, C. (1989). The *Drosophila* gene *torso* encodes a putative receptor tyrosine kinase. *Nature, London*, **338**, 478–83.

Stanojevic, D., Hoey, T. & Levine, M. (1989). Sequence specific DNA binding activities of the gap proteins encoded by *hunchback* and *Krüppel* in *Drosophila. Nature, London*, **341**, 331–5.

Stent, G.S. (1984). From probability to molecular biology. *Cell*, **36**, 567–71.

Stent, G.S. & Weisblat, D.A. (1982). The development of a simple nervous system. *Sci. Am.*, **246**, 100–11.

Stephens, L., Kitajima, T. & Wilt, F. (1989). Autonomous expression of tissue specific genes in dissociated sea urchin embryos. *Development*, **107**, 299–307.

Stern, C.D. (1990). The marginal zone and its contribution to the hypoblast and primitive streak of the chick embryo. *Development*, 109, 667–82.

Stern, C.D. & Canning, D.R. (1990). Origin of the cells that give rise to the primitive streak in the chick embryo. *Nature, London*, **343**, 273–5.

Stern, C.D., Fraser, S.E., Keynes, R.J. & Primmett, D.R.N. (1988). A cell lineage analysis of segmentation in the chick embryo. *Development*, **104**, 231–44.

Sternberg, P.W. & Horvitz, H.R. (1984). The genetic control of cell lineage during nematode development. *Ann. Rev. Genet.*, **18**, 489–524.

Steward, R., Zusman, S.B., Huang, L.H. & Schedl, P. (1988). The *dorsal* protein is distributed in a gradient in the early *Drosophila* embryo. *Cell*, **55**, 487–95.

Stevens, L.C. (1980). Teratocarcinogenesis and spontaneous parthenogenesis in mice. In *Differentiation and Neoplasia*. pp. 265–274, Springer Verlag, Berlin & Heidelberg.

Strecker, T.R., Kongsuwan, K., Lengyel, J.A. & Merriam, J.R. (1986). The zygotic mutation *tailless* affects the anterior and posterior ectodermal regions of the *Drosophila* embryo. *Dev. Biol.*, 64–76.

Strickland, S. & Mahdavi, V. (1978). The induction of differentiation in teratocarcinoma stem cells by retinoic acid. *Cell*, **15**, 393–403.

Strome, S. & Wood, W.B. (1983). Generation of asymmetry and segregation of germ line granules in early *C.elegans* embryos. *Cell*, **35**, 15–25.

Sulston, J.E, Schierenberg, E., White J.G. & Thomson, J.N. (1983). The embryonic cell lineage of the nematode *Caenorhabditis elegans*. *Dev. Biol.*, **100**, 64–119.

Sulston, J.E. & White, J.G. (1980). Regulation and cell autonomy during postembryonic development of *Caenorhabditis elegans*. *Dev. Biol.*, **78**, 577–97.

Summerbell, D., Lewis, J.H. & Wolpert, L. (1973). Positional information in chick limb morphogenesis. *Nature, London*, **224**, 492–6.

Surani, M.A.H., Barton, S.C. & Norris, M.L. (1984). Development of reconstituted mouse eggs suggests imprinting of the genome during gametogenesis. *Nature, London*, **308**, 548–50.

Surani, M.A.H. & Handyside A.H. (1983). Reassortment of cells according to position in mouse morulae. *J. Exp. Zool.*, **225**, 505–11.

Svajger, A., Levak-Svajger, B. & Skreb, N. (1986). Rat embryonic ectoderm as renal isograft. *J. Embryol. Exp. Morph.*, **94**, 1–27.

Swain, J.L., Stewart, T.A. & Leder, P. (1987). Parental legacy determines methylation and expression of an autosomal transgene: a molecular mechanism for parental imprinting. *Cell*, **50**, 719–27.

Takasaki, H. (1987). Fates and roles of the presumptive organizer region in the 32 cell embryo in normal development of *Xenopus laevis*. *Dev. Growth Diffn.*, **29**, 141–52.

Takasaki, H. & Konishi, H. (1989). Dorsal blastomeres in the equatorial region of the 32 cell *Xenopus* embryo autonomously produce progeny committed to the organizer. *Dev. Growth Diffn.*, **31**, 147–56.

Tam, P.P.L. & Beddington, R.S.P. (1987). The formation of mesodermal tissues in the mouse embryo during gastrulation and early organogenesis. *Development*, **99**, 109–26.

Tarin, D., Toivonen, S. & Saxen, L. (1973). Studies on ectodermal–mesodermal relationship in neural induction. II. Intercellular contacts. *J. Anat.*, **115**, 147–8.

Tarkowski, A.K. (1959). Experiments on the development of isolated blastomeres of mouse eggs. *Nature, London*, **184**, 1286–7.

(1961). Mouse chimeras developed from fused eggs. *Nature, London*, **190**, 857–60.

Tarkowski, A.K. & Wroblewska, J. (1967). Development of blastomeres of mouse eggs isolated at the 4 and 8 cell stage. *J. Embryol. Exp. Morph.*, **18**, 155–80.

Tautz, D. (1988). Regulation of the *Drosophila* segmentation gene *hunchback* by two maternal morphogenetic centres. *Nature, London*, **332**, 281–4.

Tautz, D., Lehmann, R., Schnürch, H., Schuh, R., Seifert, E., Kienlin, A. *et al.* (1987). Finger protein of novel structure encoded by *hunchback*, a second member of the gap class of *Drosophila* segmentation genes. *Nature, London*, **327**, 383–9.

Technau, G.M. (1987). A single cell approach to problems of cell lineage and commitment during embryogenesis of *Drosophila melanogaster. Development*, **100**, 1–12.

Thisse, B., Stoetzel, C., El Messal, M. & Perrin-Schmitt, F. (1987). Genes of the *Drosophila* maternal dorsal group control the specific expression of the zygotic gene *twist* in presumptive mesodermal cells. *Genes and Dev.*, **1**, 709–15.

Thomas, J.B., Crews, S.T. & Goodman, C.S. (1988). Molecular genetics of the *single minded* locus: a gene involved in the development of the *Drosophila* nervous system. *Cell*, **52**, 133–41.

Thomas, R. (1973). Boolean formalization of genetic control circuits. *J. Theor. Biol.*, **42**, 563–85.

Thomson, J. & Solter, D. (1988). Transgenic markers for mammalian chimeras. *Wilhelm Roux Arch. Dev. Biol.*, **197**, 63–5.

Tiedemann, H. (1978). Chemical approach to the inducing agents. In *Organizer – A Milestone of a Half Century from Spemann* ed. Nakamura O. & Toivonen S., pp. 91–117, Elsevier/N. Holland.

Titlebaum, A. (1928). Artifical production of *Janus* embryos of *Chaetopterus. Proc. Natl Acad. Sci. USA*, **14**, 245–7.

Tomlinson, C.R., Bates, W.R. & Jeffrey, W.R. (1987). Development of a muscle actin specified by maternal and zygotic mRNA in ascidian embryos. *Dev. Biol.*, **123**, 470–82.

Thomson, J. & Solter, D. (1988). Transgenic markers for mammalian chimeras. *Wilhelm Roux Arch. Dev. Biol.*, **197**, 63–5.

Turing, A.M. (1952). The chemical basis of morphogenesis. *Phil. Trans. Roy. Soc. B.*, **237**, 37–72.

Ullman, S.L. (1964). The origin and structure of the mesoderm and the formation of the Coelomic sacs in *Tenebrio molitor* L (Insecta, Coleoptera). *Phil. Trans. Roy. Soc. B.*, **248**, 245–77.

van den Biggelaar, J.A.M. (1977). Development of dorsoventral polarity and mesentoblast determination in *Patella vulgata. J. Morphol.*, **154**, 157–86.

van den Biggelaar, J.A.M. & Guerrier, P. (1979). Dorsoventral polarity and mesentoblast determination as concomitant results of cellular interactions in the mollusc *Patella vulgata. Dev. Biol.*, **68**, 462–71.

(1983). Origin of spatial organization. In *The Mollusca*, vol 3, Chap. 5, Development. Academic Press: New York.

van den Heuvel, M., Nusse, R., Johnston, P. & Lawrence P.A. (1989). Distribution of the *wingless* gene product in *Drosophila* embryos: a protein involved in cell–cell communication. *Cell*, **59**, 739–49.

Van Dongen, C.A.M. (1976). The development of *Dentalium* with special reference to the significance of the polar lobe. V. Differentiation of the cell pattern in lobeless embryos of *Dentalium vulgare* (da Costa). during late larval development. *Proc. Kon. Ned. Akad. Weterisch.*, C **79**, 245–55.

Van Dongen, C.A.M. & Geilenkirchen, W.L.M. (1974). The development of *Dentalium* with special reference to the significance of the polar lobe. I–IV. *Proc. Kon. Ned. Akad. V. Wet.*, C **77**, 57–100.

Vincent, J.P. & Gerhart, J.C. (1986). Subcortical rotation in *Xenopus* eggs: an early step in embryonic axis specification. *Dev. Biol.*, **123**, 526–39.

Vogt, W. (1929). Gestaltungsanalyse am Amphibienkeim mit ortlicher Vitalfarbung. II. Teil, Gastrulation and Mesodermbildung bei Urodelen und Anuren. *Wilhelm Roux Arch. EntwMech. Orgs.*, **120**, 384–706.

von Ubisch, L. (1938). Uber Keimverschmeltzungen an *Ascidiella aspersa. Wilhelm Roux Arch. EntwMech. Orgs.*, **138**, 18–36.

Waddington, C.H. (1952). *The Epigenetics of Birds*. Cambridge University Press.

Warner, A.E. & Gurdon, J.B. (1987). Functional gap junctions are not required for muscle gene activation by induction in *Xenopus* embryos. *J. Cell Biol.*, **104**, 554–64.

Warner, A.E., Guthrie, S.C. & Gilula, N.B. (1984). Antibodies to gap junction protein selectively disrupt junctional communication in the early amphibian embryo. *Nature, London*, **311**, 127–31.

Weeks, D.L. & Melton, D.A. (1987). A maternal messenger RNA localized to the vegetal hemisphere in *Xenopus* eggs codes for a growth factor related to TGF-ß. *Cell*, **51**, 861–7.

Weigel, D., Jürgens, G., Küttner, F., Seifert, E. & Jäckle, H. (1989). The homeotic gene *fork head* encodes a nuclear protein and is expressed in the terminal regions of the *Drosophila* embryo. *Cell*, **57**, 645–58.

Weisblat D.A. & Blair, S.S. (1984). Developmental indeterminacy in embryos of the leech *Helobdella triserialis*. *Dev. Biol.*, **101**, 326–35.

Weisblat, D.A., Kim, S.Y. & Stent, G.S. (1984). Embryonic origins of cells in the leech *Helobdella triserialis*. *Dev. Biol.*, **104**, 65–85.

Wessel, G.M. & McClay, D.R. (1985). Sequential expression of germ layer specific molecules in the sea urchin embryo. *Dev. Biol.*, **111**, 451–63.

West, J.D. (1976). Clonal development of the retinal epithelium in mouse chimeras and X inactivation mosaics. *J. Embryol. Exp. Morph.*, **35**, 445–61.

Wharton, R.P. & Struhl, G. (1989). Structure of the *Drosophila* Bicaudal-D protein and its role in localizing the posterior determinant *nanos*. *Cell*, **59**, 881–92.

Whitington, P.Mc.D. & Dixon, K.E. (1975). Quantitative studies of germ plasm and germ cells during early embryogenesis of *Xenopus laevis*. *J. Embryol. Exp. Morph.*, **33**, 57–74.

Whitman, M. & Melton, D. (1989). Induction of mesoderm by a viral oncogene in early *Xenopus* embryos. *Science*, **244**, 803–6.

Whittaker, J.R. (1973). Segregation during ascidian embryogenesis of egg cytoplasmic information for tissue specific enzyme development. *Proc. Natl Acad. Sci. USA*, **70**, 2096–100.

(1977). Segregation during cleavage of a factor determining endodermal alkaline phosphatase development in ascidian embryos. *J. Exp. Zool.*, **202**, 139–54.

(1980). Acetylcholinesterase development in extra cells by changing the distribution of myoplasm in ascidian embryos. *J. Embryol. Exp. Morph.*, **55**, 343–54.

Whittaker, J.R., Ortolani, G. & Farinella-Ferruzza, N. (1977). Autonomy of acetylcholinesterase differentiation in muscle lineage cells of ascidian embryos. *Dev. Biol.*, **55**, 196–200.

Wieschaus, E., Nüsslein-Volhard, C. & Jürgens, G. (1984). Mutations affecting the pattern of the larval cuticle in *Drosophila melanogaster*. III. Zygotic loci on the X-chromosome and fourth chromosome. *Wilhelm Roux Arch. Dev. Biol.*, **193**, 296–307.

Wieschaus, E., Nüsslein-Volhard, C. & Kluding, H. (1984). *Krüppel*, a gene whose activity is required early in the zygotic genome for normal embryonic segmentation. *Dev. Biol.*, **104**, 172–86.

Wilkinson, D.G., Bhatt, S. & Herrmann, B.G. (1990). Expression pattern of the mouse T gene and its role in mesoderm formation. *Nature, London*, **343**, 657–9.

Wilkinson, D.G., Peters, G., Dickson, C. & McMahon, A.P. (1988). Expression of the FGF related proto-oncogene *int-2* during gastrulation and neurulation in the mouse. *EMBO J.*, **7**, 691–5.

Wilkinson, D.G., Bhatt, S., Cook, M., Boncinelli, E. & Krumlauf, R. (1989). Segmental expression of *Hox-2* homoeobox containing genes in the developing mouse hindbrain. *Nature, London*, **341**, 405–9.

Wilson, E.B. (1892). The cell lineage of *Nereis*. A contribution to the cytology of the annelid body. *J. Morphol.*, **6**, 361–480.

(1904*a*). Experimental studies on germinal localization. I. The germ regions in the egg of *Dentalium*. *J. Exp. Zool.*, **1**, 1–72.

(1904*b*). Experimental studies in germinal localization. II. Experiments on the cleavage mosaic in *Patella* and *Dentalium*. *J. Exp. Zool.*, **1**, 197–268.

(1904*c*). Mosaic development in the annelid egg. *Science*, **20**, 748–50.

Wilt, F.H. (1987). Determination and morphogenesis in the sea urchin embryo. *Development*, **100**, 559–75.

Winfree, A.T. (1980). *The Geometry of Biological Time*. Springer Verlag: New York.

Winkel, G.K. & Pedersen, R.A. (1988). Fate of the inner cell mass in mouse embryos as studied by microinjection of lineage tracers. *Dev. Biol.*, **127**, 143–56.

Wolpert, L. (1969). Positional information and the spatial pattern of cellular differentiation. *J. Theor. Biol.*, **25**, 1–47.

Wolpert, L. & Lewis, J.H. (1975). Towards a theory of development. *Fed. Proc.*, **34**, 14–20.

Wolterek, R. (1904). Beitrage zür praktischen Analyse der Polygordius-Entwicklung nach dem 'Nordsee' und dem 'Mittelmeer' Typus. *Arch. EntwMech. Orgs.*, **18**, 377–403.

Wozney, J.M. (1990). Bone morphogenetic proteins. *Prog. Growth Factor Res.*, **1**, 267–80.

Wylie, C.C., Snape, A., Heasman, J. & Smith, J.C. (1987). Vegetal pole cells and their commitment to form endoderm in *X. laevis*. *Dev. Biol.*, **119**, 496–502.

Yamaguchi, Y. & Shinagawa, A. (1989). Marked alteration at midblastula transition in the effect of lithium on formation of the larval body pattern of *Xenopus laevis*. *Develop. Growth and Differ.*, **31**, 531–41.

Zalokar, M. & Sardet, C. (1984). Tracing of cell lineage in embryonic development of *Phallusia mammillata* (Ascidia) by vital staining of mitochondria. *Dev. Biol.*, **102**, 195–205.

Ziomek, C.A. & Johnson, M.H. (1980). Cell surface interaction induces polarization of mouse 8 cell blastomeres at compaction. *Cell*, **21**, 935–42.

Index

abdominal-A gene, 255, 258–60, 277
Abdominal-B gene, 255, 258–60, 277
acetylcholinesterase, 153–5
acron, 218
activation centre, 262
'activation-transformation theory', 108
actin genes, 115–16
actin (muscle type), 75, 95, 116, 152, 156
actin networks, 135, 146, 151
actinomycin, 136, 152, 156
activin, 98–100, 108, 206
adaptive characters, 3
aggregation, *see* fusion of embryos
alkaline phosphatase, 122, 124, 156
allantois, 173, 177, 195
allocation, 15–16, 31, 220
alpha-foetoprotein, 175, 188, 193
Ambystoma (axolotl), 67, 73, 109
Americans, 166
amnion, 173, 175, 195
amnioserosa, 217
amorphic mutations, 224
amphibian embryo, 67–112
animal hemisphere, 68–9
animalization, 124
annelid embryo, 133–6, 142–6
Antennapedia gene, 255–60, 277
Antennapedia complex, 230, 254–7
anterior midgut, 217
anterior system, in *Drosophila*, 236–7
anterior-posterior specification, 281
    in Crendonian snapper, 291–6
    in *Drosophila*, 236–60
    in *Xenopus*, 104–6
aphidicolin, 155
appositional induction, 27–8, 31
Arbacia, 119
archencephalic induction, 108
archenteron, 70–1, 114
area opaca, 197
area pellucida, 197
armadillo gene, 252, 276
Ascaris, 158
Ascidia, 150–1. 152, 154

ascidian embryo, 7–8, 146–57
Ascidiella, 152
assymetric cell division, 17, 128–70 *passim*,
    179
Astriclypeus, 120
attractors, dynamic, 42
axis, embryonic, 26, 31, 68, 201–3
axial determinant,
    in ascidian, 156–7
    in *Xenopus*, 84
axolotl, *see Ambystoma*

balancer chromosomes, 222–3
basin, in state space 39–40
beta galactosidase, *see* reporter constructs
Bicaudal gene, 232, 239, 271
bicoid gene, 26, 28, 30, 51, 169, 229, 236–7,
    269
Bithorax complex, 230, 257–9
Bithynia, 134, 136, 142
blastocyst, mouse, 173–4
blastomere isolation, 168
    annelids 140
    ascidians, 152–3
    molluscs 140
    mouse, 182
    sea urchin, 119–20
    Xenopus, 86, 88–9
BMPs (bone morphogenetic proteins), 98–9
body plan, 5, 296
Boltenia, 146
Bombyx, 265
border cells, 240
bottle cells, 70–1
Bruchidius, 262–3
BUdR (bromodeoxyuridine), 162

cactus gene, 231, 234–5, 267
CAT (chloramphenicol acetyl transferase),
    see reporter constructs
caudal gene, 243, 248, 273
Caenorhabditis, 7–8, 128, 157–65
cell cycle, 210
cell death, 160

cell differentiation, 1, 34, 40–1
cell lineage, *see* lineage
cell state, 32, 36–43
cell types, 36
cellular blastoderm, 217
cement gland, 108
centrifugation of eggs, 137, 167
　ascidian, 151
　leech, 143
　molluscs, 137
　sea urchin, 118–9
　Xenopus, 82
cephalic furrow, 217
cephalopharyngeal skeleton, 218
Chaetopterus, 134–5, 137
chick embryo, 7–8, 195–211
chimaeras, 181–2, 189–94 *passim*
Chironomus, 264
chorioallantoic membrane, 204
chorion, 173, 175, 195, 216
chromatin diminution, 158
Ciona, 146–57 *passim*
cleavage block,
　of ascidians, 153–5
　of *C. elegans*, 163–4
　of *Xenopus*, 95
'clock and wavefront' model, 65, 110–11, 211
clonal analysis, 13–15, 32, 88–9, 220
clonal restriction, *see* clonal analysis
cloning of genes, 227
colcemide, 148
colchicine, 82
collagen, type IV, 174
commitment, 18–20, 32
compaction, 174, 185
compartment, developmental, 16, 32, 88, 220, 226–7, 249
competence, 27, 32
Concanavalin A, 174, 183–4
convergence movements, 72
cortical granules, 112
Crepidula, 133–4, 136, 137
cubitus interruptus gene, 252, 276
Cumingia, 137
CyIIIa gene, 116, 122, 125
cyclic AMP, 107
cyclopia, 105
cytochalasin, 95, 135, 137, 148, 152, 163, 185
cytokines, 29, 98
cytoplasmic determinants, 6–7, 23, 26–7, 32, 112, 168–9, 279, 285
　in ascidian, 154–7
　in *C. elegans*, 163–4
　in *Drosophila*, 230–240
　in leech, 143
　in mollusc, 138–42
　in mouse, 183–5

in *Xenopus*, 80–4, 87, 111
cytological labels, 13
cytoplasmic localization, *see* cytoplasmic determinants

D lineage of molluscs and annelids, 137–42
DAG (diacylglycerol), *see* inositol lipid cycle
D_2O (deuterium oxide), 85
decidual swelling, 174
defect regulation, 24, 46
　of ascidian unfertilized egg, 151
　of chick blastoderm, 204
　of molluscs 138, 141
　of sea urchin egg, 118
　of sea urchin morula, 121
Deformed gene, 59, 255–6, 259–60, 276
decapentaplegic gene, 99, 235, 269
Dentalium, 133, 136, 137, 140
denticle belts, 218
determination, 13, 18–19, 32, 40–41
deuterencephalic induction, 108
developmental pathway, 32
dexiotropic cleavage, 132, 136
DIA (differentiation inhibitory activity) 190
differentiation, 32
diffusion, of morphogen, 43–57
DNA methylation, 59–60, 186
dorsal gene, 28, 30, 229–36, 267
dorsal lip, 70–1, 101–2
dorsalization, 92, 96–8, 101–4
dorsoventral specification
　in Crendonian snapper, 285, 288–9
　in *Drosophila*, 230–6
　in sea urchin, 125
　in *Xenopus*, 80–5, 101–4
Drosophila, 7–8, 213–77
duplications, 46, 49, 82, 102, 135, 137, 239, 263–4, 288, 295
dynamical systems theory, 35–43

E-cadherin, 174, 185, 199
easter gene, 231, 234, 267
EC (embryonal carcinoma) cells, *see* teratocarcinoma
echinus rudiment, 115
ecto-5 antigen, 115, 122
ectoplacental cone, 173, 175
EGF (epidermal growth factor), 162
egg cylinder, 173–4
embryoid bodies, 189–90
endo-1 antigen, 115, 122
endoderm regionalization, 111
engrailed gene, 230, 248–54, 264, 275
enhancer trap, 177, 224
entactin, 174
epiblast, 197
epigenetic coding, 31, 32
equivalence, developmental, 31, 32

equivalence group, 25, 32, 164–5
ES (embryonic stem) cells, 189–91
Euscelis, 262–3
even-skipped gene, 59, 230, 245–8, 274
evolution, 3
extension movements, 72
extracellular matrix, 64
extraembryonic structures, 51
exuperantia gene, 237, 270

F9 cells 191–2
fate, embryonic, 32
fate map, 10–16, 32
    of ascidian, 148, 150–1
    of *C. elegans*, 160
    of chick, 200–1
    of *Drosophila*, 219–20
    of mollusc, 132–3
    of mouse, 178–81
    of *Platycnemis*, 261–2
    of sea urchin, 117–8
    of *Xenopus*, 76–80
FDA (fluorescein dextran amine), 77–9,
    117, 143, 178
female sterile mutations, 222
FGFs (fibroblast growth factors), 99–101,
    108, 177, 192
FGF receptor, 100–1
Fick's second law, 44
ficoll, 80
field, embryonic, 25, 32, 46
Filzkörper, 218, 232
follicle cells, in *Drosophila* ovary, 215, 231–
    3, 240
fork head gene, 259, 277
fused gene, 252, 276
fushi tarazu gene, 59, 230, 246–8, 274
fusion of embryos, 23–4
    amphibian, 93
    ascidian, 151
    chick, 202
    mouse, 183
    sea urchin, 120

gain of function mutations, 30, 225
gap genes, 240–3, 254, 289
gap junctions, 29–30, 71, 98, 142, 149, 174,
    185
gap phenomenon, 262–3
gastrulation,
    ascidian, 149–50
    C. elegans, 158
    chick, 197
    Crendonian snapper, 285, 288
    Drosophila, 217
    mollusc, 133
    mouse, 175
    sea urchin, 114
    Xenopus, 70–1

gastrulation defective gene, 268
gene trap, 178
genetic labels, 12–13
genetic mosaics, 181–2, 225–7
genetics, developmental, 160–3, 221–8,
    279–80
germ cells, 87, 158, 220
germinal bands, of leech, 143
giant cells, 174–5
giant gene, 243, 273
Gierer-Meinhardt models, *see* GM1, GM2
glp-1 gene, 162, 165
GM1 model, 53–7, 205, 206
GM2 model, 62–3
GPI (glucose phosphate isomerase), 178–80
gnathocephalon, 214
'gold rush', 106
gooseberry gene, 251, 275
gradient, *see* morphogen gradient
gravity, effects of, 81–2
growth, 16–17, 32
growth factors, *see* cytokines
growth zone, 59, 261, 289–91
gurken gene, 231, 233, 268
gynandromorph, 225

Halocynthia, 146–57
hairy gene, 230, 244–8, 273
halteres, 214
haploinsufficient mutations, 224
head process, 173, 175, 197
heavy water, *see* D$_2$O
hedgehog gene, 252, 276
Hensen's node, 197–211 *passim*
Helobdella, 134, 142–6
Hemicentrotus, 117–18, 124
hemimetabolous insects, 215, 260–5
herkunftsgemäss development, 19, 32
heterochronic grafts, 95
heterochronic mutations, 161
hierarchy, developmental, 5–6, 19–20
histoblasts, abdominal, 215, 218
HNK-1 antibody, 205
holometabolous insects 215
HOM complex, 264
homeobox genes, 75, 176, 192–3, 228
homeogenetic induction, 60–1, 101, 108
homeotic genes, 30–1, 32, 160–2, 254–9, 264
homologous recombination, 194
homology, 3
HRP (horseradish peroxidase), 12, 77–8,
    133, 143, 178–81, 218–220
huckerbein gene, 240, 243, 273
human embryo, 1, 176, 189
human teratocarcinoma, 192
hunchback gene, 230, 239–41, 272
hyaline layer, 112
hypoblast, 197
hypomorphic mutation, 224

I-antigen, 175
ICM (inner cell mass), 173–4
IGF-II (insulin-like growth factor II), 192
Ilyanassa, 7–8, 129–41
imaginal discs, 215, 218
imprinting, 185–7
induction, 6, 8, 27–30, 32, 43–57, 60–1, 112, 279
  interactions between segment polarity genes, 253–4
  of brain in ascidians, 152
  of mesoderm in *Xenopus*, 93–101
  of mesodermal structures by the organizer, 96–8
  of muscle in ascidians, 154
  of neural plate by archenteron roof: in *Xenopus*, 106–10; in chick, 207
  of parts by *bicoid* gradient, 237
  of pharynx in *C. elegans*, 165
  of polar trophectoderm by ICM, 187–8
  of primitive streak by hypoblast, 205–7
  of somites by Hensen's node, 207
  of structures by D lineage, 139–41; (*see also* appositional, homeogenetic, instructive, permissive induction; morphogen gradient)
infraabdominal genes, 258
inositol lipid cycle, 85, 124, 126
instructive induction, 27, 32
int-1 gene, *see* wnt-1
int-2 gene, 99, 175, 177
intermediate germ insects, 261
IP$_3$ (inositol triphosphate), *see* inositol lipid cycle

Janus-larva, 135

K10 gene, 233, 268
Keilen's organs, 218
knirps gene, 230, 239, 242, 272
Krox-20 gene, 177
Krüppel gene, 242, 272

L-CAM (cell adhesion molecule), *see* E-cadherin
labels, cell, 10–13, 33
labial gene, 255, 259, 276
laevotropic cleavage, 132, 136
laminin, 174
Lanice, 140, 141
laser ablation, 164–5
law of mass action 36
leech embryo, 7, 142–6
lin-12 gene, 162
limb specification, in Crendonian snapper, 292–4
lineage, cell, 20–1, 32, 33, 166–7
  of ascidian, 148, 150–1
  of *C.elegans*, 158–60

of leech, 143
of mollusc, 132
lineage labels, *see* HRP, FDA
lithium, 84–5, 100, 104, 105, 124, 126
long germ insects, 261
loss of function mutations, 30, 224
LSDS (localized source, dispersed sink) model, 45–51, 123, 205, 237
Lucifer Yellow, 117, 142
Lymnaea, 134, 136, 142
Lytechinus, 122

macromeres,
  of sea urchins, 113–14
  of molluscs, 131
malic enzyme, cytoplasmic, 188
Malpighian tubules, 218
markers, of cell state, 33
marginal zone, 70–2, 197
maternal inheritance, 9–10, 33, 136, 221–4, 231–4, 236–40
maternal mRNA, 74, 114, 116, 136, 231–4, 236–9
mating type, of yeast, 60
mechanical instabilities, 64–5
memory, of cell state, 59
merogones, 35
mesentoblast, 132
meso-1 antigen, 115
mesoderm induction, 93–101, 111
mesomeres, 114
mesothorax, 214
messenger RNA complexity 42, 116
metaplasia, 31
metathorax, 214
methylation, see DNA methylation
micromeres,
  of molluscs, 131
  of sea urchins, 113–14
microtubule array, 82, 85
mid-blastula transition, 71
MIFs (mesoderm-inducing factors), 98–101
mix-1 gene, 75, 100
mollusc embryo, 129–42
molluscan cross, 132
morphogen, 28, 33
morphogen gradient, 27–8, 32, 43–57, 97, 122–3, 125–6, 236, 237, 247
morphogenesis, 1, 33, 64–5
mosaic behaviour, 20–1, 33, 128–9, 167–8
mosaic analysis 225–7, 231, 243
mouse embryo, 7–8, 171–95
multiple wing hairs gene, 226
mural trophectoderm, 173–4
mutagenesis, 222–4
mutation, loss of function, 30
  gain of function, 30
Myo-D gene, 75
myoplasm, 146, 152–7

N-CAM (cell adhesion molecule), 75, 199
naked gene, 252, 276
nanos gene, 169, 229, 238–9, 270
Nasrat gene, 271
Nereis, 128, 134, 140
neural induction, 106–10, 152
neural plate, 18, 73, 175, 197–9
neuroblasts, 217
normal development, 9–17, 33
notochord, 73, 147–9, 175, 197–9
nuclear transplantation,
    in *Drosophila*, 219
    in mouse, 185–6
    in *Xenopus*, 87
nudel gene, 233, 268
null mutations, 224
nurse cells, 215

odd-paired gene, 274
odd-skipped gene, 275
oestrogen receptor, 59
ooplasmic segregation, 134, 146, 167
organizer, 33, 46, 87, 101–4
orthodenticle gene, 237, 243, 273
ortsgemäss development, 19
oskar gene, 238, 270
overexpression experiments, 75, 193, 256, 294–5

P elements, 221, 227
P granules, 158
pair-rule genes, 244–8
paired gene, 246, 274
par genes, 163
Paracentrotus, 117, 123
parallel processing, 283
parasegments, 217, 250–60 *passim*
parietal endoderm, 173–4, 188
patched gene, 251, 276
Patella, 133–4, 140, 141–2
pattern formation, 33
pelle gene, 234, 267
permissive induction, 29, 33
Phallusia, 146, 151
pharynx, of *C. elegans*, 165
PIP_2 (phosphatidyl inositol bisphosphate), *see* inositol lipid cycle
pipe gene, 233, 268
Platycnemis, 261–2
Pleurodeles, 72, 73
pluteus larva, 113–4
polar lobe 129–30, 136, 138–40
polar trophectoderm, 173–14
polarity, 25–6, 33
    of chick blastoderm, 201–2
    of *Drosophila* (dorsoventral), 234
    of mouse blastomeres, 183–5
    of sea urchin, 125
    of *Xenopus*, 80–5
    see also duplications

polarization, *see* polarity
pole cells, 215
pole plasm, of annelids, 134
    of *Drosophila*, 216, 220, 237–9
polehole gene, 271
polyclone 25, 33
Polygordius, 128
polytene chromosomes, 221
positional information, 33
posterior midgut, 217
posterior system, in *Drosophila*, 237–9
potency, 19, 27, 33, 40–1
pp60 c-src, 101
primary hypoblast 197
primary mesenchyme, 113–14, 124
primary pair-rule genes, 230, 246–7
primordial cell numbers, 14–15
primitive endoderm, 173–4, 187
primitive streak, 175, 197–9
prism stage of sea urchin, 113–14
pristine characters, 2
proboscopedia gene, 255
procephalon, 214
program, developmental, 282–96
progress zone, 59
proportion regulation, 22–4, 49–50, 55–6, 126
    annelids, 141
    ascidian egg, 151, 155
    chick embryo, 202
    molluscs, 138
    mouse embryo, 182–3
    sea urchin egg, 118, 120
    Xenopus, 92, 101
prospective region, *see* fate map
protein kinase A, 107, 108
protein kinase C, 107, 108
prothorax, 214
prototroch, 133
pseudocleavage, of *C. elegans*, 158–9
pumilio gene, 238, 270

quail nucleolar label, 13, 205

Rana, 67, 73, 87, 111
RDA (rhodamine dextran amine), 209–10
reaction–diffusion models, 61–4
reduction gradient, 116–17
regional specification, 1–2, 33
regulation, embryonic, 21–5, 33
    for specific examples, *see under* defect regulation, twinning, fusion, proportion regulation
Reichert's membrane, 174
repeating patterns, 61–6
    see also segmentation, somitogenesis
reporter constructs, 122, 228
retinoic acid, 29, 108, 192
rhabditin granules, 158, 164
rhombomeres, 177

Ring-X chromosome 225
runt gene, 230, 245–8, 274

Sabellaria, 140
Schistocerca, 264
Schizosaccharomyces, 161
SDS (sodium dodecyl sulphate), 124
sea urchin embryo, 7–8, 112–26
secondary hypoblast, 197
secondary mesenchyme, 114
secondary pair-rule genes, 230, 246–7
segmentation, 61–6
    of amphibian embryo somites, 110–11
    of chick embryo somites 209–11
    of *Drosophila*, 244–54
    of leech embryo, 142–6
segment polarity genes, 248–54, 264
selector genes, *see* homeotic genes
separatrix, 39–40
serendipity gene, 226
serial homology, 61
Sex combs reduced gene, 255–6, 259–60,
    277
sex lethal gene, 226
shell gland, 133
shell, of chick, 197
short germ insects, 261
shrew gene, 269
size regulation, *see* proportion regulation
sinestral form of *Lymnaea*, 136
single-minded gene, 229, 234, 269
sloppy-paired gene, 274
SM50 antigen, 115, 122, 124
Smittia, 264
snail gene, 235, 268
snake gene, 234, 267
snapper, Crendonian, 282–96
somatoblast, first and second, 132
somites, 73, 110–11, 175, 197–9, 209–11
somitogenesis, 65, 110–11, 209–11
somitomeres, 198
spalt gene, 259, 277
spätzle gene, 268
Spec genes, 115, 122, 124
specification, 18, 33, 40–1
    of ascidian blastomeres, 152–3
    of mollusc blastomeres, 140–1
    of parts in chick blastoderm, 202–4
    of parts in rat embryo, 189
    of parts in sea urchin morula, 120–2
    of parts in *Xenopus* blastula, 89–92
    of sea urchin blastomeres, 119–20, 122
spinocaudal induction, 108
spiracles, 218
spiral cleavage, 132, 136
state, *see* cell state
state space, 37–43
states, stable, 38–9
statocysts, 133
staufen gene, 271

stem cells, 16–17, 33, 133, 135, 156, 163
    see also ES cells
stereocilia, 113–14
steroid-thyroid receptors, 242
Strongylocentrotus, 112–26
Styela, 128, 146–57
subgerminal cavity, 196–7
suramin, 105
swallow gene, 237, 270
symmetry breaking processes, 50, 52–7,
    202
syncytial blastoderm, 217

Tachycines, 264–5
tailless gene, 240, 242–3, 273
teloblasts, of molluscs, 133
    of annelids, 134–5, 143–5
teloplasm, 134, 143
temperature shock, 110
telson, 218
Tenebrio, 214
teratocarcinoma cells, 191–2
terminal system, in *Drosophila*, 239–40
TGFβ (transforming growth factor β), 98–
    100, 235
three-signal model, 97–8
threshold, 33, 46, 48, 57–61
Toll gene, 169, 231, 234–6, 267
tolloid gene, 269
torpedo gene, 231, 233, 268
torso gene, 26, 30, 169, 229, 240, 271
torsolike gene, 240, 271
totipotency, 19
TPA (12–O–tetradecanoylphorbol-13-
    acetate), 107
trajectories, in state space, 37–43
transferrin, 175
transfilter induction, 98, 106
transgenic mice, 192–4
trefoil, 129–30
Tribolium, 254, 264
triploids, 93
tritiated thymidine, 12, 178, 180, 183, 200,
    207, 210
Triturus, 108
trochophore larva, 133–5
trophectoderm, 173–4, 187–8
trophoblast, 174
trunk gene, 271
tube gene, 267
Tubifex, 135, 141
tudor gene, 270
twinning, 22–3, 126
    of ascidian, 151
    of chick, 202–3
    of insects, 264–5
    of mammals, 182, 189
    of sea urchins, 119–20
    of *Xenopus*, 82, 85, 86
twist gene, 75, 229, 268

twisted gastrulation gene, 269
tyrosine kinase, 4, 101, 240

ultraviolet irradiation, 83–4, 104, 157,
    263–4
ultraviolet microbeam, 218, 261
Ultrabithorax gene, 255, 257–60, 277
universality of mechanisms, 3, 125–6, 211–
    12, 278–81
urodeles, 67–112 *passim*
uvomorulin, *see* E-cadherin

valois gene, 270
vasa gene, 220, 270
vegetal hemisphere, 68–9
vegetalization, 124–5
vg-1 gene, 74, 99
Veitch matrix, 283
veliger larva, 129, 133
velum, 133
ventral furrow, 217
vimentin, 175
visceral endoderm, 173–4, 188

vital stains, 12, 76–7, 200–1
vitelline membrane, 68, 112, 216

windbeutel gene, 233, 268
wingless gene, 230, 248–54, 275
wnt-1 gene, 194

X-chromosome inactivation 181–2
X-ray induced somatic mutation, 225–7
Xenopus, 7–8, 67–112
xhox-1A gene, 111
xhox-3 gene, 75, 100

yellow crescent, 147–8
yellow gene, 226

zebrafish, 13–15
zerknüllt gene, 229, 235, 269
zinc, as animalizing agent, 124
zinc fingers, 177, 241
zoologists, 2
zona pellucida, 172
zygotic character, 10, 33

Printed in the United States
By Bookmasters